Latest Developments in Geographic Information Systems

Volume I

Latest Developments in Geographic Information Systems
Volume I

Edited by **James Cook**

𝒞𝓁 LANRYE
INTERNATIONAL

New Jersey

Published by Clanrye International,
55 Van Reypen Street,
Jersey City, NJ 07306, USA
www.clanryeinternational.com

Latest Developments in Geographic Information Systems
Volume I
Edited by James Cook

International Standard Book Number: 978-1-63240-325-4 (Hardback)

Printed in the United States of America.

Contents

Preface   VII

Chapter 1   **Understanding the Urban Sprawl in the Mid-Size Latin American Cities through the Urban Form: Analysis of the Concepción Metropolitan Area (Chile)**   1
Carolina Rojas, Iván Muñiz, Joan Pino

Chapter 2   **A Proposal for a Geospatial Database to Support Emergency Management**   14
Ivan Frigerio, Stefano Roverato, Mattia De Amicis

Chapter 3   **Using a GIS to Assessment the Load-Carrying Capacity of Soil Case of Berhoum Area, Hodna Basin, (Eastern Algeria)**   22
Amar Guettouche, Farid Kaoua

Chapter 4   **A Multidisciplinary Approach to Mapping Potential Urban Development Zones in Sinai Peninsula, Egypt Using Remote Sensing and GIS**   28
Hala A. Effat, Mohamed N. Hegazy

Chapter 5   **Cartographic-Environmental Analysis of the Landscape in Natural Protected Parks for His Management Using GIS: Application to the Natural Parks of the "Las Batuecas-Sierra de Francia" and "Quilamas" (Central System, Spain)**   45
Antonio Miguel Martínez-Graña, Jose Luis Goy, Caridad Zazo

Chapter 6   **Extraction of Urban Vegetation in Highly Dense Urban Environment with Application to Measure Inhabitants' Satisfaction of Urban Green Space**   60
Fatwa Ramdani

Chapter 7   **Simulation Models and GIS Technology in Environmental Planning and Landscape Management**   66
Giuliana Lauro

Chapter 8   **A Web-Based Cancer Atlas of Saudi Arabia**   77
Khalid Al-Ahmadi, Ali Al-Zahrani, Atiq Al-Dossari

Chapter 9   **Accuracy of Stream Habitat Interpolations Across Spatial Scales**   92
Kenneth R. Sheehan, Stuart A. Welsh

Chapter 10   **The Accuracy of GIS Tools for Transforming Assumed Total Station Surveys to Real World Coordinates**   103
Ragab Khalil

Chapter 11   **Using GIS Data to Build Informed Virtual Geographic Environments (IVGE)**   109
Mehdi Mekni

Chapter 12   **Location of Large-Scale Concentrating Solar Power Plants in Northeast Brazil**   120
Verônica Wilma Bezerra Azevedo, Chigueru Tiba

Chapter 13   **The Contribution of the Geospatial Information to the Hydrological Modelling of a Watershed with Reservoirs: Case of Low Oum Er Rbiaa Basin (Morocco)**   139
Youness Kharchaf, Hassan Rhinane, Abdelhadi Kaoukaya, Abdelhamid Fadil

Chapter 14   **Application of LiDAR Data for Hydrologic Assessments of Low-Gradient Coastal Watershed Drainage Characteristics**   150
Devendra Amatya, Carl Trettin, Sudhanshu Panda, Herbert Ssegane

Chapter 15   **Modelling of Environment Vulnerability to Forests Fires and Assessment by GIS Application on the Forests of Djelfa (Algeria)**   167
Mohamed Said Guettouche, Ammar Derias

Chapter 16   **GIS as an Efficient Tool to Manage Educational Services and Infrastructure in Kuwait**   176
Khalid Al-Rasheed, Hamdy I. El-Gamily

Chapter 17   **Geospatial Evaluation for Ecological Watershed Management: A Case Study of Some Chesapeake Bay Sub-Watersheds in Maryland USA**   188
Isoken T. Aighewi, Osarodion K. Nosakhare

Permissions

List of Contributors

Preface

Geographic Information Systems (GIS) is a sub-category of a broader academic discipline called Geoinformatics. The origin of this term can be traced to the year 1968, when Roger Tomlinson, also fondly nicknamed as the 'Father of GIS', used it on one of his research papers. Today, GIS has emerged as a powerful tool, which can organize a complex spatial environment with tabular equations.

In a nutshell, GIS is a computer system for capturing, storing, and displaying data related to spatial locations on earth's surface. It uses the spatio-temporal location as the key index variable for providing all other information. What makes the GIS a smart tool is the fact that it can show many different kinds of data on one map. From population, income, or education level to information on different kinds of vegetation and soil, the GIS can provide all sorts of data. GIS derives its data sets from precise navigation and imaging satellites, aircrafts and transactional databases.

GIS can use any information that includes location, which can be expressed in many different forms such as latitude and longitude, address, or ZIP code. The GIS combines the information from different sources in such a way that it all has the same scale.

I'd like to thank all our contributors, who've shared their studies on GIS with us.

Editor

Understanding the Urban Sprawl in the Mid-Size Latin American Cities through the Urban Form: Analysis of the Concepción Metropolitan Area (Chile)

Carolina Rojas[1], Iván Muñiz[2], Joan Pino[3]

[1]Department of Geography, Faculty of Architecture, Urbanism and Geography, University of Concepción, Center for Sustainaible Urban Development—Chile, Concepción, Chile
[2]Department of Applied Economics, Autonomous University of Barcelona, Barcelona, Spain
[3]CREAF (Center for Ecological Research and Forestry Applications), Autonomous University of Barcelona, Barcelona, Spain

ABSTRACT

Latin American cities, like those from North America and Europe, experience problems of urban sprawl. However, few studies have dedicated exclusively to this phenomenon in specific cities, and this omission is particularly noticeable regarding cities not considered among the megalopolis of the continent. The present work analyzes urban sprawl through an urban form in the Concepción Metropolitan Area, Chile, between 1990 and 2009, considering local aspects that may have played a role in the process. The main empirical results obtained from this study reveal a metropolitan area that has expanded intensely over a 20-year period, growing from 9000 hectares to more than 17,000 ha for a 96% increment in the built-up area. The new urban surfaces consolidate a central conurbation that strengthens the role of the main downtowns, with less-intense occupation towards the sub-centers but in a structure that follows the transportation infrastructure. Over the last 20 years, the distance between the shapes has grown progressively by around 2 km, increasing the size of the ellipse by more than 1000 km^2. In particular the complexity of the urbanized surfaces has grown, becoming more irregular in shape and less compact as they come to occupy larger areas. So our principal findings include: an increment of nearly 100% in the urban surface, the importance of a polycentric urban structure in the process of consolidation as a support for analyzing different spatial dynamics, and the growing morphological irregularity of the territory of the sprawl.

Keywords: Urban Sprawl; Latin American Cities; Metropolitan Areas; Geographical Information System

1. Introduction

In Europe and especially the United States, urban sprawl has been the subject of a consolidated line of research since the 1960s. However, in Latin America, this phenomenon only very recently has become a point of interest for the academic community.

In general, Latin American literature has focused on contextualizing the phenomenon of sprawl within a broader framework, taking urban sprawl, polycentrism, functional specialization, network economies, and the growing privatization of the urban space to be reflections of a new type of globalized city that grows and changes according to the logic of the market and improvements in the fields of transport and telecommunications. This contextualization has allowed elaborating a coherent tale regarding the changes that have taken place. However, there is a clear lack of symmetry in terms of the quality and sophistication of the theoretical tale and the scant empirical evidence on which it is based.

Urban sprawl increases the cost of supplying public services; occupies spaces with elevated agricultural, ecological, or landscape value; and leads to a model of mobility based on the automobile, which is expensive, unequal in the distribution of opportunities, and environmentally unsustainable [1-3]. Furthermore, urban sprawl often accompanies or accentuates characteristic problems of Latin American cities such as phenomena of social segregation, difficult access to employment for the most disadvantaged groups, or the geological vulnerability of some urbanized zones [4-6]. Bearing in mind the economic, social, and environmental importance of urban sprawl, a better understanding of this phenomenon is indispensable.

This work is intended to characterize the phenomenon of urban sprawl that has occurred in Concepción during the last 20 years. In order to achieve this, the rhythm of land consumption and its relation to the existing spatial structure—Concepción is a Medium size Chilean city, polycentric and with an interesting particularity, it counts on two CBDs, almost twins and very close each other. Then, two indices of urban form are now calculated [7], in order to calculate the evolution of the urban form adopted by the dimension of the sprawl [8]. As indicated in the specialized literature. A drop of the compactness index and an increase in the complexity index imply a greater sprawl. The information obtained will allow to determinate the rhythm of land consumption, its preferential location, proximity among urban zones and irregularity (Complexity index).

The rest of this article is organized as follows: Section two is dedicated to the recent transformations of Latin American cities and various visions of urban sprawl. Section three deals with the concept of urban sprawl, and section four presents the methodological aspects related to processing satellite images, extracting urbanized surfaces, and calculating the metrics of the forms and ellipse associated with the measurement of sprawl in terms of the distance between the shapes of the expansion. In section five, the case study is characterized, and the results are presented according to the proposed objectives, detailing the urban growth according to the structure of the downtowns that make up the Concepción metropolitan area (CMA), the new spatial shapes, and the sprawl of the urbanized surfaces. Finally, section six contains the main conclusions of the study.

2. Recent Transformations in Latin American Cities

Many cities are currently growing faster than what their planning can accommodate, and urban growth is becoming a worrisome global reality that increasingly affects the balance between humans and ecosystems. In this sense, cities are good examples of complex systems that require enormous effort to maintain their equilibrium [9]. Understanding urban growth is a fundamental planning concern. Environmental and social challenges require transforming rapid urban growth into sustainable growth, with an internal balance between economic activity, population growth, infrastructure, pollution, waste, and noise [10].

Urban growth is a worrisome reality that affects global urban ecosystems and is even considered to be a type of global environmental change [11]. Such growth is seen as a potential threat to ensuring adequate housing, sanitation, health, and transportation services in a sustainable urban environment, especially in the cities of less developed countries like those in Latin America.

As a whole, Latin America, including Mexico, constitutes one of the most urbanized regions of the world, with an urban population that grew rapidly from 69 million in 1950 to 448 million in 2007. By 2025, 575 million people are expected to be living in Latin American cities. This notable increment (78%) is higher than that one for Europe, Africa, and Asia for the same dates [12]. The intensity, speed, and impact on natural systems of this accelerated urban growth in Latin American cities, especially considering the ever-more-segregated distribution of the population groups, has awoken interest in the matter.

The rapid urbanization in Latin America has been largely explained by its industrial development between the 1950s and 1980s, and the changing role of private capital in territorial organization beyond its competencies in the economic model. In this sense, all Latin American countries are experiencing transformations due to globalization as well as endogenous factors. Although their origins are debatable, these factors are related to urban effects such as new modalities of metropolitan expansion, suburbanization, polycentralization, social polarization, residential segregation, and the fragmentation of the urban structure [13]. Three types of regions can be distinguished within the context of this new, quickly growing urban configuration: first are countries with more than 80% of the population living in urban areas (Argentina, Chile, Uruguay, Venezuela); second are countries with 50% to 80% of the population in urban centers (Mexico, Brazil, Ecuador, Colombia, Cuba, Bolivia, Peru); and third are countries with less than 50% of the population in cities (Paraguay) [14]. In terms of urban development, Mexico City, Sao Paulo, Rio de Janeiro, and Buenos Aires are world class examples of large cities with extensive growth of the urbanized surface, whereas Santiago de Chile stands out as a metropolitan area whose growth is primarily based on the creation of new downtowns and the use of agricultural land.

The large metropolitan areas of Latin America are complex, competitive, and dynamic hubs in which these tendencies are notorious. Although most cities in Chile are mid-sized, they, too, reveal certain restructuring processes that should be mentioned. Chile is highly urbanized, with more than 80% (13,090,113 inhabitants) of its population living in urban areas of over 5000 inhabitants. However, this country is characterized by mostly mid- to small-sized cities. Three metropolitan areas (Santiago, Concepción, Valparaíso), all located in the macro-central zone, reflect spatially concentrated development, with more than 6 million inhabitants, which represents

Understanding the Urban Sprawl in the Mid-Size Latin American Cities through the Urban Form: Analysis of the Concepción Metropolitan Area (Chile)

3

48.3% of the population. Chile is centralized in the sense that more than 5 million individuals live in the city of Santiago alone.

The changes experienced by Chilean cities, like those of the rest of Latin America, are described as being tied to globalization, particularly since Chile began to apply the neoliberal model as early as the 1970s. One primary transformation is the shift from relatively compact structures associated with the foundational model to diffuse models present in peripheral areas or the places towards which the growth of these cities tends [15].

3. Urban Sprawl in Latin America

The repercussions of the urban changes resulting from the effects of globalization and the neoliberal model can be seen in new expressions and spatial forms of growth that have led mainly to an unequal distribution of the population based on the social condition and ever more disperse urban structures. The most important phenomenon is known as "peripheralization". This is an "oil stain" type of growth process at the edges of the city, with peripheral settlements typically being located in areas where the land value and residential quality are low [14]. Peripheralization has favored the development of disperse sprawl, fragmented spaces, polycentralities, residential segregation, and landscapes dominated by large commercial centers and highways. As noted by [16], it seems incredible how, over the course of a few decades of urbanization, the neoliberal model has acted as one more agent in the territorial ordering, redefining the relation between the public and the private, with a clear preference for the latter.

No single tale explains the causes of urban sprawl. Rather, at least three approaches have been taken into account to this end, and the theoretical roots of these can be clearly differentiated. First, according to the natural growth approach, urban sprawl is the logical result of an increase in the per capita income, declining transportation costs, and innovations in the field of telecommunications [17]. Greater per capita incomes lead to higher demands for single-family housing; lower transportation costs allow financing housing farther from the workplace; and innovations in telecommunications (e.g., telecommuting) make physical proximity increasingly less important. Second, the economic restructuring approach emphasizes the spatial repercussions of a new economic logic marked by the globalization of capitalism. From this perspective, the emergence of new sub-centers, the appearance of commercial centers or business parks on the periphery, and the phenomenon of sprawl are the main features of a new urban model stemming from globalization in a context in which the private sector tends to impose itself over the public sector. Finally, the

path-dependency approach breaks with the economic-technological determinism of the two previous approaches, indicating that forces that are global in nature feed sprawl, but that these forces interact with the local context, where history, geography, and urban planning are important. In this sense, sprawl is understood to be a problem of poor territorial planning due to poor housing policies [18,19].

In general, the study of the causes of urban sprawl in the USA has followed the first approach (natural growth), with special emphasis on the role of technological change. In Latin America and the United Kingdom, the second approach (economic restructuring) has been more popular, associating urban sprawl with the growing liberalization of the economy that has occurred in Latin America since the 1980s. Finally, in continental Europe, the third approach (path-dependency) is the most common since not all European countries and cities experience processes of sprawl. The countries that have experienced the most intense sprawl are precisely those in which the public sector has had a waning presence in the housing policy and urbanism.

In Latin America, the study of urban sprawl is first characterized by the fact that this topic is not generally the subject of specific analysis. Research on sprawl tends to form part of broader studies of the spatial transformations of the city that includes aspects such as the decentralization of the population and activity, the formation of sub-centers, the emergence of a new geography of consumption and leisure, spatial segregation or the constitution of networks, and new patterns of mobility. Although all these phenomena are related to a certain degree, specific studies should be proposed to delve further into each one of them. Secondly, the economic and social costs of sprawl are those that have been granted the most attention. Therefore, studies that tackle the environmental aspects of sprawl, especially in relation to the occupation of areas with elevated ecological and/or landscape value, are lacking. In the third place, most studies have examined the causes of sprawl through the lens of economic restructuring. The problem with this approach is that it lacks a certain economic determinism in the same way that the natural growth approach lacks a certain technological determinism. The path-dependency approach allows a more detailed look at the economic-social circumstances—from land and housing policies to the power of local agents in planning or the policy of inflation—that have interacted with the global tendencies feeding the process of sprawl.

One of the main reasons by which sprawl in Latin American cities has not been considered an issue of preferred analysis is that it has been refused until recently. In general, other labels were preferred rather than alternatives to "dispersion". The fundamental reason behind this refusal is that in both USA and Europe is only given within a context of population stagnation. This implies the reorganization of the residential and productive space whose net balance implies a higher consumption of land. If on the contrary a population growing is given as well as the land consumption, the term "urban growing" becomes more usual and if what id given is the functional annexation of peripheral area, the term "metropolization" should be used. These are two processes that certainly have determined the evolution of Latin American cities. However, it must be recognized that the forms adopted by some Latin American metropolis are very similar to United States cities qualified as dispersed. On the other hand it also must be taken into account that in some Latin American countries the rural-urban migratory processes have been stopped and the population growing of some metropolis has been clearly below the land consumption rates. For all this, it seems appropriate to start analyzing the evolution of the shape of the Latin American cities under the perspective of dispersion.

Empirical works that have addressed dispersion of Latin American cities by mean of indices of urban form are scarce. Among them, the index by Huang et al. (2007) is noteworthy [20]. This index allows testing for the presence of spatial urban dispersion patterns only in major Latin American cities such as Buenos Aires, Córdoba, Porto Alegre, Rio de Janeiro, Sao Paulo, Santiago, San Salvador, Tegucigalpa, Guadalajara, Ciudad de México, Monterrey, Managua, Montevideo, Bogotá, Caracas, Quito and Guatemala. Even without considering measurement in metropolitan areas of medium size, urbanization trends reflect the maintenance of more compact and dense shapes, mainly in cities that have manifested a radial growing influenced by a more European foundational planning, in contrast to the general trend that indicated a greater similarity towards the patterns of the US cities [20].

In the case of Chilean cities, urban expansion has been characterized by the dominance of the private automobile, the emergence of new downtowns and the installation of commercial shopping centers. This type of urban sprawl has been specifically characterized by low-density peripheral expansion; new centralities; new more fragmented, heterogeneous spaces; and residential segregation [15]. When a closer look is given at these transformations, the case of Santiago de Chile immediately stands out. This city of more than 5 million inhabitants is closely tied to the American process, experiencing accelerated sprawl and being cataloged as a suburbanized, extremely segregated, fragmented, polycentric region with imprecise limits whose expansive dynamics have incorporated nearby urban and rural lands, configuring a vast periurban area [21]. Surprisingly, not only are these processes notorious in the capital, but they are also present in

the rest of the metropolitan areas (Concepción, Valparaíso) and regional capitals with more than 100,000 inhabitants (known as "regional cities") in varying phases of the metropolitan life cycle [15].

4. Methodology

By means of a GIS-based approach to urban form, we first selected the major built-up polygons of urban growth in the CMA from Landsat images and then calculated a set of spatial indices concerning the size and shape of these built-up areas. In order to understand the spatial pattern of urban sprawl and the alterations in urban form over time, the changes in these indices from 1990 to 2009 were tracked in terms of the size and spatial position of the corresponding urban areas using linear models.

The proposal of this methodology is to apply a set of spatial indices in order to understand the evolution of urban form of CMA, even though it is not designed to this study area, they are perfectly applicable to other metropolitan areas of Latin America. The measures could be useful to understand the link between land use pattern and urban sprawl.

4.1. Image Classification and Selection of Built-Up Areas

The remote sensing techniques used here constitute powerful tools for studying urban growth and sprawl because they facilitate the identification of different land uses and covers. These techniques can be applied to satellite images to identify built-up areas. The main methods used to identify urban and built up areas are supervised classification [20,22,23] and principal component analysis [24, 25]. In this study, we applied supervised classification using maximum likelihood with over 70% accuracy.

We used six Landsat orthorectified images that completely covered the CMA: two images from 1990 (path 001/row 085: 22 January 1990; path 001/row 086: 22 January 1990), two from 2000-2001 (path 001/row 085: 18 January 2000; path 001/row 086: 22 December 2001), and two from 2009 (path 001/row 085: 18 January 2009; path 001/row 086: 18 January 2009). The 1990 and 2009 images were obtained with the Thematic Mapper (TM) sensor at a spatial resolution of 30 m, and the 2000 and 2001 images with the Enhanced Thematic Mapper (ETM +) sensor at a resolution of 30 m. The images from path 001/row 085 and path 001/row 086 were joined, generating three mosaics: one for 1990, one for 2000, and one for 2009. The spatial reference system in the study was UTM (WGS84, zone 18S).

The built-up area was identified with supervised classification and principal component analysis. The first component retained more information about the built-up area, the second about vegetation, and the third about water. Subsequently, using these three bands, a supervised maximum likelihood classification was performed and the images were reclassified into either urban (built-up) or non-urban areas (two classes).

4.2. Landscape Metrics of Built-Up Areas

There are numerous studies that have analyzed urban dispersion through the calculation of urban form indices, making an intense GIS use. The three more utilized indices are: compactness index, complexity index and to a lesser extent, the porosity index [20,23,26-29].

In this study, we calculated two indicators using spatial metrics based on the area (A) and perimeter (P) of the built-up areas in 1990, 2000, and 2009. These indicators are:

Compactness of the largest path index (CLPI): This index measures landscape fragmentation (Equation (1)). When the shape is more regular and the patch number is smaller, the index value is larger [20] (Equation (1)):

$$\text{CLPI} = \frac{2\pi\sqrt{A/\pi}}{P} \tag{1}$$

Complexity index (CO): This index measures the irregularity of shapes. It is a medium weighting of the fractal dimension index (Equation (2)). The higher values indicate greater complexity. The fractal dimension is discussed in [7].

$$\text{CO} = \frac{\sum\limits_{i=1}^{i=n} \dfrac{2\ln 0,25P_i}{\ln A_i}}{N} * \frac{A_i}{\sum\limits_{i=1}^{i=N} A_i} \tag{2}$$

Compactness of forms is considered an antonym of the sprawl [29] and therefore, an urban expansion associated to more compact zones. Furthermore, it relates areas of greater population density closer to the services centers and jobs, promoting a lesser land consumption. It also has been related to less fragmented forms, by considering the number of forms of a greater urban zone [29]. Complexity indicator associated to a more disperse growing is related to a more irregular edge expansion and therefore, to more complex spaces in its integration and urban connection.

4.3. Directional Distribution (Standard Deviational Ellipse)

This technique comes from the ecological study of the extension of animal habitats and consists basically of obtaining a bivariate confidence interval corresponding to the coordinates X and Y. These define the major and minor axes of an ellipse with the smallest possible area. Then the standard distance (Euclidian) is calculated be-

tween the locations X and Y of growing zones. In this study, the ellipse is used as a measure of the sprawl from the urban areas or built-up areas.

The size of the confidence ellipse depends on the number of standard deviations of the distance of the urban areas with respect to downtown metropolitan Concepción. In this case, we use two deviations, covering 95% of the urban forms.

The interpretation is interesting to analyze because of that area resulting as an indicator of sprawl related with the distance to central business center, and if the evolution to urban growth tends to extended, it increases the territory could be interpreted as a spatial process of urban sprawl or first steps of this phenomenon. Therefore from the points in space and the angle of the main axis of the ellipse, as an indicator of the geographic orientation of the urban growth and interpret in which areas with another pattern it is experimenting or starting a process of urban sprawl.

The increment of the distances of the new urban forms along to the loss of proximity to the center reflects a more deficient model for connecting urban forms and the dependence of the existence of highway of transportation networks. Finally, related to a technical point, the tool Directional Distribution of ArcGIS (Standard Deviational Ellipse) is used.

5. Study Area

The CMA is located in central-southern Chile (36°35'S - 37°00'S and 72°45'W - 73°15'W). This is a coastal territory or the internal margin of a trench with tectonic origins that contains modeled plains and terraces of fluvial origins [30]. According to the delimitation proposed by the Concepción Metropolitan Regulatory Urban Plan (2003), the city is defined as a functional, hierarchical territory made up of 11 apparently closely related townships (municipalities): Chiguayante, Concepción, Coronel, Hualqui, Hualpen, Lota, Penco, San Pedro de la Paz, Santa Juana, Talcahuano, and Tomé. These municipalities congregate a population of over 900,000 inhabitants and a density of 318.9 inhabitants per km^2 [31]. This metropolitan area stretches along 60 km of coastline from the township of Tomé (the northern limit) to that of Lota (the southern limit), including the entirety of the townships: Chiguayante, Concepción, Coronel, Hualqui, Lota, Penco, San Pedro de la Paz, Santa Juana, Talcahuano, and Tomé.

The CMA covers a surface of 2830.40 km^2, or 7.63% of the regional surface, and its population constitutes 48.49% of the total population and 57.31% of the urban population of the Biobío Region. The inhabitants reside mainly in urban zones with low slopes (97%) located on the coastal plain. These intermediate cities are located on the shores of the Biobío River and coexist with complex, conditioning geographic elements (beaches, dunes, rocky cliffs, marsh areas, river mouths, wetlands, bays, peninsulas, islands, gulfs, and the coastal mountain range). Given its complex geography and climate, this territory is heavily exposed to recurring natural phenomena (floods) and, environmentally, is very vulnerable to urbanization.

The urban structure (**Figure 1**) is heavily influenced by the downtowns of Concepción and Talcahuano, the urban articulators of this eminently industrial space that has experienced heavy urban growth. It is important to note that this is the second urban concentration in Chile, with a traditional presence of industrial and service activities. Recent socio-spatial studies carried out regarding this territory have revealed certain dynamics that are related to the aforementioned rapid urban growth of the last decade (2000-2010). The CMA constitutes a unit of greater scale and size, characterized by the existence of flows of goods, services, and persons; productive and economic specialization; and an increasingly hierarchical economic

and urban organization of the cities [32]. In 1975, the urbanized surface was 5219.6 ha, concentrated in downtown Concepción and Talcahuano; in 1990, this area grew up to 9012.2 ha, conforming the central conurbation around the transport routes that unite these two downtowns; and in 2001, the urbanized surface reached 12,000 ha, experiencing an increase of 33% [33].

Although earlier information justified the population growth and, therefore, that of the urbanized surface in terms of migration [34], at present, this situation is more complex and also has to do with other spatial factors that may be natural or biophysical in nature (altitude, slope, distance to forests, natural spaces and water bodies) as well as anthropogenic factors such as the tendency for new growth to be concentrated in areas that may be near socioeconomic points of interest (downtown, universities, shopping centers, etc.) and highway networks. These areas are environmentally vulnerable given the geographic restrictions of the emplacement of the most urbanized zone of the CMA [35].

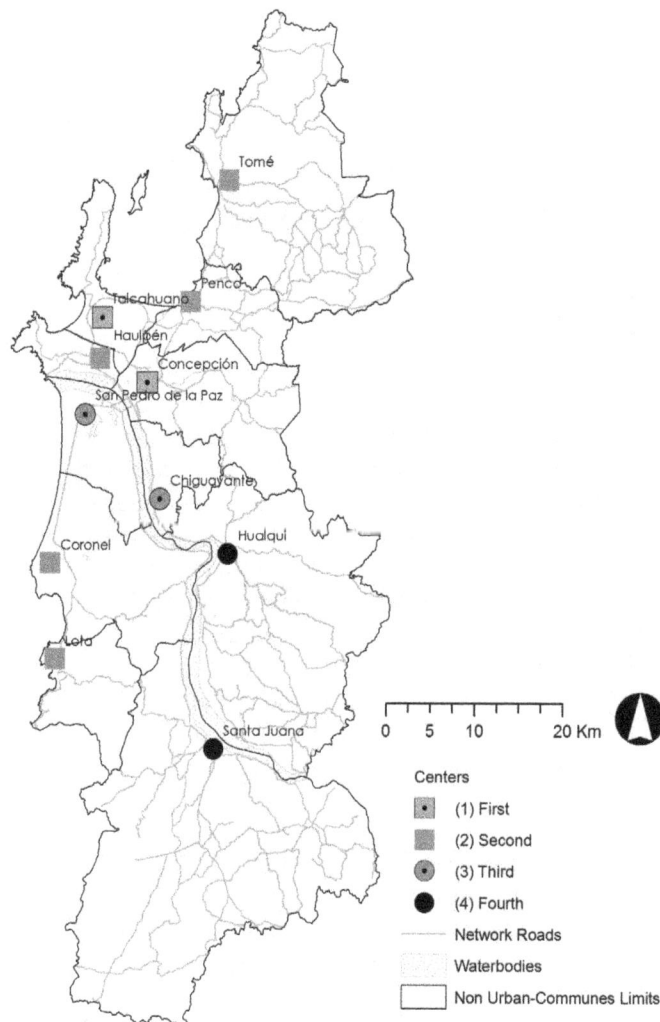

Figure 1. Metropolitan centers in CMA.

Understanding the Urban Sprawl in the Mid-Size Latin American Cities through the Urban Form: Analysis of the Concepción Metropolitan Area (Chile)

7

The organization of recent urban growth attempted to increase the density of certain central spaces of renewal. However, this growth has spilled over into the periphery under a model of low densities, decreasing the compactness of the city. Functionally, the settlements are organized in a restricted polycentric model, with Concepción and Talcahuano acting as main downtowns and the remaining townships acting more as sub-centers with a much lower spatial influence than the two main downtown areas; these sub-centers are also more localized in terms of their surroundings [32]. The morphology of the sub-centers is completely dominated by a central conurbation; that is, an oil stain phenomenon that spreads over the peripheral limits of the historically consolidated spaces. Some of the main factors that have favored this representation include the distance to the urban downtown (taken to be the civic downtown of the city of Concepción) and the distance to highways and the transport infrastructure. These factors directly affect the formation of a tentacular growth structure. Finally, the recent urban growth structures have led to some complex and conflictive consequences. One such outcome is socio-spatial segregation related to the location of social housing, its concentration, and the value of the land on which it is built [36]. Another consequence is the impact on swamplands: over the last three decades, 23% (1.734 ha) of the area occupied by swamplands has been lost due to the expansion of the city, leading to swamplands with different degrees of habitat fragmentation, loss, and alteration [37].

6. Results

6.1. Urban Growth and Urban Spatial Structure

Built-up areas in the CMA grew from 9021 ha in 1990 to 12,007 ha in 2000 and 17,683 ha in 2009 (**Table 1** and **Figure 2**). Built-up areas increased by 28% and 47% in the first (1990-2000) and second (2000-2009) time periods; that is an annual rate of urban growth of about 639 ha for the overall time period. The population, however, grew 12% and 7% for the first and second periods, respectively (**Table 1**). Comparison of growth rates of population and land, much favorable for the second one, strongly validates the convenience of studying the spatial evolution of Concepción under the perspective of the urban sprawl.

The built-up areas increased differently at different places within the CMA. According to the hierarchy of the centers proposed by Rojas, Muñiz and García (2009), the first-order downtowns or central places (Concepción and Talcahuano, including Hualpén) consolidated and concentrated the central urban area of the CMA [32]. Concepción and Talcahuano are clearly the largest

Table 1. Built-up areas and population (1990-2000-2009).

Years	Built-up areas (ha)	Built-up areas (Km²)	Population
1990	9392	9392	821,694
2000	12,047	1204	922,651
2009	17,683	1768	994,781

Source Population: National Statistics Institute (2009) www.ine.cl.

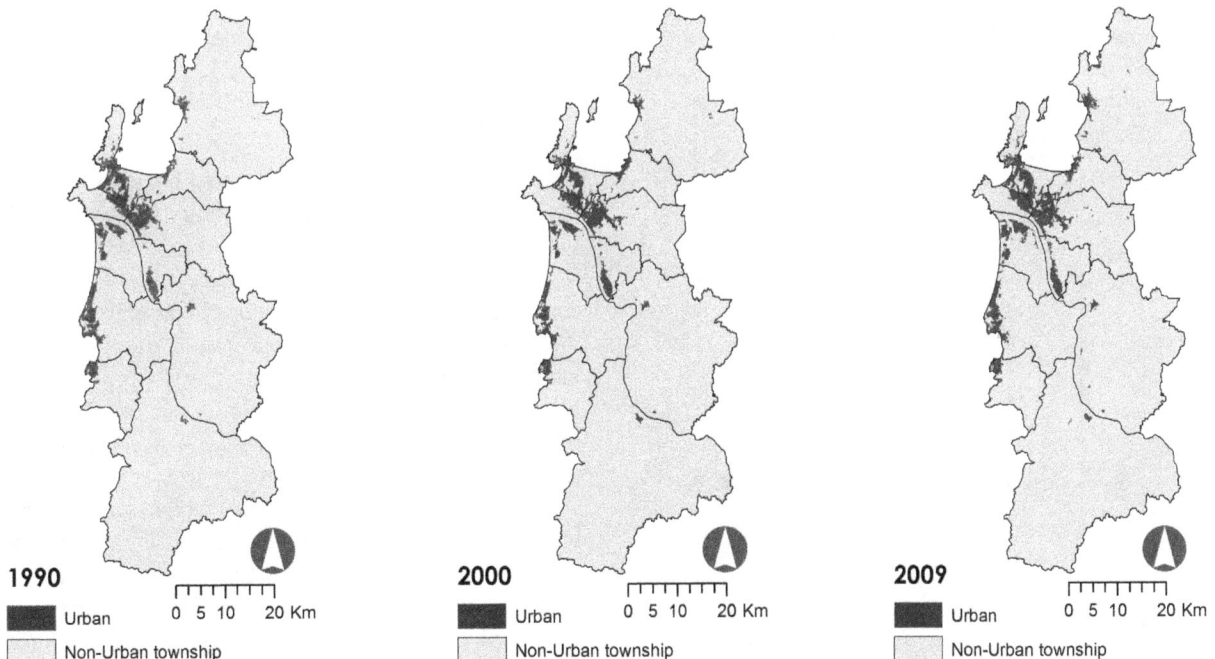

Figure 2. Urban area is equal to built-up areas in the CMA in 1990, 2000, and 2009.

townships, with more than 3000 ha. In 2009, a trend to consolidate was observed in the urban centers of Concepción and Talcahuano, with an increment of built-up areas exceeding 3000 ha.

Third order or bedroom cities (Chiguayante and San Pedro de la Paz) have been added to the central conurbation following an oil stain model. The growth of the emerging center of San Pedro has been impressively fast, and the built-up area now exceeds 2000 ha. This central conurbation concentrated 71% of the total built area in the CMA in 1990, 74% in 2000, and 71% in 2009. Thus, the tendency over time is for the central conurbation to maintain about 70% of the CMA population. However, new urbanizations began to emerge in the townships of Chiguayante, Hualpén, and San Pedro de la Paz since 2000. As a result, each of these townships now accounts for nearly 1000 ha of built-up areas.

In second-order downtowns or integrated cities (Penco, Tomé, Lota, and Coronel), urban occupation has been particularly intense around the transportation network, particularly in Penco and Tomé along the northern coast of the region, with more than 500 ha, below the 1000 ha. Growth in Coronel has been greater than 500 ha, but Lota has shown negligible growth. Growth was also minimal in fourth-order or rural downtowns with large rural areas (Hualqui and Santa Juana), and these built-up areas remained below 200 ha (**Figure 3**).

6.2. Shape Patterns of Urban Sprawl

Changes in these landscape metrics over time reveal a drop in compactness (from 0.65 to 0.56) in the first study period and an increment (from 0.56 to 0.59) in the second period. Conversely, complexity increased (0.009 to 0.016) in the first period and decreased in the second period, although values in 2009 were still higher than those in 1990 (0.016 to 0.012) (**Table 2**).

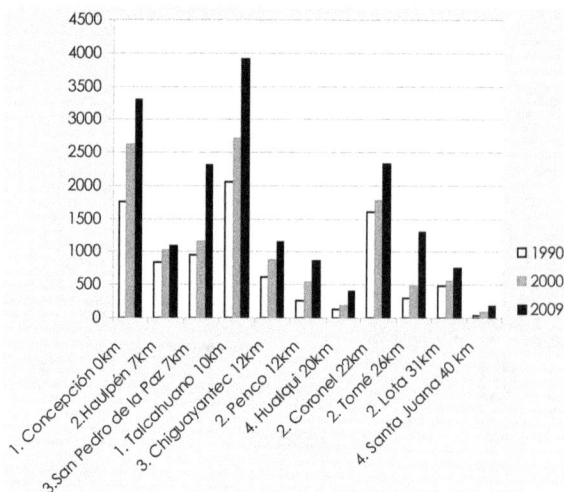

Figure 3. Increase of built-up area (ha) by township.

Table 2. Descriptive statistics of shape index (N = number of patches).

Indices	1990 (N = 118)		2000 (N = 66)		2009 (N = 96)	
	Mean	S.D.	Mean	S.D.	Mean	S.D.
Compactness (CLPI)	0.65	0.17	0.56	0.17	0.59	0.15
Complexity (COI)	0.009	0.0003	0.016	0.0007	0.012	0.004

The results regarding the central conurbation (**Table 2**) are also interesting. In 1990, the CLPI was 0.23 and the COI 0.009; in 2000, the CLPI was 0.16 and the COI 0.01; and finally in 2009, the CLPI was 0.17 and the COI 0.01.These values indicate that the evolution of urban form is more clear in the most urbanized area (central conurbation) and that sprawl tends to less regular in compactness and more irregular in complexity.

The shape of the built-up polygons became less compact from 2000 to 2009 and at the same time that the complexity of these tended to increase (**Table 2**). Although our data do not permit a deeper analysis, the non-parametric Kruskall-Wallis test confirmed these trends in compactness and complexity at $p < 0.05$.

In general, non-significant or weak linear associations were found between the selected landscape metrics and the area and distance to the CMA centroid. The only significant relation took place between compactness and area, although the association was weak. The relation between these variables was negative; *i.e.*, compactness decreased when the size of the built-up area increased (**Figure 4**).

6.3. Urban Sprawl in the Concepción Metropolitan Area

The possible causes of urban sprawl began to appear in the second half of the twentieth century and are related to industrialization and the migratory processes favored by the road for the development of scattered cores at flat and coastal sites. Due to the limitations imposed by the relief, the plains offered the best location, despite environmental restrictions, especially in the Concepción-Talcahuano conurbation. Given the development of the transport infrastructure, this conurbation also took on a star and tentacular shape.

The low-density model, another factor causing sprawl, has been present since the 1960s, that is, at the demographic and industrial peak, but paradoxically with scant investment in roadway infrastructure, which is absolutely necessary for industrial expansion. Although the urban planning of those years intended for the development to be concentrated and decentralized, the plans were not

Understanding the Urban Sprawl in the Mid-Size Latin American Cities through the Urban Form: Analysis of
the Concepción Metropolitan Area (Chile)

9

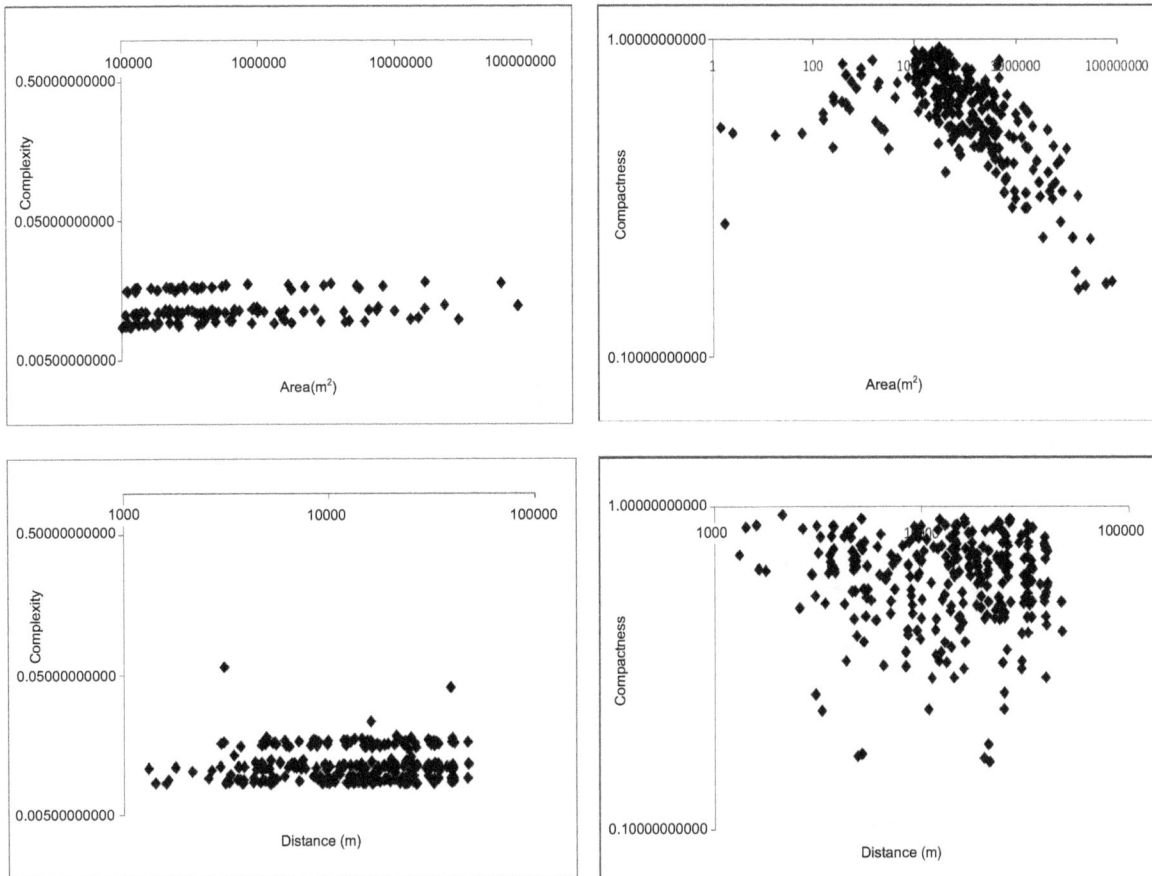

Figure 4. Scatterplot comparing compactness and complexity of built-up areas with respect to their area and distance from Concepción.

able to resolve the poorly distributed concentrations of the economically active population [38].

Urban planning and the current standards were another factor forcing sprawl in the 1970s (National Urban Development Policy). Both strengthened urban expansion, increasing the urban limit. However, a natural growth model, regulated by real estate speculation and industrial expansion, was adopted. In the 1980s, the market influence was maintained with overtones of urban renewal and infrastructure.

Results indicate that even the internal growing processes of the CMA demonstrate that the capacity is reducing as the complexity increases; it is still far from reaching the spatial patterns of major cities [20]. When compared to the Santiago metropolitan area, CMA is more compact and less complex than the Chilean capital city, which presents a CLPI of 0.038 and a CO of 1.44 [20].

The results indicate that in 1990, the area of the ellipse was 2657 km² (265,723 ha), increasing to 3379 km² (337,915 ha) in 2000 (an increment of 722 km²) and to 4330 km² (433,019 ha) in 2009 (an increment of 951 km² or 28%). This indicates that the distance between the

urban surfaces grew considerably, broadening the size of the ellipse. This evidence of the quantitative a result is related to the consolidation of the causes of sprawl in the 1990s, especially in the cities of the conurbation that are classified as spatially dispersed and disarticulated [39].

The angles of orientation indicate the distribution and direction of the urban growth. In this sense, the orientation in 1990 (8.9°) was more centralized, with a more concentrated extension in the conurbation (Talcahuano-Concepción). However, in 2000, urban growth was directed towards San Pedro and Chiguayante, moving westward at 13.7°, a tendency that was maintained in 2009, with an angle of 14.8°. According to the results and the factors forcing sprawl in the CMA, the urban expansion that began to increase in the 1990s found an ideal scenario for intensely disperse and agglomerated growth on the coastal plain (**Figure 5**).

7. Conclusions

The present research contributes to the study of urban sprawl in Latin American cities, specifically in mid-sized metropolitan areas with intense expansion. The methodology allows a thorough examination of the empirical

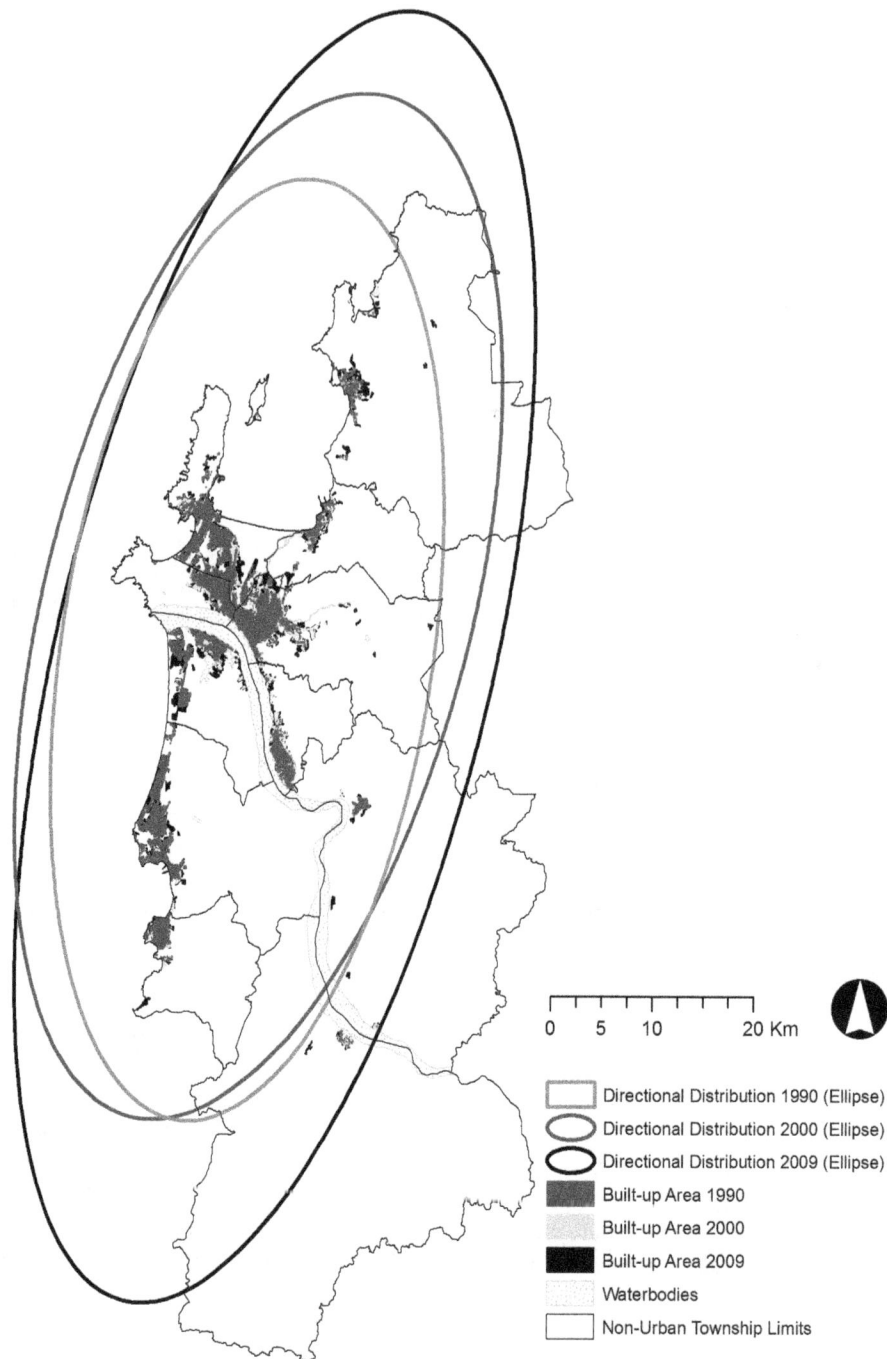

Figure 5. Standard deviational ellipses map.

evidence of urban land use, its compactness and complexity, and the spatial distance between new urbanized surfaces to central area.

Thus, the theoretical context linked to the influence of the neoliberal system and global economic tendencies in the process of urbanization can be related to the spatial dynamic of the CMA over the last 20 years. This process is not new, because it manifested in other medium size cities of the Biobío region such as Los Ángeles, redefin-

ing the urban expansion in more polarized and fragmented spaces [40].

Our results about CMA indicate that the trend for horizontal and peripheral growth has been maintained since the 1990s, coinciding with a strengthening of the neoliberal model in Chile. This horizontal trend, associated with peripheral urbanization and influenced by the land market, has favored disperse, agglomerated, and complex urban expansion. In the case of CMA, the

Understanding the Urban Sprawl in the Mid-Size Latin American Cities through the Urban Form: Analysis of the Concepción Metropolitan Area (Chile)

11

increase in the distances among new peripheral urban areas and the center can be evidences as an element of urban sprawl.

The case of CMA the urban sprawl related with the number of forms is not based on a fragmentation model that is not evident and is not highly influenced by the role of metropolitan centers. In this case, the hypothesis of urban sprawl indicates that areas of greater built-up will be accompanied by greater complexity but declining compactness, especially in the most urbanized and largest areas such as the central conurbations.

Although we do not note an intense fragmentation process, we do see progressive growth in the distance between the new forms, increasing the size of the ellipse by more than 1000 Km^2. The use of confidence ellipses is generally inconvenient given their high dependence on the points farthest from the downtown as this, in general, generates very large areas. However, this "disadvantage" is a boon to the present analysis as it is directly related to the phenomenon to be measured: the sprawl of urbanized forms.

Based on our results, we can conclude that the CMA is evolving into a mixed urban model, with first elements of sprawl occurring mainly on the periphery and near the largest population centers or in the polygon on the central area (conurbation) that concentrates infrastructure facilities and public services. The phase of spatial evolution of the CMA can be categorized as "agglomerated dispersion". In other words, cities that keep their size compact, but being dispersed in the periphery, making their forms more complex and increasing the distance among urbanized areas.

In CMA this process is closely tied to the social and political changes that occurred in Chile in the 1990s. However, these spatial tendencies towards more dispersed forms are far below of the means of cities from developed and developing countries. Latin American capitals are therefore within an international comparison. CMA is presented as a compact city without further complexities, in a context where most Latin American cities can be distinguished by their moderate centrality, density, complexity, centrality and a moderately low level of open space [20].

Eventhough authors consider the proposal by Huang et al. (2007) valuable, but a proposal that simplifies the complexity of Latin American cities; North American and European cities are not comparable with Latin American cities with the same indices or metrics, because Latin American cities take on more complex and concentrated recent forms. Agglomerated dispersion is the outcome of urban development consisting of a continuous, horizontal, extensive, and not evidently fragmented built-up area that loses compactness and gains complexity. The urban form tends to join and regroup in a dispersed

oil stain. Compact areas are small, and they are located near the sea and in rural areas; urban growth is connected through fewer polygons or urban forms. Furthermore, the less compact, more complex forms are located in larger urban areas such as the central conurbation.

The specific spatial growth patterns for the built-up area of the CMA continued growing in the main center. The new sprawl consolidates the idea of a central urban conglomerate expanding along the banks of the Biobío River and presenting concentric or radial growth that is dispersed in the new built-up areas around the main centers in the periphery of the cities (Hualpén, Penco, Hualqui, and Santa Juana). Concepción and Talcahuano, the main centers, show tentacular growth along the highways.

The measures also suggest some ideas about the sustainability of the compact structures found in the urban centers versus scattered suburban growth in the periphery. Scattered urbanizations are located in the periphery, that is, in the areas between Concepción and Talcahuano and in the residential area of San Pedro. One main effect of this is that of being located in spaces exposed to natural risks (landslides and flooding). The disperse growth has been intense and temporally very fast, in complete contradiction with the geographic limitations that determine the territory in zones that are little apt environmentally for human settlements.

The CMA, with its complex built-up area, faces constant urban pressure whose dispersive trend is not the most appropriate given the goal of sustainable urban development, especially when considering the environmental and geographic factors that constrain horizontal expansion: the coastline, rivers, small lakes, and surrounding hills. For this reason, many new urban areas are developing in unsuitable directions, generally increasing exposure to natural risks [41]. Consequently, the recent urban sprawl observed in the CMA does not reflect a pattern of smart growth.

As can be seen, this is a functionally very dynamic and changing area, belonging to a metropolitan space, attractive for residential and renewal purposes. Thus, further studies are required to monitor the changes and to elucidate the factors that explain them. Another important idea is the process of polycentrism that runs parallel to sprawl and should be distinguished from this since these are two alternative processes. On the one hand, the area has grown in a disperse manner towards the peripheries and, on the other, the townships have retained average densities around their main downtown areas. In general, the densities of the periphery have not increased. Future research lines should aim to update the monitoring of urban sprawl and its relationship with polycentrism. It would be appropriate to detect whether the new peripheral growth has given rise to or seen the emergence of density-attracting downtowns.

Finally, the measures that help understanding the first degree or evidences of urban sprawl (GIS techniques, landscape metrics, and statistical analysis) have demonstrated their application and effectiveness in the modeling of dynamic phenomena in a mid-sized metropolitan area of Chile.

8. Acknowledgements

This research was sponsored by the projects FONDE-CYT No. 11090163: "Valoración del Territorio Metropolitano. Aproximaciones desde su Sostenibilidad y Evaluación Ambiental Estratégica" (Evaluation of the metropolitan territory by means of sustainable approaches and strategic environmental assessment). CONICYT/ FONDAP/15110020.

REFERENCES

[1]   J. K. Brueckner, "Urban Sprawl: Lessons from Urban Economics," Brookings-Wharton Papers on Urban Affairs, 2001, pp. 65-97.

[2]   P. W. G. Newman and J. R. Kenworthy, "Cities and Automobile Dependence: A Sourcebook," Gower Technical, Aldershot, 1989.

[3]   I. Muñiz and A. Galindo, "Urban Form and the Ecological Footprint of Commuting. The Case of Barcelona," Ecological Economics, Vol. 55, No. 4, 2005, pp. 499-514.

[4]   M. Munro, "Homo-Economicus in the City: Towards an Urban Socio-Economic Research Agenda," Urban Studies, Vol. 32, No. 10, 1995, pp. 1609-1621.

[5]   P. Wilson, "Building Social Capital: A Learning Agenda for the Twenty-First Century," Urban Studies, Vol. 34, No. 5-6, 1997, pp. 745-760.

[6]   M. C. Derloo and S. Musterd, "Ethnic Clusters in Amsterdam, 1994-1996. A Micro area Analysis," Urban Studies, Vol. 35, No. 3, 1998, pp. 385-396.

[7]   I. Thomas, P. Frankhauser and C. Biernacki, "The Morphology of Built-Up Landscapes in Wallonia (Belgium): A Classication Using Fractal Indices," Landscape and Urban Planning, Vol. 84, No. 2, 2008, pp. 99-115.

[8]   G. Galster, R. Hanson, M. Ratcliffe, M. Wolman, S. Coleman and J. Freihage, "Wrestling Sprawl to the Ground: Defining and Measuring an Elusive Concept," Housing Policy Debate, Vol. 12, No. 4, 2001, pp. 681-717.

[9]   M. Batty, "The Size, Scale, and Shape of Cities," Science, Vol. 319, No. 5864, 2008, pp. 769-771.

[10] J. Barredo and L. Demicheli, "Urban Sustainability in Developing Countries Megacities: Modelling and Predicting Future Urban Growth in Lagos," Cities, Vol. 20, No. 5, 2003, pp. 297-310.

[11] N. B. Grimm, S. H. Faeth, N. E. Golubiewski, C. L. Redman, J. Wu, X. Bai, et al., "Global Change and the Ecology of Cities," Science, Vol. 319, No. 5864, 2008, pp. 756-760.

[12] United Nations, "World Urbanization Prospects," Department of Economic and Social Affairs, 2007. http://esa.un.org/unpd/wup/index.htm.

[13] C. de Mattos, "Santiago de Chile, Globalization and Metropolitan Expansion: What Was There Still Is," Latin American Regional Urban Studies Review, Vol. 25, No. 76, 1999, pp. 29-56.

[14] J. X. Barros, "Urban Growth in Latin American Cities Exploring Urban Dynamics through Agent-Based Simulation," Ph.D. Thesis, University of London, London, 2004.

[15] R. Hidalgo and F. Arenas, "From Urban Country to Metropolitan. Recent Transformation in the Chilean Cities," In: R. Hidalgo, C. De Mattos and F. Arenas, Eds., From Urban country to Metropolitan, GEOlibro, Pontifical Catholic University of Chile, Santiago, 2009, pp. 9-29.

[16] P. Ciccolella, "Recent Transformation in the Latin American Metropolis," In: F. M. Victoria and R. Gurevich, Eds., Geography: New Topics, New Questions. An Agenda for Teaching, Buenos Aires, Biblos, 2007, pp. 17-38.

[17] P. Mieszkowski and E. Mills, "The Causes of Metropolitan Suburbanization," Journal of Economic Perspectives, Vol. 7, No. 3, 1993, pp. 135-147.

[18] A. G. Champion, "Population Change and Migration in Britain Since 1981: Evidence for Continuing Deconcentration," Environment and Planning A, Vol. 26, No. 10, 1994, pp. 1501-1520.

[19] P. C. Cheshire and D. G. Hay, "Urban Problems in Western Europe: An Economic Analysis," Unwin Hyman, London, 1989.

[20] J. Huang, X. Lu and J. M. Sellers, "A Global Comparative Analysis of Urban Form: Applying Spatial Metrics and Remote Sensing," Landscape and Urban Planning, Vol. 82, No. 4, 2007, pp. 184-197.

[21] C. de Mattos, "Transformación de las Ciudadess Latinoamericanas Impactos de la Globalización?" Estudios latinoamericanos Urbanos Regionales (EURE), Vol. 28, No. 85, 2002, pp. 5-10.

[22] M. K. Jat, P. K. Garg and D. Khare, "Monitoring and Modeling of Urban Sprawl Using Remote Sensing and GIS Techniques," International Journal of Applied Earth Observation and Geoinformation, Vol. 10, No. 1, 2008, pp. 26-43.

[23] W. Ji, M. Jia, R. Wahad and K. Underhill, "Characterizing Urban Sprawl Using Multi-Stage Remote Sensing Images and Landscape Metrics," Computers. Environment and Urban Systems, Vol. 30, No. 6, 2006, pp. 861-879.

[24] H. S. Chae, S. J. Kim and J. A. Ryu, "A Classification of

Understanding the Urban Sprawl in the Mid-Size Latin American Cities through the Urban Form: Analysis of
the Concepción Metropolitan Area (Chile)

13

Multitemporal Landsat TM Data Using Principal Component Analysis and Artificial Neural Network," *Geosciences and Remote Sensing*, Vol. 1, No. 1, 1997, pp. 517-520.

[25] X. Li and A. Yeh, "Principal Component Analysis of Stacked Multi-Temporal Images for the Monitoring of Rapid Urban Expansion in the Pearl River Delta," *International Journal Remote Sensing*, Vol. 19, No. 8, 1998, pp. 1501-1518.

[26] E. Irwin and N. Bockstael, "The Evolution of Urban Sprawl: Evidence of Spatial Heterogeneity and Increasing Land Fragmentation," *PNAS*, Vol. 104, No. 52, 2007, pp. 20672-20677.

[27] G. Shen, "Fractal Dimension and Fractal Growth of Urbanized Areas," *International Journal of Geographical Science*, Vol. 16, No. 5, 2002, pp. 419-437.

[28] M. Zhu, N. Jiang, J. Li, J. Xu and Y. Fan, "The Effects of Sensor Spatial Resolution and Changing Grain Size on Fragmentation Indices in Urban Landscape," *International Journal of Remote Sensin,g* Vol. 27, No. 21, 2006, pp. 4791-4805.

[29] S. Mubareka, E. Koomen, C. Estreguil and C. Lavalle, "Development of a Composite Index of Urban Compactness for Land Use Modelling Applications," *Landscape and Urban Planning*, Vol. 103, No. 3-4, 2011, pp. 303-317.

[30] M. Mardones and V. Vidal, "Zonning and Assesment of Natural Geomorphological Risk. A Instrument of Urban Planning in the City of Concepción," *Latin American Urban Planning Review*, Vol. 27, No. 81, 2004, pp. 97-122.

[31] SEREMI—MINVU (Secretaría Regional Ministerial de Vivienda y Urbanismo Región del Bío-Bío), Report of Urban Metropolitan Planning of Concepción, Concepción, 2003.

[32] C. Rojas, I. Muñiz and M. A. García-López, "Estructura urbana y Policentrismo en el Área Metropolitana de Concepción," *Latin American Regional Urban Studies*, Vol. 35, No. 105, 2009, pp. 47-70.

[33] C. Rojas, S. Opazo and E. Jaque, "Growth Dynamics and Patterns in the Concepción Metropolitan Area. Last Decades Trends," In: R. Hidalgo, C. De Mattos and F. Arenas, Eds., *From Urban Country to the Metropolitan Country*, GEOlibros—Instituto de Geografía Pontificia Universidad Católica de Santiago, 2011, pp. 257-268.

[34] L. Pérez and E. Salinas, "Urban Growth and Globalization: Transformations of the Metropolitan Area of Concepción, Chile, 1992-2002, Scripta Nov," *Geography and Social Science Electronic Review*, Vol. XI, No. 51, 2007. http://www.ub.edu/geocrit/sn/sn-251.htm

[35] C. Rojas and W. Plata, "Metropolitan Area of Concepción: Driving Factors of Urban Growth (2001-2009) by a Spatial Logistic Regression Model," *International Congress of Territorial Planning and Geographical Information Technology*, Tegucigalpa, Honduras, 2010.

[36] F. Machiavello and R. Hidalgo, "Socio-Spatial Consequenses of the Social Housing in the Metropolitan Area of Concepción: Integrated Neighborhoods or Urban Spaces?" In: L. Pérez and R. Hidalgo, Eds., *Metropolitan Area of Concepción (MAC), Evolution and Challenges*, Geolibros, Institute of Geography Pontifical Catholic, University of Chile, Santiago, 2010, pp. 153-170.

[37] A. Pauchard, M. Aguayo, E. Peña and R. Urrutia, "Multiple Effects of Urbanization on the Biodiversity of Developing Countries: The Case of a Fast-Growing Metropolitan Area (Concepción, Chile)," *Biological Conservation*, Vol. 127, No. 3, 2006, pp. 272-281.

[38] L. Perez and P. Fuentes, "The Intercommunal Urban Plan of Concepción in 1963. Beginning of the Metropolitan Planning," In: L. Pérez and R. Hidalgo, Eds., *Metropolitan Area of Concepción (MAC), Evolution and Challenges*, Geolibros, Institute of Geography Pontifical Catholic, University of Chile, Santiago, 2011, pp. 45-58.

[39] I. Gysling and A. Hoffmann. "Intercommunal and Communal Planning of the Metropolitan Area of Concepción," In: L. Pérez and R. Hidalgo, Eds., *Metropolitan Area of Concepción (MAC), Evolution and Challenges*, Geolibros, Institute of Geography Pontifical Catholic, University of Chile, Santiago, 2010, pp. 59-82.

[40] G. Azócar, H. Romero, R. Sanhueza, C. Vega, M. Aguayo and M. Muñoz, "Urbanization Patterns and Their Impacts on Social Restructuring of Urban Space in Chilean Mid-Cities: The Case of Los Angeles, Central Chile," *Land Use Policy*, Vol. 24, No. 1, 2007, pp. 199-211.

[41] C. Rojas, J. Pino and E. Jaque, "Strategic Environmental Assessment in Latin America: A Methodological Proposal for Urban Planning in the Metropolitan Area of Concepción (Chile)," *Land Use Policy*, Vol. 30, No. 1, pp. 519-527.

2

A Proposal for a Geospatial Database to Support Emergency Management

Ivan Frigerio[*]**, Stefano Roverato, Mattia De Amicis**

Department of Earth and Environmental Sciences, University of Milano Bicocca, Milan, Italy

ABSTRACT

The basic procedure of the Italian Civil Protection Department aims at reducing disaster losses by giving prominence to a proactive strategy, focusing on prevision and prevention of hazard events rather than postdisaster activities. Italian law commits municipalities to produce Emergency Plans that include risk scenarios as well as all data required for emergency management, such as structures, infrastructures and human resources. However the law in the matter of Civil Protection does not supply information about how to produce and archive necessary data for emergency planning and management. For this reason, we propose a standard methodology to create a geodatabase using GIS software, to collect all data that could be used by municipalities to create Emergency Plans. The resulting geodatabase provides a tool for hazard mitigation planning, allowing not only the identification of areas at risk, but also the structures, infrastructures and resources needed to overcome a crisis, thus improving all strategies of risk reduction and the resilience of the system [1].

Keywords: Emergency Management; Geospatial Database; Civil Protection; Emergency Planning

1. Introduction

Italy is considered a vulnerable country due to the high occurrence of hazardous events and its peculiar geological, geomorphological and climatic conditions. Excessive human activities, illegal constructions and poor environmental maintenance are still increasing natural hazards such as earthquakes, landslides and floods in residential areas. In addition, the presence of holdings that use and produce dangerous substances in industrialized areas, exposes both population and the environment to industrial risk. An accurate knowledge of the incidence of these phenomena (both natural and anthropogenic) is the key to reduce risk, increase resilience and minimize anthropic and environmental damages [2].

The National Civil Protection Department has already developed prevision and prevention plans as well an instruments to identify priority actions in case of necessity: emergency operations linked to hazard events, vulnerable areas and financial resources availability. The role of the Department is to develop Emergency Plans for "expected" events, which may require the intervention of the central organs of the government, whereas the regions

are responsible for identifying guidelines of Municipal Emergency Plans. This includes mapping the risks in the considered area and identifying the available structures, infrastructures and resources for effective emergency management [3].

As so far, the only document available in Italian country is the "Manuale Operativo" (an Emergency Operations Manual) [4], which represents guidelines to provide Emergency Plans. However it does not specify which mapping elements to consider and how these should be represented in cartography. Consequently, each region adopts its own methodology and it is a lack of a valid standard nationally model.

However, currently, it is quite difficult to create a valid standard model, as all data needed are produced at different institutional levels (national, regional, provincial, municipal) and different scales. On one hand there are policies that provide guidelines to produce data to define maps for environmental planning and in the same time, there are specific directives in order to define the Civil Protection System specifying the methods and the organizations involved in emergency management.

For all the reasons mentioned is strictly necessary to provide national standard, valid over the whole Italian

---

[*]Corresponding author.

territory, in order to obtain precise rules that ensure uniformity of the data to be mapped [5].

2. Guidelines and Policy of Italian Civil Protection

The main Italian Civil Protection law (225/92) [6] classify the Civil Protection's tasks in four types: 1) prevision, 2) prevention, 3) rescue and 4) emergency overcome. Among these actions, this law focuses on the development of prevision and prevention activities. This approach is a new cultural orientation based on a systematic and widespread risk analysis in order to reduce the consequences of hazardous events in anthropic areas [7]. In this context, it is necessary to guarantee continuous data exchange between both emergency management and urban planning.

Within the prevention task, the determination of Emergency Plan is the main activity to complete. That is considered as "a set of operating procedures to deal with any expected disaster in a given territory". The aim of Emergency Plan is define event scenarios and elaborate a database to efficiently support emergencies. Thus, the first step in preparing an Emergency Plan is the collection and analysis of spatial data and their mapping at different scales. In this way it is possible to allow not only an overview of the area but also provide a detailed vision on a possible impact of hazardous events on vulnerabilities [8]. So, it is initially necessary to identify and map structures and infrastructures within every municipality, such as roads and strategic buildings (schools, hospital, etc.). Successively, strategic building, parking places and other public constructions have to be classified among three types of emergency areas: 1) rescue and resource areas, 2) population waiting areas and 3) recaption gathering area of the population; even in this case there are thematic symbols for their cartographic representation. For each class, events scenarios must be identified: in particular, possible areas of impact on vulnerable elements must be derived with a detailed scale of at least 1:10,000 [9].

In Emergency Plans is also necessary consider, for each risk scenarios identified (hydraulic and hydrogeological, seismic, volcanic, industrial and bushfire), the presence of human resources, material, equipment, as well as operational and decisional capabilities, in order to react very quickly to minimize the damage caused by the occurrence of an hazardous event [10]. A support database for emergency management must therefore contain not only maps, but also all available resources within the area and responsible people for the intervention operations. The expected risk scenarios must then be linked to a precise model of intervention functional to the considered risk, assigning responsibilities for decision-making at different levels, using all resources rationaly and de-

fining a communication system that allows a constant information exchange [11]. The plan must be flexible enough to be used in all emergency situations (either expected or unexpected), and simple, in order to be rapidly operative. It also needs to be an easily upgradable document, as it must consider environmental and urban planning changes, as well as the necessity of modifying the extent of an expected risk scenario [12]. Maps are also very important in emergency planning: in fact, the cartography allows a faster and more intuitive knowledge of territory, as well as a better management, both during planning and in the operative phase [13]. In recent years, also the normative recognised the importance digital mapping tools (GIS) for this analysis, in support to traditional paper maps. Even if, as said before, the Plan realized using GIS software is a useful and dynamic tool [14], producing documents that are easily upgradeable [15]. Currently, the use of these technologies is not required by the law, but it is a discretion of individual municipalities.

3. Regional Contest

As evidenced by the 225/92 law, each region has developed its own guidelines to assist their municipalities in the preparation of Emergency Plans. In order to understand the state of the art of each region guidelines, a literature search in the archives of 19 Italian regions and Autonomous Provinces of Trento and Bolzano has been carried out. In particular, all local regulations have been considered in order to propose a standard methodology that does not come in conflict with any other existing regional law and that could be easily replicable throughout Italy.

The results show that more than 50% of the Italian regions are not prepared enough Emergency Planning, thus they leave to municipalities and provinces the freedom to refer only to national guidelines (**Table 1**).

As evidenced in **Table 1**, only two Italian regions elaborated in detail Civil Protection laws: Lombardia and Emilia Romagna. In particular, Emilia Romagna Region provided that every municipality and province should organize a own database to support emergency management using Geographic Information Systems. In fact, the main objective of this legislation is creation of digital standardized and geo-referenced database, which all municipalities must take into account when preparing Emergency Plans. The law also evidences the scale at which data should be represented and the topographic maps to be used. It also specifies to use: 1) a point theme to represent the same items shown in the "Manuale Operativo" (Emergency Area and Coordination Centres), 2) a linear theme to represent administrative boundaries, transport infrastructure and technological networks and 3) a polygon theme to represent the event scenarios.

Table 1. List of provisions of low about emergency planning for each Italian region.

Region	Regulation
Abruzzo	National
Basilicata	National
Calabria	D.g.r. 472/2007
Campania	National
Emilia Romagna	**D.g.r. 1166/2004**
Friuli Venezia Giulia	National
Lazio	National
Liguria	D.g.r. 746/2007
Lombardia	**L.r. 16/2004**
Marche	National
Molise	National
Piemonte	D.g.r. 42/2004
Puglia	National
Sardegna	National
Sicilia	D.g.r. 2/2011
Toscana	D.g.r. 26/2000
Umbria	National
Valle d'Aosta	National
Veneto	National
Prov. Trento	L.p. 9/2011
Prov. Bolzano	Guidelines 2009

Table 2. Example of the category risk areas that represent the event scenarios. The field "type of risk" describes the typology of risk that must be mapped within the municipality. Instead, the field "type of hazardous event" contains a description of the type of risk considered. In addition, each item is identified by a unique identification code.

Type of risk		Type of hazardous event	
Code	Description	Code	Description
0	Other	−1	Value to be assigned as the default
1	Hydrogeological risk	0	Other
		1	Landslide surface
		2	Topple
		3	Rockfall
		4	Debris flow
		5	Bank erosion of hydrographic network
		6	Flooding of the minor rivers
		7	Flooding of the major rivers (PAI zones)
		8	Flooding of the lakes
		9	Avalanche
2	Seismic risk	0	Other
		1	Earthquake
3	Bushfire risk	0	Other
		1	Bushfire on forested area
		2	Bushfire on urbanized area
		3	Bushfire on infrastructure
4	Industrial risk	0	Other
		1	Productive plant
		2	Burst-productive plant explosion
		3	Gaseous emissions into the atmosphere
		4	Dispersion of toxic or harmful liquid
		5	Emission of radioactive, toxic or harmful materials
		6	Incident to transport dangerous substances
5	Environmental risk	0	Other
		1	Tornado
		2	Hailstorm
		3	Water crisis

On the other hand, the guidelines issued by the Lombardy Region through the DGR n.8/4732 ("Direttiva Regionale per la pianificazione di emergenza degli enti locali" of 16 may 2007) [16], discipline how to draw up Emergency Plans, but it does not constrain to use informatics tools. In particular, it indicates how to map these elements, from spatial data (such as structures and infrastructures) to event scenarios and expects the use of GIS software to generate the Emergency Plan [17]. In this perspective, the Lombardy Region developed a standard web oriented methodology, called PEWEB, which aims to create a regional spatial database in order to share data of every municipality and therefore to effectively manage emergency planning. The PEWEB system requires spatial data in shapefile format to be loaded instead tabular data has to send in XLM format. Data are divided into 5 categories: 1) risk areas, 2) strategic structures, 3) strategic areas, 4) point of accessibility and 5) road network infrastructure. Each category is composed by a unique geometry and are identified by specific codes. **Table 2** show the example of the category "risk areas". All categories have point geometry, with the exception of the first one ("risk areas"), that is a polygonal one.

The category "strategic structure" is referred to buildings or built-up areas, while "strategic area" identifies open areas that can be used as logistic bases for the rescuers, resources and materials, or as zones able to receive a great number of people in case of emergency. The category "point of accessibility" identifies structures finalized to the movement of vehicles, materials and people such as railway stations and airports. Finally, the category "road network infrastructure" is referred to significant infrastructures for the viability such as bridges, viaducts and overpasses [18].

4. Proposal for a Geo-Referenced Spatial Database to Support Emergency Management

The classification adopted by Lombardy Region through the PEWEB system is the most complete and systematic among those analyzed, but it is not entirely exhaustive because it does not include all elements the National Operating Manual recommends for Emergency Plans mapping. However, it was decided to use this system, with the required changes and additions, as the basis for the elaboration of a new classifying proposal of the geo-referenced spatial database. In fact, the PEWEB system has a well good organized and items classification clearly reflect guidelines. For this purpose, it was necessary to define a new taxonomy at national level, to be used as the basis for the creation of the territorial geo-referenced database that should consider the elements to be mapped in a logical and functional way. This item was grouped into eight categories with a specific geometry (**Table 3**):

Each category also contains several classes with different objects, refered to the final elements to be mapped. Among the objects of each class it was included the item "Other" to denote generic elements that can not be classified by the current items. For example, in **Table 4**, the

Table 3. The eight category of the proposed taxonomy.

Category	Type of geometry
Area at risk	Polygon
Strategical surface	
Generic facility	
Operative strategic structure	
Non operative strategic structure	Point
Road network infrastructure	
Point of access	
Technological networks and infrastructure	Line

Table 4. Classification proposal: category operative strategic structure. The classes that bring together the different elements to be mapped are identified by a unique code: 11: Institutional head office, 12: Head of operating structure, 13: Head of the emergency management center, 14: Emergency facility.

Class		Objects	
Code	Description	Code	Description
11	Institutional head office	0	Other
		1	Municipality
		2	Prefecture
		3	Province
		4	Region
		5	Consortium Park Authority
		6	Mountain Communities
12	Head of operating structure	0	Other
		1	Fire Department
		2	SSUEM-118
		3	Red Cross
		4	Military
		5	Police District Department
		6	Voluntary of Civil Protection
		7	Emergency Multipurpose Centers
		8	Municipal Warehouses
		9	State Forest Management
		10	Carabinieri District Department
		11	Police State
		12	Alpine and Speleological Rescue
13	Head of the emergency management center	0	Other
		1	Rescue Coordination Center
		2	COM
		3	COC
		4	UCL
14	Emergency facility	0	Other
		1	Reception Center and Shelter
		2	Suitable Structure for Operational Centers
		3	Health Care Facility

category "Operative Strategic Structure", identified by a punctual geometry, includes four classes that represent the basic structures to be activated in case of emergency (**Figure 1**). In total, in all eight categories studied, 138 items (grouped in 38 classes) were identified, with the exclusion of items classified as "Other".

Once processed this new taxonomy, the next step was to create a digital geo-referenced database. It was decided to organize the data using a relational database, in which tables without duplicate rows represent all data: therefore, a suitable key uniquely identifies a row.

In this structure, the user is also able to query the system with complex requests, crossing data from different tables. It is however necessary to include a common key field ("key") in all tables allowing the combining of information from one table to another. Therefore, it is possible to consider the same object at the same time from different points of view (e.g. chronological, regulatory, typology, etc.); instead, in hierarchical database only one type of interdependence between objects is possible [19]. It was also created the domain, an outstanding part of the geodatabase, which contains all possible values assigned to a field. The domain has been applied to the field "Code" that specifies the type of identified object. With this structure, the user will be constrained to choose

among different elements of a list: their priority is specified in the guidelines of the Operations Manual. For example, in the class "Emergency facility", the user can choose among reception centers and shelters, suitable structures for operational centers or health care facilities. It is therefore guaranteed the representation of the classes identified, avoiding the possibility of mapping structures at will, leaving so the proposed standard of the Operations Manual.

It should be remembered, however, that for each class there is always the generic field "Other" that allows the classification of not provided items.

In this project, the geodatabase has been completed using ESRI ArcGIS 10.1, but it is also possible to use other database format such as PostgreSQL with PostGIS extension for geospatial data. Afterwards, the geodatabase has been divided into feature dataset, each containing the feature classes identified in the proposed taxonomy. For each feature class, descriptive fields (the same provided in the proposed taxonomy) were then assigned. These fields will be the same that appear in the representation of the elements in the cartography. It was used "short integer" format for integer numbers, "float" for decimal numbers and "text" for alphanumeric strings (**Table 5**). The field "text" was also used for the fields

Figure 1. Example of strategic operative structure in Castellanza town (in the province of Varese, Italy).

Table 5. Fields about feature class of "emergency facility".

Field name	Data type
Shape	Geometry
Code	Short integer
Address	Text
Municipality Code	Text
Number of floors	Short integer
Surface area	Float
Number of sleeping accomodation	Short integer
Type of use	Text
Vulnerable	Text

Table 6. Classes of the structure, type and numbers of elements present in Castellanza town.

Structure	Number and type of objects in Castellanza town
Crowd aggregation center	3 trade center and 3 sport centre
School	1 university, 3 nursery school and 8 schools
Industrial plant	215 production
Industrial plant at risk	4 plant of the following industries: Agrolinz, Melamin, Cisalpina and Perstrorp
Place of worship	3 churches
Health facility	2 hospital, 4 pharmacies and 1 private hospital
Accommodation	2 hotels
Road network infrastructure	4 viaducts and 13 crossing roadway
Point of access	2 rail station and 1 helicopter landing pad
Power plant	84 high pylons and 52 electrical box (other)
Institutional head office	Municipality
Emergency facility	2 health care facility and 18 rescue and resource structure
Head of operating structure	Carabinieri
Head of the emergency management center	3 COM
Framework for telecommunications	9 antenna tower for mobile
Materials warehouse	14 gas station
Structure of bushfire interest	29 point of water supply, 199 hydrants and 6 well

true/false, associating it to a domain in which two of the values "0—No", "1—Yes" were included, for example the field "Vulnerable". In addition, to improve the interoperability of the data, coordinates in the geographic reference system WGS84 were assigned to all feature classes. In the future, the use of a geo-database to compile Civil Protection Plans will be easier and more convenient in order to organize all information needed. The adoption of a standard method will also allow faster information exchange between various administrations, considering not only different Regions, but entire Italy: this would lead to planning simplification and allow to easily manage emergencies.

5. Test the Geo-Database on a Real Case

The geo-database has been tested on a real case, using data of Castellanza town (province of Varese), in order to assess the effectiveness of the developed database.

First of all, the spatial data in shapefile format were prepared for import, to be compatible with the new adopted taxonomy. In particular, a numeric field called "Code" was added to each attribute table of the shapefile: this has the same values used in the taxonomy table, part of which is shown in **Table 6**. It was then possible to import the records of the shapefile representing Castellanza territorial data in the feature classes, organized according to considered fields. This operation has been performed for all classes of structures identified and the results were loaded into the geodatabase, as shown in **Table 6**. During this operation, emerged a critical situation regarding "Strategical surface" feature class: in fact, the geodatabase, which was structured to use polygonal geometries, conflicted with Castellanza available punctual data. This prevented data loading. This problem emphasizes the need of a standard methodology that could bring the users to collect and store all data in polygon shapefile format.

With relation to the risks, this municipality is subject to three different types of events: 1) flooding risk (of the river Olona), 2) buschfire risk and 3) industrial risk (**Figures 2** and **3**). These data have been loaded in the geodatabase using the same procedure described for the structures. The linear shapefile, referred to both "infrastructure networks" and "technological networks". This refers to provincial and state roads, railways, mains supply, electricity and sewer distribution over the Castellanza territory were lastly loaded.

6. Conclusion

The analysis described in this paper was conducted to respond to the absence of Italian standard methodology in order to identify, analyze and archive data to be used for the development of local Emergency Plans. The results of the first part of this work emphasize the frag-

Figure 2. Hydrogeological and industrial risk scenarios.

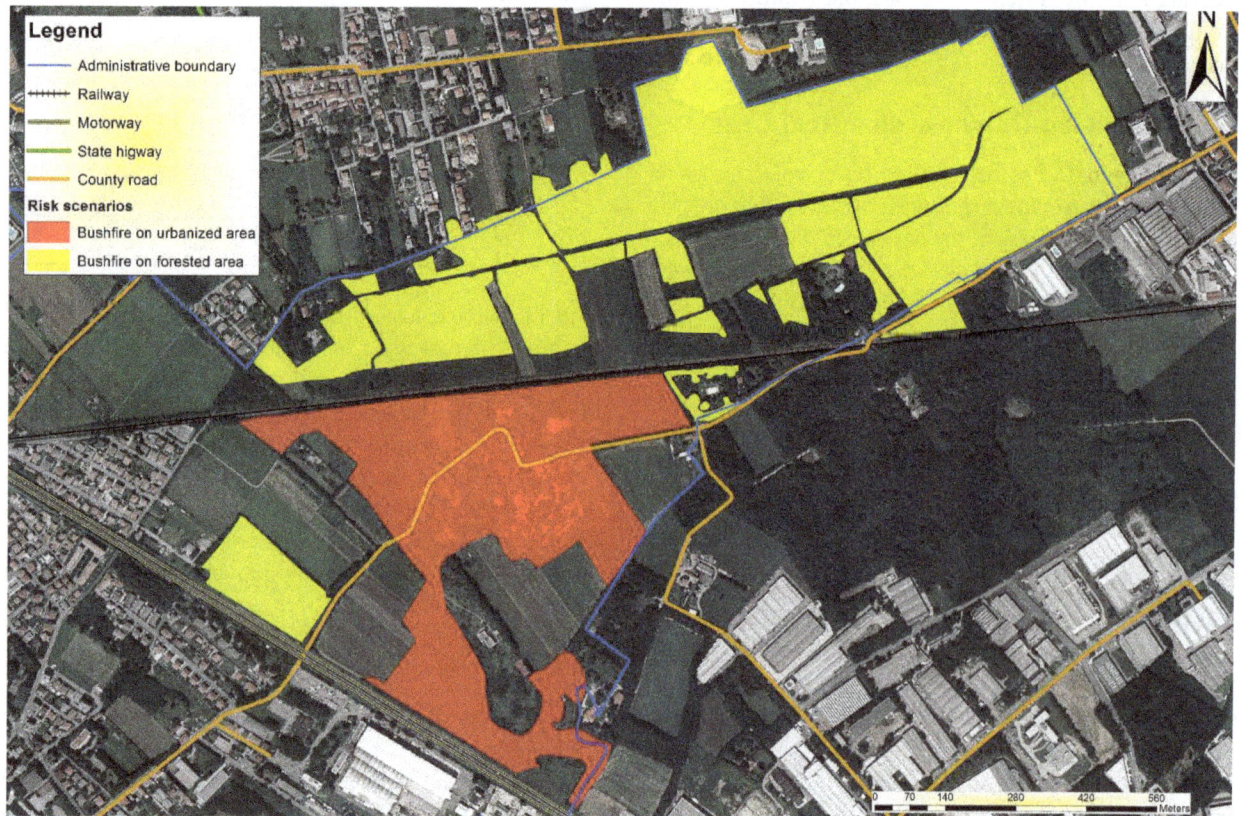

Figure 3. Bushfire risk scenarios.

mented framework of Italy. The Regions that show interest in this issue are few: only Lombardy and Emilia Romagna Regions, in fact, implemented solutions to overcome this problem by adopting a specific system, as the PEWEB system for Lombardy. This problem is due to the absence of a national law that imposes to municipalities which data use for Emergency Plans and where to find them. The result is that, actual needed data are mainly produced either by urban planning or by specific directives linked to the definition of the Civil Protection System which refers to emergency planning, and these two systems do not interact. The presented analysis attempts to close this gap, proposing a standard methodology to uniform the input data so as to increase their interoperability. The GIS techniques, implemented to create geodatabase, are not overly complex and required input data is easily available to the most public administrations. The development of these procedures made a breakthrough towards the unification of the systems of emergency management of Civil Protection on a national scale. The described procedure revealed to be highly flexible and simple, two features those are at the base of emergency planning. The system was tested only in one municipality (Castellanza), from which some critical points emerged, even if the global evaluation of the proposed methodology is highly positive. This test underscores the necessity, in the near future, to test this geodatabase on a great number of municipalities in order to highlight other gaps or problems, to develop a precise and efficient methodology, improving therefore emergency planning and management throughout the whole Italian country.

REFERENCES

[1] S. L. Cutter, "GI Science, Disaster and Emergency Management," *Transactions in GIS*, Vol. 7, No. 4, 2003, pp. 439-446.

[2] J. Birkmann, "Measuring Vulnerability to Natural Hazards: Towards Disaster Resilient Societies," United Nations University Press, New York, 2006.

[3] S. L. Cutter, C. T. Emrich, B. J. Adams, C. K. Huyck and R. T. Eguchi, "New Information Technologies in Emergency Management," In: W. I. Waugh and K. Tierney, Eds., *Emergency Management: Priciples and Practice for Local Governement*, International City/Country Management Association (ICMA) Press, 2007.

[4] Dipartimento di Protezione Civile, "Manuale Operativo per la Predisposizione di un Piano Comunale o Intercomunale di Protezione Civile," 2007.

[5] D. Alexander, "Towards the Development of a Standard in Emergency Planning," *Disaster Prevention and Management*, Vol. 14, No. 2, 2005, pp. 158-175.

[6] Law 225/92, "Istituzione del Servizio Nazionale di Protezione Civile".

[7] S. L. Cutter, E. Tate and M. Berry, "Integrated Multiazard Mapping," *Environmental and Planning B: Planning and Design*, Vol. 37, No. 4, 2010, pp. 646-663.

[8] J. Blauth, I. Poretti, M. De Amicis and S. Sterlacchini, "Database of Geo-Hydrological Disaster for Civil Protection Purposes," *Natural Hazard*, Vol. 60, No. 3, 2012, pp. 1065-1083.

[9] T. J. Cova, "GIS in emergency management," 1999.

[10] J. R. Jensen and M. E. Hodgson, "Remote Sensing of Natural and Man-Made Hazards and Disasters," In: M. K. Ridd and J. D. Hipple, Eds., *Manual of Remote Sensing: Remote Sensing of Human Settlements*, American Society for Photogrammetry and Remote Sensing, 2006, pp. 401-429.

[11] A. Cavallin, S. Sterlacchini, I. Frigerio and S. Frigerio, "GIS Techniques and Decision Support System to Reduce Landslide Risk: The Case Study of Corvara in Badia, Northern Italy," *Geografia Fisica e Dinamica Quaternaria*, Vol. 34, 2011, pp. 81-88.

[12] D. Alexander, "Principles of Emergency Planning and Management," Terra Publishing, 2002.

[13] P. A. Longley, M. F. Goodchild, D. J. Maguire and D. W. Rhind, "Geographic Information System and Science," John Wiley and Sons, Hoboken, 2005.

[14] National Research Council, "Successful Response Starts with a Map: Improving Geospatial Support for Disaster Management," National Academies Press, Washington DC, 2007.

[15] M. Konecny, S. Zlatanova and T. L. Bandrova, "Geographic Information and Cartography for Risk and Crisis Management," Springer, Berlin, 2010.

[16] Regione Lombardia, "Direttiva Regionale per la Pianificazione dell'Emergenza Degli Enti Locali," Bollettino Ufficiale della Regione Lombardia, Milano, 2007.

[17] Dipartimento di Protezione Civile—Presidenza del Consiglio dei Ministri, "Metodo Augustus," DPC Informa, No. 12, 1997.

[18] Regione Lombardia, "GEODB Prevenzione e Sicurezza, Sintesi Progettazione Esterna e Specifiche Funzionali. Architettura Informativa. Mosaico Piani di Emergenza," Bollettino Ufficiale della Regione Lombardia, Milano, 2011.

[19] S. Sumathi and S. Esakkirajan, "Fundamentals of Relational Database Management System," *Springer Science and Buisness Media*, Vol. 47, 2007.

3

Using a GIS to Assessment the Load-Carrying Capacity of Soil Case of Berhoum Area, Hodna Basin, (Eastern Algeria)

Amar Guettouche, Farid Kaoua

Faculty of Civil Engineering, University of Sciences and Technology Houari Boumediene, Algiers, Algeria

ABSTRACT

The concept of load-carrying capacity of the soil can be evaluated by two main components: permissible stress and permissible depth; and therefore, running it begins its assessment that allows an outline of exploitation. Nevertheless, the assessment of the load-carrying capacity made the object of several works of research and many models, based on the multi-criteria analysis, have been established. This work examines the contribution of GIS approach to assessment load-carrying capacity of the soil. This one has been finished in two practicums: 1) Assessment of the capacity of soil by a multi-criteria approach, using the Weighted Sum Model (WSM); 2) It brought to use the GIS approach to evaluate and spatialize degree of soil bearing stresses resulting from the buildings, as well as load distribution. The method has been applied to the Berhoum area of Hodna Basin, in eastern Algeria, where each is characterized by its various natural properties and density of equipment. Final results are better in the classification of the degree of load-carrying capacity possible in each site. This results in allowing exploiters to program their optimal designs for the rational management of the area.

Keywords: Load-Carrying Capacity; WSM; GIS; Berhoum; Hodna Basin's; Algeria

1. Introduction

Some Algerian middle areas are witnessing difficulties in construction caused by problems related to soil, especially the instability of foundations. This is due to the weakness of knowledge of soil characteristics like: its ability to bear the load of buildings and geological composition. Conversely, these areas are characterized by population density and speed of urban growth.

The danger is caused by these problems which are annoyed and still disturbing the population life progress. However, it leads to negative effects on working life in these areas and thus, evacuating them. So, the preventive methods remain the best way to overcome these problems or reduce them.

One of the means that allows the adoption of effective preventive methods is to understand the phenomenon and predict the places where it occurs. And this is what matters in our research, which aims to assess the load-carrying capacity of soil mathematically using the intersection approach relying on the calibration analysis and to calibrate it according to field measurements.

2. The Study Area

The area that has been selected as a study area is Berhoum, (previously Souk Ouled Nedjaa), in El Hodna Basin, Eastern Algeria which is characterized by its various natural properties and density of equipment in it. It is located at a distance of 300 km south of Algiers and 50 km to the east of the city of M'sila (**Figure 1**).

3. Methodology

It is known that the geotechnical mapping, including Multi evaluation of load-carrying capacity of the soil, was the subject of a lot of research scientific work, including: [1-7].

Many relationships and mathematical models that have been developed either partial; depend on the standard one or several criteria mostly "geographical" and not "Geotechnical".

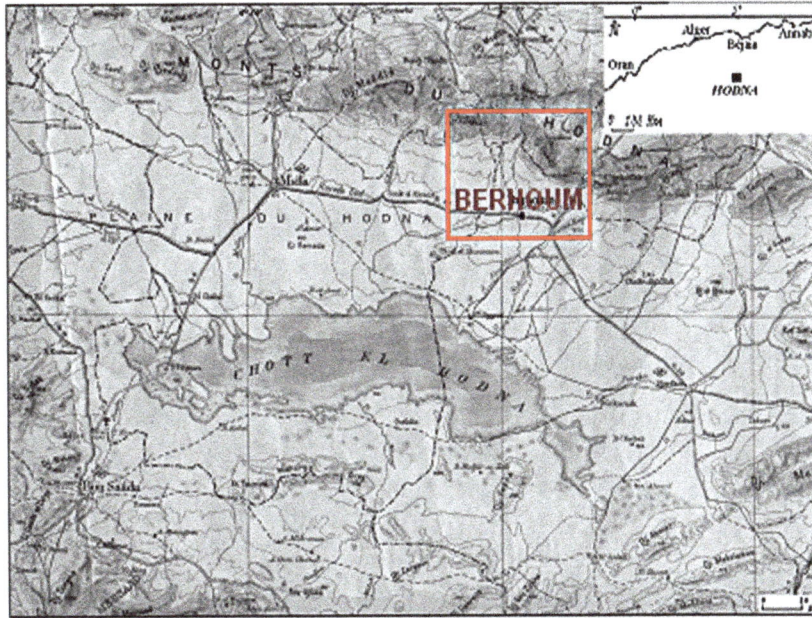

Figure 1. Location of the study area.

On the other hand, the endurance or the load-carrying capacity of soil is known as the maximum pressure that can be borne by the soil before the collapse happens in the construction ground [8,9].

Based on this definition, the load-carrying capacity of soil is associated with two basic components: the soil allowable pressure and the depth at which the pressure is applied. And by identifying the factors that control the load-carrying capacity of the construction soil, a mathematical model has been adopted. We review it as follows.

3.1. Hypotheses

We pose, in this occasion, two basic assumptions:

The load-carrying capacity of soil is a phenomenon that can be evaluated by geotechnical factors related to soil.

The possibility of the phenomenon occurrence is expressed by the total weighted factors leading to it.

3.2. Modeling Load-Carrying Capacity

Load-carrying capacity of soil (P) is the sum of the determinants of the phenomenon, but each factor has its weight (WSM) and it is expressed in the following relationship:

$$P = \sum_{l}^{n} w_i \cdot C_i$$

where:

w_i: gravity or weight specified for each worker.

C_i: determinants.

For the load-carrying capacity of soil; the first stage of

modeling begins by choosing the geotechnical criteria that intervene or help in the process of its occurrence, its spread, its intensity, or all combined. These criteria are in fact interrelated and therefore their intersection does the phenomenon. Consequently, the load-carrying capacity of soil can be found depending on WSM for both stress and depth. This scientific explanation can be expressed as follows:

$$P = 0.6\sigma_{perm} + 0.4H_{perm} \tag{1}$$

where:

P: the load-carrying capacity of soil.

σ_{perm}: permissible stress of the soil.

H_{perm}: the permissible depth of the soil (depth at which applied stress).

In this method, the weight values, and classification is based on expert opinion.

3.2.1. Permissible Stress of the Soil

The permissible stress is one of the main factors that govern the stability of the foundations and it has been prepared (σ_{perm}) taking into account the highest recorded stresses.

Taking into account the lithology map, and geotechnical surveys available, the permissible stresses for each lithologic facies collected then the permissible stress of the entire region is determined.

The permissible stress values are divided into four main categories (**Table 1**).

3.2.2. Permissible Depth of Soil

The permissible depth (H_{perm}) has been prepared accounting the largest registered depths which agreed for

Table 1. Classification and grade of the criteria specifying permissible stress.

Permissible stress	Grade	Classification
$\sigma_{perm} < 1$ bar	1	Low
1 bar $\leq \sigma_{perm} < 2$ bars	2	Medium
2 bars $\leq \sigma_{perm} < 3$ bars	3	Strong
$\sigma_{perm} \geq 3$ bars	4	Very strong

Table 2. Classification and grade of the criteria specifying permissible depth.

Permissible depth	Grade	Classification
$H_{perm} \geq 2$ m	1	Low
1.5 m $\leq H_{perm} < 2$ m	2	Medium
1 m $\leq H_{perm} < 1.5$ m	3	Strong
$H_{perm} < 1$ m	4	Very strong

Table 3. Order of weights and values of the criteria.

Criteria	Weights	Classification	Values
σ_{perm}	0.6	Low	0.6
		Medium	1.2
		Strong	1.8
		Very strong	2.4
H_{perm}	0.4	Low	0.4
		Medium	0.8
		Strong	1.2
		Very strong	1.6

Table 4. Degree of the load-carrying capacity of soil.

σ_{perm} \ H_{perm}	0.6	1.2	1.8	2.4
0.4	1	1.6	2.2	2.8
0.8	1.4	2	2.6	3.2
1.2	1.8	2.4	3	3.6
1.6	2.2	2.8	3.4	4

Colours reflect the degree of load-carrying capacity (P).

the recorded permissible stresses. And depending on the lithology map and the available geotechnical surveys; the permissible depths for each lithological face have been collected then they have been distributed in the entire region.

The permissible depth values have been divided into four main categories (**Table 2**).

3.3. Intersection between the Standards and the Completion of Tables

In order to be able to use the proposed model to assess the load-carrying capacity of soil, we first identify the factors controlling the phenomenon and classify them then digitize them by the degree of impact of each factor. Relying on quantitative data, to assess the load-carrying capacity of soil, we used a Weight Somme Model (WSM) where the rank of weights and values are in **Table 3**.

Intersection matrix is given as **Table 4**.

Intersection between stress and depth permissible they give the strapless soil foundations as shown in **Table 4**.

4. Spatialization by GIS Approach: Application in the Study Area

The approach we have adopted for the spatial distribution map preparation to assess the load-carrying capacity of soil is based on the geographic information systems in which we used two types of data:

-The first is related to the lithological map of the area (rocks' map),

-And the second is related to data of the field survey and geotechnical drilling.

These data obtained from the rocks' map or from the field survey and geotechnical drilling, have been implemented in GIS program (MapInfo 8) for the primary maps drawing.

After the building of a database of the geographic reference, organizing and structuring it by the use of programs, a map of the load-carrying capacity of soil that shows the best areas for building.

4.1. Lithological Map of Study Area

The rocky map of the area is derived from the geological map "Souk Ouled Nedjaa" (map No. 169, scale 1:50.000), prepared by the R. Guiraud [10]. It has been prepared

after the geological map numeration, then building the rocky facets' database. After the topical analysis, the following major rock, units has been identified (**Figure 2**).

4.2. Field Survey Data and Geotechnical Borehole

To complete the database we have adopted central values of all the criteria selected for this study. Data used in this study were collected from 49 study geotechnical surveys, well field investigation based on direct observations, as well as recordings and field sampling (GPS, images, etc.).

Geotechnical studies prepared soil laboratories: LHCC-M'sila, LCTP-M'sila, LIEG-M'sila, AICH.GEOSOL-Setif, EAST.LCTP-Setif, and LBEMA-Bordj Bou Arreridj.

These data were consolidated and exploit digitization units have been analysed and evaluated these data for my map permissible stress (σ_{perm}) and permissible depth (H_{perm}) (**Figures 3** and **4**).

4.3. Identification and Classification of Load-Carrying Capacity

Depending on the relationship (1) that have been entered into of GIS program guide, strapless been assessing of capacity of the soil (*P*) according to of proposed classification previously, She of output of map shown in **Figure 5**.

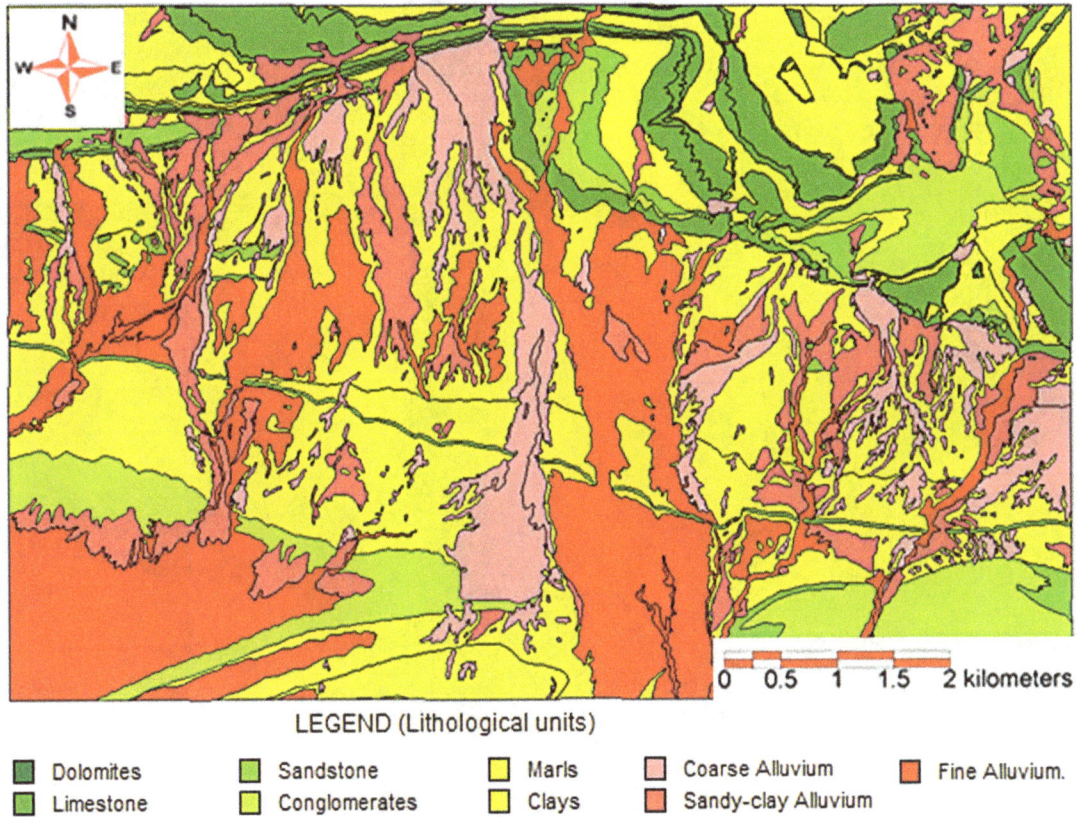

LEGEND (Lithological units)

◼ Dolomites	◼ Sandstone	◻ Marls	◻ Coarse Alluvium	◼ Fine Alluvium.
◼ Limestone	◼ Conglomerates	◻ Clays	◼ Sandy-clay Alluvium	

Figure 2. The main rock units of the study area.

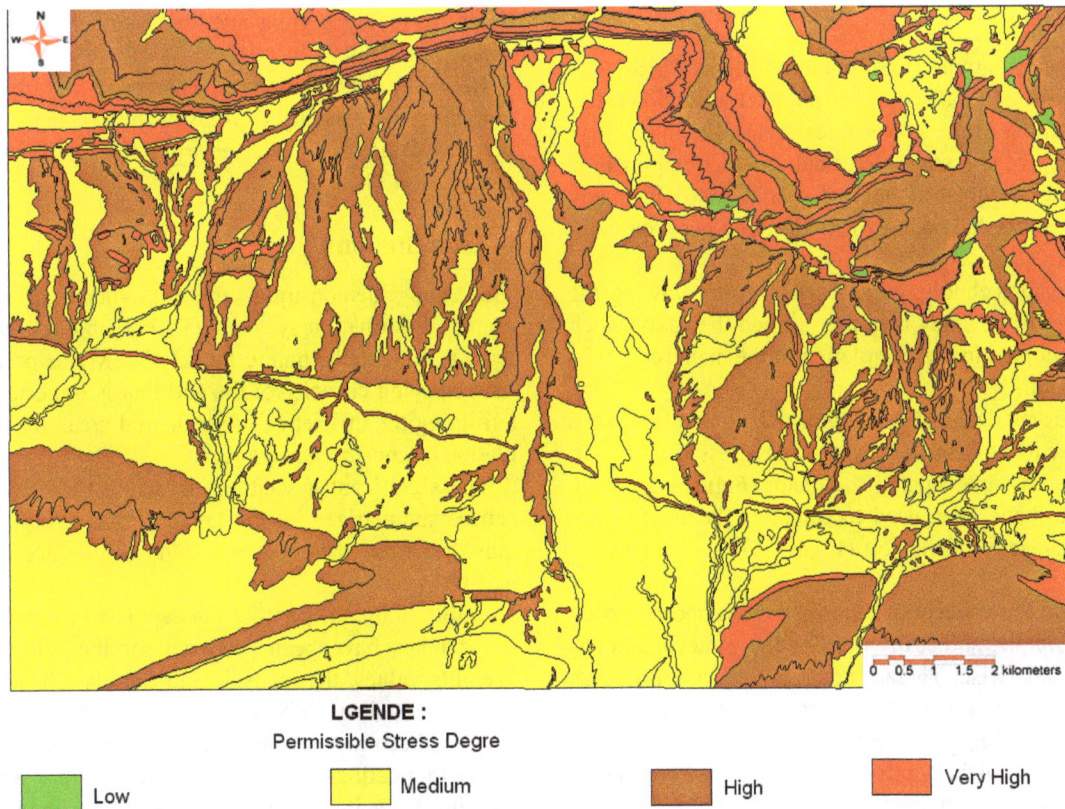

LGENDE :

Permissible Stress Degre

◼ Low	◻ Medium	◼ High	◼ Very High

Figure 3. The permissible stress map of study area.

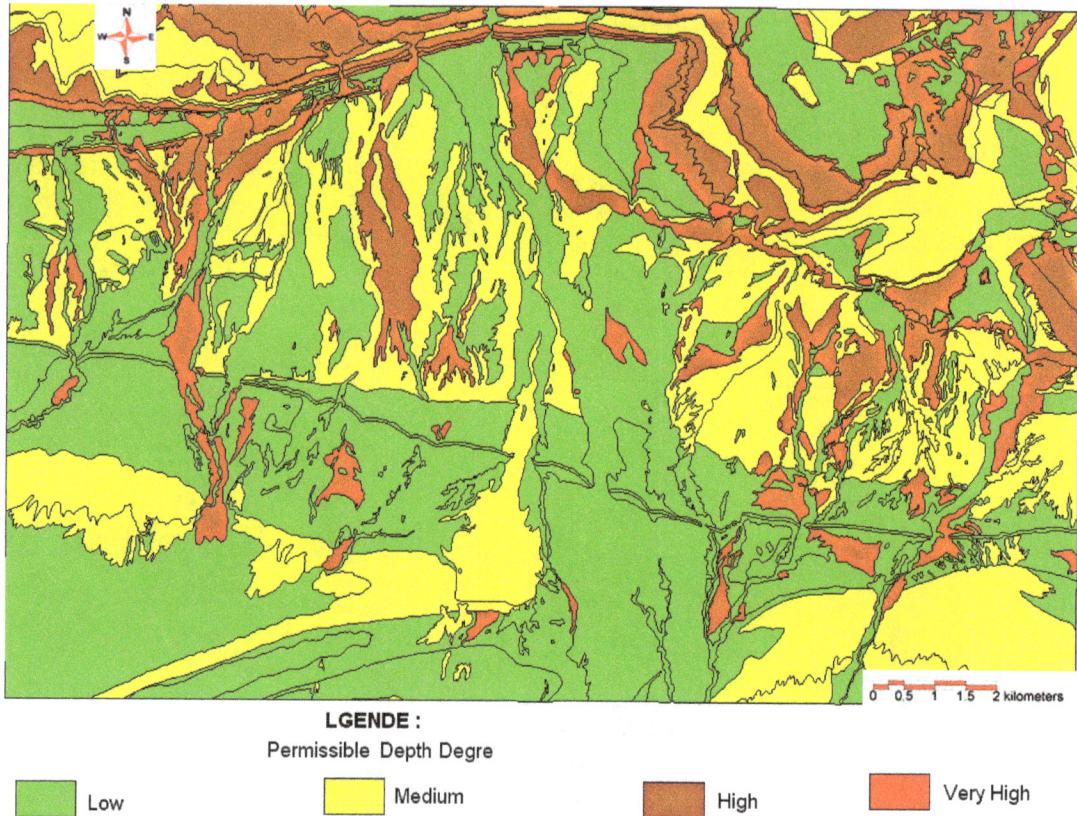

LGENDE :
Permissible Depth Degre

Low   Medium   High   Very High

Figure 4. The permissible depth map of study area.

5. Discussion

The use of geographic information systems (GIS) with the adoption of (WSM) model allowed and allows fast the preparation of the geotechnical maps, the load carrying capacity of soil map in particular; which requires the combination of great amount of spatial data with many different geographic data, experts data and decision-makers details.

The digital technology has been used in drawing the rocky map of the study area that formed the basis which we have adopted in the completion of the main rock units of that region.

Field survey data and geotechnical drilling allowed the data collection to get the permissible stress (σ_{perm}) map and the permissible depth (H_{perm}) maps (**Figures 2** and **3**). Thees data have been used to calculate the field' load-carrying capacity of the soil (**Figure 4**) by the relationship (1).

Based on the concept of the intersection, the proposed model is simple, and easily used by the geotechnical engineer or otherwise. Thread model is very interesting as it is being built on a small number of geotechnical charcteristics according to experts' judgment. Similarly, this model can be easily applied to other similar regions in terms of the environmental landscape and climate. This model is not static; however, it can be changed some-

times by giving a great attention to one factor and decrease it to another.

Because of the obvious advantages of this method, the load-carrying capacity of the soil map of the study area (**Figure 4**) has been selected to be the final map for the study.

6. Conclusions

This study demonstrates the importance of the geographic information systems (GIS) techniques which use with a Weighted Somme model (WSM) to prepare the load carrying capacity of the soil that is important in determining the efficiency of residential areas for the completion of construction projects.

The significant advantages of using these techniques can be summarized in: the low cost, the easily use data, quickly update data and the ability to produce new scenarios.

In this study, the permissible depth and the permissible stress maps have been prepared for the selected area. Weights values, the analysis and the classification have been assigned by (WSM) model. As a result, the study area has been divided into four different areas according to the foundations efficiency: 1) areas with a weak load carrying capacity of soil; 2) areas with a medium load carrying capacity of soil; 3) areas with a strong load car-

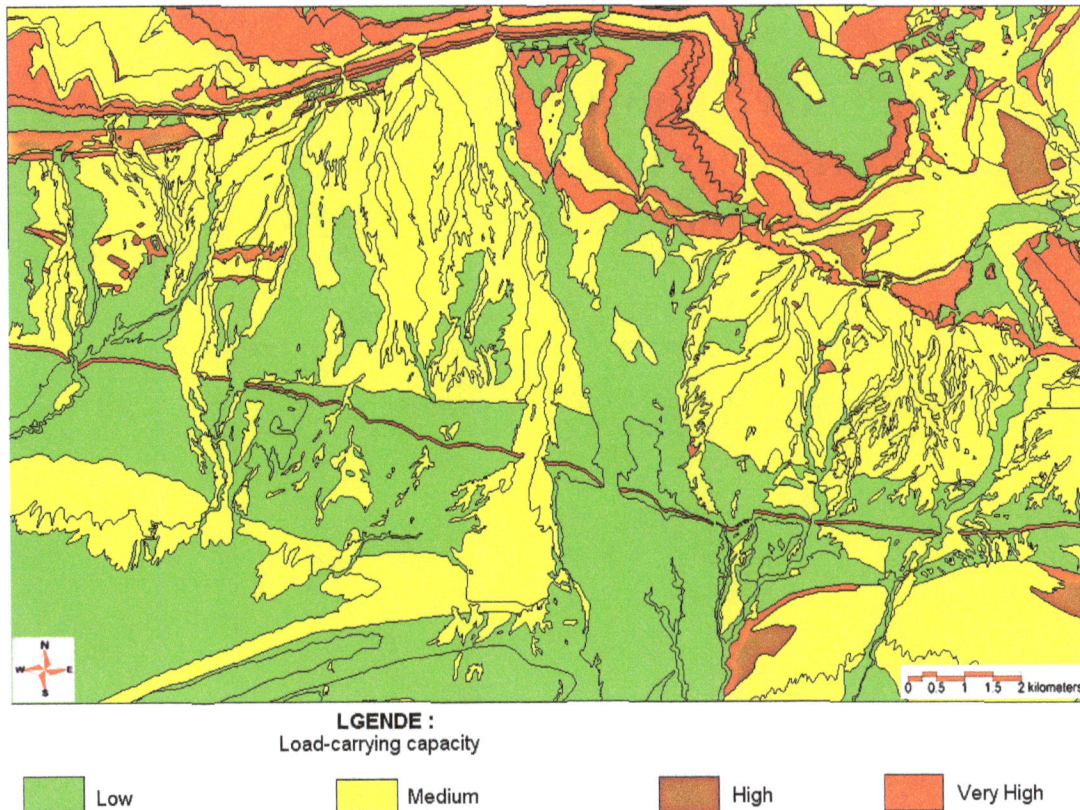

LGENDE :
Load-carrying capacity

Low   Medium   High   Very High

Figure 5. Load-carrying capacity of soil in study area.

rying capacity of soil; and 4) areas with very strong load carrying capacity of soil. According to this map, most of the study area was identified suitable for construction.

The methodology used for the selected area in this study can be applied to other places and other procedures for the selection of the site and be used to rebuild the necessary standards appropriately.

REFERENCES

[1]   C. Ayday, M. Altan, H. A. Nefeslioğlu, A. Canigür, S. Yerel and M. Tün, "Preparation of Engineering Geological Map of Eskişehir Urban Area," Research Institute of Satellite and Space Sciences, Anadolu University (in Turkish), 2001.

[2]   S. J. Carver, "Integrating Multicriteria Evaluation with Geographical Information Systems," *International Journal of Geographical Information Systems*, Vol. 5, No. 3, 1991, pp. 321-339.

[3]   J. T. Diamond and J. R. Wright, "Design of an Integrated Spatial Information System for Multiobjective Land Use Planning," *Environment and Planning B*, Vol. 15, No. 2, 1988, pp. 205-214.

[4]   J. R. Eastman, W. Jin, P. A. K. Kyem and J. Toledano, "Raster Procedures for Multicriteria/Multiobjective Decisions," *Photogrammetric Engineering and Remote Sensing*, Vol. 61, No. 5, 1995, pp. 539-547.

[5]   P. Jankowski, "Integrating Geographical Information Systems and Multiple Criteria Decision Making Methods," *International Journal of Geographical Information Systems*, Vol. 9, No. 3, 1995, pp. 251-273.

[6]   C. P. Keller, "Decision Making Using Multiple Criteria," NCGIA Core Curriculum, Unit 57, Santa Barbara, National Center for Geographic Information and Analysis, 1996. http://www.geog.ubc.ca/courses/klink/

[7]   J. Malczewski, "GIS and Multicriteria Decision Analysis," John Willey and Sons Inc., Hoboken, 1999, 392 p.

[8]   K. Meftah, "Cours Mécanique des Sols," 2008, pp. 70-72. http://pf-mh.uvt.rnu.tn/

[9]   G. Philipponnat and B. Hubert, "Fondations et Ouvrages en terre, Troisième Tirage," Eyrolles, 2002, p. 548.

[10] R. Guiraud, "Notice Explicative de la Carte Géologique au 1/50.000è de Souk Ouled Nedja," Publication du Service Géologique de l'Algérie, Pl. Hors-texte, 1971, 35 p.

4

A Multidisciplinary Approach to Mapping Potential Urban Development Zones in Sinai Peninsula, Egypt Using Remote Sensing and GIS

Hala A. Effat[1], Mohamed N. Hegazy[2]

[1]Environmental Studies and Land Use Department, National Authority for Remote Sensing and Space Sciences, NARSS, Cairo, Egypt

[2]Division of Geological Applications and Mineral Resources, National Authority for Remote Sensing and Space Sciences, NARSS, Cairo, Egypt

ABSTRACT

One of the main concerns of physical planning is the proper designation of suitable sites for feasible and sustainable land use. A main importance of such issue is that it withdraws attention to the necessity of adopting a multidisciplinary approach to the zoning and site selection problem. Egypt has a top priority objective to develop Sinai Peninsula and to create new sustainable and attracting communities that should ensure a stable, economic and sustainable environment in vast desert zones. Due to the difficulty in solving a zoning problem in a desert, the use of remote sensing and Geographic Information System (GIS) was to explore the desert potentials in the region. Five sub-models were created for five themes using Spatial Multicriteria Analysis (SMCA) and used as inputs to the final suitability model. These themes are: land resources, land stability, accessibility, cost of construction and land protection. A GIS-based model was designed following a sustainable development approach. Economic, social and environmental factors were introduced in the model to identify and map land suitable zones for urban development using Analytical Hierarchy Process (AHP). The suitability index map for urban development was produced by weighted overlay of the five sub-models themes. The most suitable zones for urban development in Sinai Peninsula amounted to 5327 square kilometers representing 17% of total area, whereas high suitable zones reached 40% indicating a high suitability of Sinai Peninsula lands for residing new urban communities.

Keywords: Urban Development; Remote Sensing; GIS; Site Selection; Themes; Analytical Hierarchy Process; Sinai; Egypt

1. Introduction

Sustainable development is becoming a popular concept among planners and researchers because it guides resource use in a way that aims to meet the needs of today's populations without compromising the ability for future populations to meet their needs [1]. Using the principle of sustainable development, the environmental, social and economic impacts can be managed to maximize positive impacts while minimizing negative effects. Urban planners use Geographic Information Systems (GIS) in sustainable development research and decision-making [2].

Land use suitability assessment is an important fundamental work in land use planning. The use of remote sensing and GIS technology in land suitability evaluation is a new technology and a new method in urban-rural planning. Such techniques provide a quantitative analysis. The rapid development of remote sensing technology and gradually maturing of GIS technology applications provides the foundation for urban-rural planning from the qualitative analysis to quantitative analysis. The core technology of the urban-rural planning is land use suitability with comprehensive evaluation [3]. Spatial analysis combined with multi-criteria evaluation methods was proven to be useful for both facing the main issues relating to land consumption and minimizing environmental impacts of spatial planning [4]. One of the most useful applications of GIS for planning and management is the landuse suitability mapping and analysis [5-8]. Broadly defined, land-use suitability analysis aims at identifying the most appropriate spatial pattern for future land uses according to specify requirements, preferences, or predictors of some activity [6,8]. The GIS-based land-use suitability analysis has been applied in a wide variety of situations including ecological approaches for defining land suitability/habitat for animal and plant species [9, 10], geological favorability [11], suitability of land for agricultural activities [12,13], landscape evaluation and planning [14], environmental impact assessment [15] selecting the best site for the public and private sector facilities [16,17], and regional planning [18].

In the present study, remote sensing and GIS techniques have been applied to explore the potentials of Sinai Peninsula, Egypt for urban development. The main objective of this work is to identify and delineate the optimum locations for developing new urban communities in such vast desert area. To achieve this objective, several factors influencing the suitability of the land for the required development are investigated, and the main five effective factors include: accessibility, land stability, natural resources, costs of construction, and protection of natural protectorates and archaeological sites. For each of these themes, a special sub-model was prepared based on analysis and interpretation of high resolution satellite images. Then, all of the sub-models were integrated, as layers, in a GIS environment, in order to create an overall model to depict the most suitable zones for urban development in Sinai Peninsula.

2. Description of the Study Area

Sinai is a triangular peninsula covering an area of 61,000 sq km in the northeastern Egypt and joining the great continental land masses of Africa and Asia within the geographic location falling between latitudes 27°43' and 31°19' North and longitude 32°19' and 34°54' East The Peninsula is situated between the Gulf of Aqaba and Gulf of Suez, and is bounded from north by the Mediterranean Sea. (**Figure 1(a)**). It comprises two administrative governorates, North Sinai covering an area of about 27,564.0 square kilometers and South Sinai covering an area of about 31,272.0 square kilometers. North and South Sinai Governorates population reach 395,271 and 159,029 respectively as of 2012 estimates [19] (Central Agency for Public Mobilization and Statistics). The Peninsula also covers portions of three governorates; namely Port Saied, Ismailya and Suez Governorates (**Figure 1(b)**). The physical geography includes desert plains, sand dunes and sea coasts, plateaus and mountainous zones. The Mediterranean Sea borders the Peninsula from the north with a shoreline reaching 205 km. The region is rich in mines, where kaolin, manganese, zircon, coal and feld-

spar exist. Quarrying activities such as gypsum, glass sand, marble, granite, dolomite and limestone are being extracted [20] (Center of Housing and Building Researches, 2007). The Peninsula is popular for its unique protectorates and historical and religious sites such as St. Catherine Monastery and Mount Moses. Despite its rich resources, the peninsula is among the least governorates in population density in Egypt. The Egyptian Government has put Sinai's development plan in its top priorities in the previous years.

3. Materials

3.1. Remotely Sensed Data

Remote sensing is an important data acquisition means, it plays a vital role in land use suitability assessment. In the present study, land use, cultivated land, urban distribution, water bodies, ecological elements are all obtained from the analysis and interpretation from satellite images including:

1) Landsat ETM satellite images, acquired in 2013, for regional investigation of the whole Sinai Peninsula.

2) SPOT4 satellite images acquired in 2011, for more de tailed analysis of some particular areas.

3) Shuttle Radar Topography Mission (SRTM) [21] data for extraction of Digital Elevation Model (DEM), slope, aspect and stream network.

The data were digitally processed in ESRI Spatial Analyst.

3.2. Maps

Various thematic maps covering Sinai Peninsula at different scales have been collected from the specialized agencies. They include the following:

1) Topographic maps at scale 1:500,000 published by the Egyptian Military Survey Department, 1995 [22].

2) Geologic maps scale 1:500,000 obtained from the Egyptian general Petroleum Corporation, 1987 [23].

3) The protected areas map was obtained from the Egyptian Environmental Affairs Agency, [24].

4) The Hydro-geologic map, scale 1:200,000 was obtained from the Research Institute for Groundwater (RIGW) [25].

5) The mineral resources map was obtained from the Center of Housing and Building Researches, 2007 [20].

All maps were obtained in hard copies, scanned, rectified, digitized in ESRI ArcGIS10 and saved as feature classes in a geographic database for further analysis.

4. Methodology

The selection of suitable sites for specific land uses must be based upon a set of local criteria to ensure that the maximum benefit and least cost for a selected site/zone

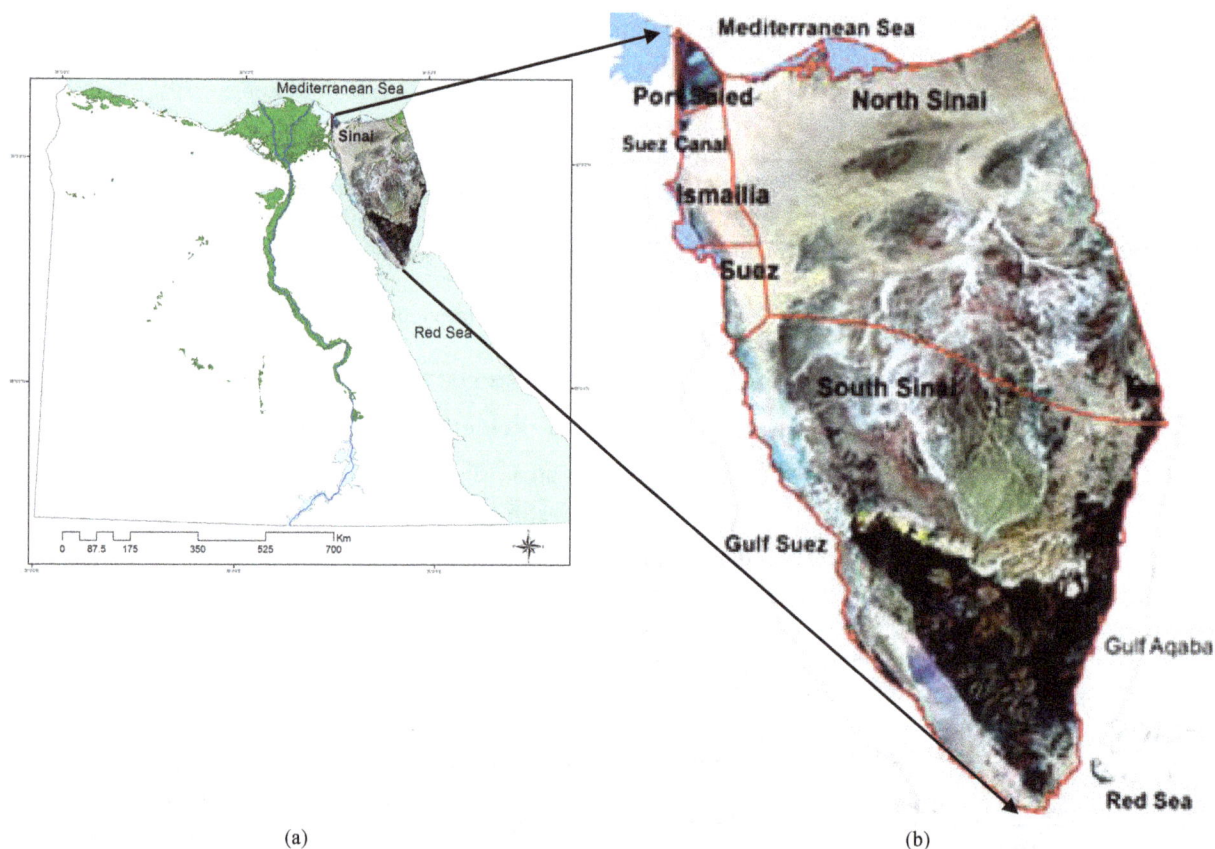

Figure 1. (a) Location of Sinai Peninsula; (b) Administrative divisions of Sinai Peninsula.

are attained. The following general overlay analysis steps were followed based on ESRI Spatial Analyst, [26]:

1) Define the problem.
2) Break the problem into sub-models.
3) Determine significant layers.
4) Transform the data within a layer into a common scale (normalize).
5) Weight the input layers.
6) Combine the layers.
7) Analyze.

A conceptual diagram of the applied analysis is shown in **Figure 2**.

4.1. Define the Problem

In this step, the overall objective was identified. All aspects of the remaining steps of the overlay modeling process must contribute to this overall objective. Ensuring geological safety, resources, least costs, accessibility and protection of natural and archeological assets are the objectives of this study.

4.2. Break the Problem into Sub-Models

Most overlay problems are complex, and it is recommended to break them down into sub-models for clarity, to organize thoughts, and to more effectively solve the

overlay problem. For this study, several factors that contribute to the suitability of the lands for accommodating new urban communities were identified. Such factors were then grouped into five main sets or rather themes. For each of the five themes a Multicriteria Evaluation sub-model was created These themes are: 1) accessibility, 2) natural resources, 3) land stability and 4) construction costs and 5) natural and cultural values protection.

4.3. Determine Significant Layers and Defining the Related Themes

The significant attributes were defined and the related layers were created in this step. Factors used to model the urban development site suitability model; include social, economic and environmental themes, in addition to a land safety theme and a natural and cultural values protection theme. For each of such themes an independent sub-model was created. The attributes and layers that affect each sub-model was identified. Each factor describes a component of the phenomena a sub-model is defining. Such factor contributes to the goals of the sub-model, and each sub-model contributes to the overall goal of the overlay model. (ESRI spatial analyst) [27]. The themes used to create the sub-models are described in the following section:

A Multidisciplinary Approach to Mapping Potential Urban Development Zones in Sinai Peninsula, Egypt Using Remote Sensing and GIS

31

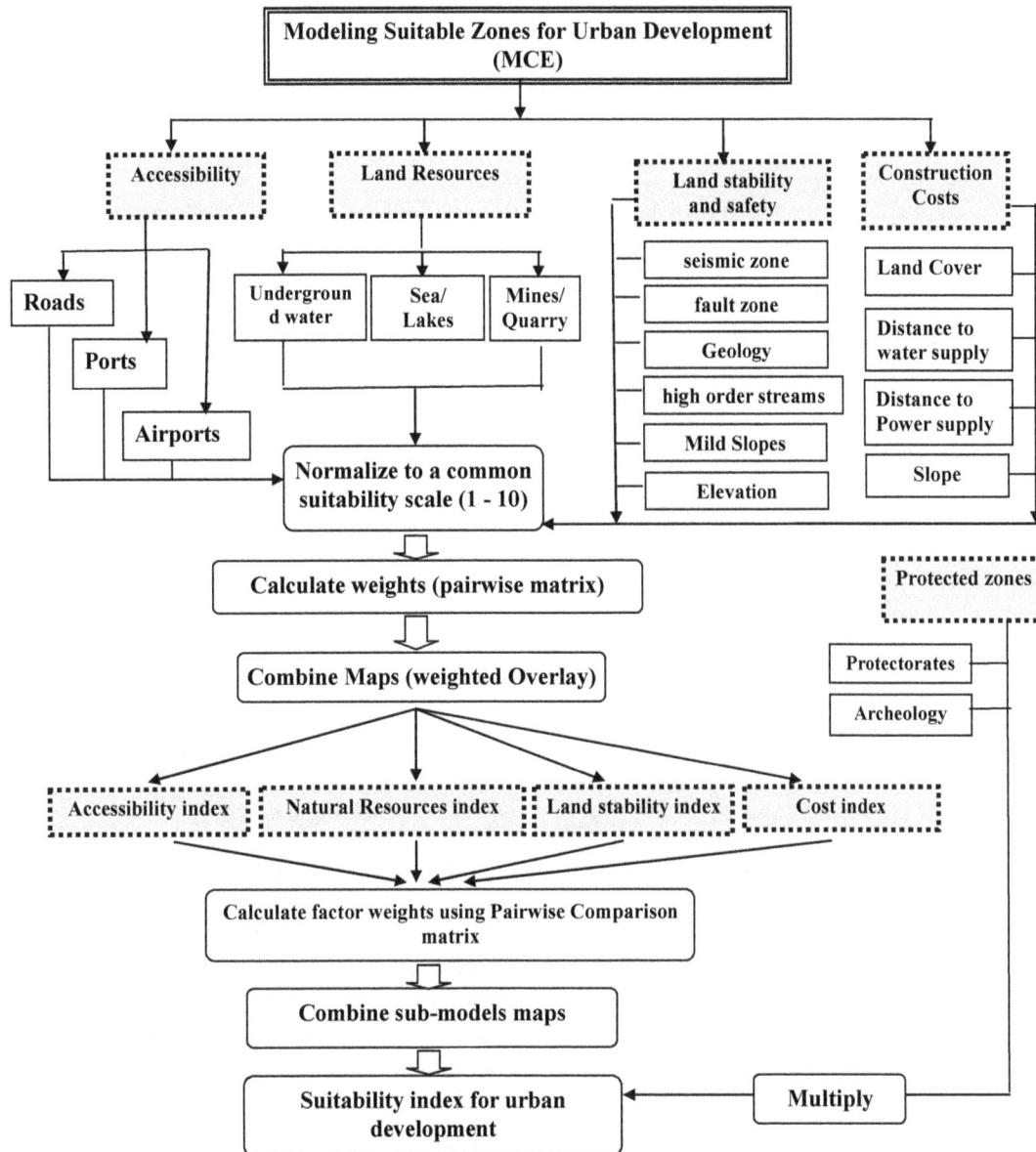

Figure 2. A conceptual model for the applied methodology.

4.3.1. Natural Resources Theme

Potential land resources are the backbone for job creation and therefore it should take first priority and consideration in development of new urban zones. Factor maps used for modeling the land resources (**Table 1**) theme include the following: groundwater, sea/lakes and mineral resources.

1) Ground Water

Groundwater is one of the most important natural resources for the development in such areas of severe aridity as North Sinai. There are two main types of groundwater; the Quaternary aquifer in wadi alluvium and coastal sand dunes, and the deep fracture zone aquifers in Lower Cretaceous sandstone formations Hydro-geologic units were digitized from the available hydro-geologic

map (Institute for groundwater, 1999) [24].

2) Sea/Lakes

Sea water can be desalinated and provide the water needs of urban communities in Sinai Peninsula. Lake Bardawil provides an aqua culture resource and fisheries for the natives. Proximity of a site from a shoreline is considering a benefit factor. Shorelines were digitized from the Landsat ETM image acquired in 2013.

3) Mineral Resources

The availability of mines and quarries is an essential factor for industrial activities. Sinai encounters several occurrences of various types of mineral resources including: coal, black sands, glass sand, kaolin, as well as building and construction raw materials such as marl, dolomite, sand, gravel and marble and granite. Large

Table 1. Standardization of suitability factors: natural resources.

Suitability scale	Factors for resources			
	Ground water (aquifers)	Distance to sea shorelines (m)	Distance to mines (m)	Distance to quarry (m)
0		0 - 500	0 - 1000	0 - 2000
1	Non-aquifers clays and shale.	106,013 - 117,791	100,895 - 112,105	93,382 - 103,756
2	Local groundwater occurrences in fissured and weathered zone in hard rocks	94,234 - 106,012	89,685 - 100,894	83,006 - 93,381
3	-	82,454 - 94,233	78,474 - 89,684	72,630 - 83,005
4	Local and moderately to low productive.	70,675 - 82,453	67,264 - 78,473	62,255 - 72,629
5	Local and highly to moderately productive.	58,896 - 70,674	56,053 - 67,263	51,879 - 62,254
6	Extensive and moderately to low productive.	47,117 - 58,895	44,843 - 56,052	41,504 - 51,878
7	-	35,338 - 47,116	33,632 - 44,842	31,128 - 41,503
8	Extensive and highly to moderately productive.	23,559 - 35,337	22,422 - 33,631	20,752 - 31,127
9	-	11,780 - 23,558	11,211 - 22,421	10,377 - 20,751
10	Extensive and highly productive, continuous.	500 - 11,779	1000 - 11,210	2000 - 10,376

amount of raw materials for cement industry (limestone, shale and gypsum) are also available. Mineral resources map was obtained as a shape file from the Center of Housing and Building Researches, 2007 [25].

4.3.2. Land Stability Theme

Land stability sub-model aims at avoiding the land collapse vulnerable lands. Avoiding such vulnerable zones is crucial in selecting lands that would be developed as urban residential and new communities. For this purpose, a vulnerability map for land collapse hazards in Sinai was created. Such map identifies the highly vulnerable lands that should be avoided from the selection for new urban settlement. To produce the land stability map, the inverse of such map was used.

Land stability index map = 1/land vulnerability index map

The vulnerability index map was modeled using the following factors:

1) Elevation

Higher elevation zones are more vulnerable to land collapse due to the force of gravity. Thus the highest elevation sites are least suitable.

2) Slope

Lands having steeper slope angles are more vulnerable to land collapse. Slope greater than 25 degree has been classified as unsuitable.

3) Lithology

Different rock types have different response to land collapse. The lithology was reclassified according to the suitability scale

4) Seismic Intensity Zones

Zones prone to high magnitude seismic activity are more vulnerable to land collapse.

5) Stream Density

The higher the stream density, the more vulnerable to

land collapse.

6) Active Faults Zones

Zones with active faults are more vulnerable to land collapse.

7) Faults Density

Zones with high density of faults are vulnerable.

4.3.3. Accessibility Theme

Accessibility is a main factor that defines the decisions of people; accessibility to a site can be a main factor for polarization for a new location. Factors used to model the accessibility theme are defined as follows:

1) Roads

Roads are the main arteries for a development site. They provide accessibility and link remote areas, constituting the main arteries for residential needs. Main corridors, paved roads and desert tracks were considered. All roads were digitized from the topographic map, updated from SPOT-4 images and a road distance map was created.

2) Ports

The proximity to existing ports is an important factor for exchanging goods and providing jobs. A distance function was used to create a distance to ports map.

3) Airports

Airports are crucial in serving a site providing accessibility and linking its remote areas. The airports sites were digitized from the topographic map and a distance map was created.

4.3.4. Costs of Construction Theme

The cost of construction is a main economic factor for decision making in site selection. Factors used to model the cost of construction theme are defined as follows:

1) Land Cover/Land Use

Land cover-land use is the main basis for urban plan-

A Multidisciplinary Approach to Mapping Potential Urban Development Zones in Sinai Peninsula, Egypt Using Remote Sensing and GIS

33

ning; the distribution of various land-use types gives considerable constraints to urban planning. For example natural protectorates, cultivated lands, urban areas are not liable to change. The land cover layer was derived by supervised classification of SPOT4 images. Six classes were identified namely: sabkha and wetlands, water bodies, cultivated land, natural vegetation, desert, urban and roads.

2) Water Supply

Provision of a water resource is a main factor for developing an area. Underground water and sea water (for desalinization) were considered a possible water resource. Proximity from sea shore and underground water layers were produced using the distance module.

3) Power Supply

The proximity of the site to a power supply facility such as high electric lines or power stations is an economic factor. The high electric lines and power stations were mapped from the topographic map.

4) Slope

The slope was delineated using SRTM digital elevation model (DEM) in ESRI ArcGIS 9.2 spatial analyst.

4.3.5. Land Protection Theme

The investigated is unique in ecological and archaeological values. The protectorates and the archeological sites should be protected from any changes in land use. The protected zones were obtained from the Egyptian Environmental Affairs Agency, EEAA, 2007 [24]. The archaeology sites were obtained from the Egyptian General Organization for Physical Planning, GOPP, 2007 [27]. A binary function was used to classify the lands giving a zero value for the national protectorates and 500 meters buffer zone around the archaeological sites was given a zero value and masked out.

4.4. Transformation of Criteria Attributes to a Common Scale (Normalization)

Criteria attributes have different measuring scales. In order to perform analysis, a standardization has to be performed through transformation of attributes into a common suitability. For each factor, the attributes were rated in reference to a common scale. Thus for each submodel, the criteria attributes were transformed from the original values to a common suitability scale ranging from 1 - 10. The higher value being more favorable. A value of zero was given to unsuitable pixels (**Tables 1-6**).

4.5. Combining the Theme Factors (Layers)

Equal weights were used for the factors within a theme. The theme maps were created by combining such factors using the following equation:

$$S = \sum \left(W_j \cdot X_{ij} \right) \qquad (1) \ [29]$$

where: S = composite suitability score, W_j = weights assigned to each factor j, X_{ij} = attribute score i of factor j.

4.6. Calculation of the Theme Weights

Certain factors may be more important to the overall goal than others. If this is the case, the factors can be weighted based on their relative importance. Analytical Hierarchy Process [30] was used to assign weights to each criterion, factor, and thus determine their relative importance in the final decision adopted within the model. The method is based on pair-wise comparison within a reciprocal matrix, in which the number of rows and columns is defined by the number of criteria. Accordingly, it is necessary to establish a comparison matrix between pairs of criteria, contrasting the importance of each pair with all the others. Subsequently, a priority vector is computed to establish weights (W_j). These weights are a quantitative measure of the consistency of the value judgments between pairs of factors [31]. Satty's scale of measurement is used as follows:

S = {1/9, 1/8, 1/7, 1/6, 1/5, 1/4, 1/3, 1/2, 1, 2, 3, 4, 5, 6, 7, 8, 9}.

A pairwise comparison matrix was designed. The comparison ratings are provided on a nine-point continuous scale, which was proposed by (Eastman, 1995) [29]. The comparisons ratings and factors were discussed with experts and a pairwise comparison matrix was constructed based on (**Table 3**). If we call that weight aij, and use that scale of comparison and if the relative weighting is a23 = 3/1, then the relative importance of attribute 3 with regard to 2 is its reciprocal a32 = 1/3. This process generated an auxiliary matrix in which the value in each cell is the result of the division of each value judgment (aij) by the sum of the corresponding column. Finally, the average of the normalized values of rows was obtained, which corresponds to the priority vector (W_j). This was normalized by dividing each vector value by n (the number of vectors), thus obtaining the normalized overall priority vector, representing all factor weights (W_j).

1) Determination of the weighted sum vector by multiplying matrix of comparisons on the right by the vector of priorities to get a new column vector. Then divide first component of new column vector by the first component of priorities vector, the second component of new column vector by the second component of priorities vector, and so on. Finally, sum these values over the rows.

2) Determination of consistency vector by dividing the weighted sum vector by the criterion weights. Once the consistency vector is calculated it is required to compute values for two more terms, lambda (λ) and the consistency index (CI). The value for lambda is simply the average value of the consistency vector.

The calculation of CI is based on the observation that λ is always greater than or equal to the number of criteria under consideration (n) for positive, reciprocal matrices and $\lambda = n$, if the pairwise comparison matrix is consistent matrix. Accordingly, λ-n can be considered as a measure of the degree of inconsistency. This measure can be normalized as follows:

$$CI = (\lambda - n)/(n-1) \qquad (2)$$

The term CI, referred to as consistency index, provides a measure of departure from consistency. To determine the goodness of CI. The Analytical Hierarchy Process (AHP) compares it by Random Index (RI), and the result is what we call CR, which can be defined as:

$$CR = CI/RI \qquad (3)$$

Random Index (RI) is the CI of a randomly generated pairwise comparison matrix of order 1 to 10 obtained by

approximating random indices using a sample size of 500 [31], **Table 3** shows the value of RI sorted by the order of matrix.

4.7. Combining the Four Themes

In overlay analysis, it is desirable to establish the relationship of all the input factors together in order to identify the desirable locations that meet the goals of the model. If a weighted summation is used, the higher the value on the resulting output raster, the more desirable the location will be. A weighted overlay was used to combine the criteria maps and to produce the final suitability index for suitable urban zones.

$$S = \sum (W_i \cdot X_{ij}) * \Pi c_l \qquad (4) \text{ [28,29]}$$

where: S = composite suitability score, W_j = weights assigned to each factor j, X_{ij} = attribute score i for factor j,

Table 2. Saaty's nine-point weighting scale [31].

Intensity of importance	Description	Suitability class
1	Equal importance	Lowest suitability
2	Equal to moderate importance	Very low suitability
3	Moderate importance	Low suitability
4	Moderate to strong importance	Moderately low suitability
5	Strong importance	Moderate suitability
6	Strong to very strong importance	Moderate high suitability
7	Very strong importance	High suitability
8	Very to extremely strong importance	Very high suitability
9	Extremely importance	Highest suitability

Table 3. Random index (RI) (Saaty, 1980) [32].

Order Matrix	1	2	3	4	5	6	7	8	9	10
R.I.	0.00	0.00	0.58	0.9	1.12	1.24	1.32	1.41	1.45	1.49

Table 4. Standardization of the suitability factors; land stability.

Suitability scale	Factors for stability of lands					
	Seismic intensity	Faults density	Elevation (meters)	Slope (degrees)	Stream density	Rock type
0	≥7	0.25 - 0.27				Sabkha
1		0.23 - 0.24	1209 - 1342	68 - 74	0.29 - 0.38	Clay and sand clay
2	6 - 7	0.20 - 0.22	1075 - 1208	60 - 68	0.24 - 0.29	Wadi deposits
3	5 - 6	0.17 - 0.19	941 - 1074	53 - 60	0.20 - 0.24	
4		0.15 - 0.16	807 - 940	45 - 53	0.17 - 0.20	
5	4 - 5	0.12 - 0.14	673 - 806	38 - 45	0.15 - 0.17	Limestone and chalky limestone
6	3 - 4	0.09 - 0.11	538 - 672	30 - 38	0.13 - 0.15	Limestone and Marly limestone
7		0.06 - 0.08	403 - 537	22 - 30	0.09 - 0.12	Dolomitic Limestone clastic deposits.
8	2 - 3	0.04 - 0.05	269 - 402	16 - 22	0.06 - 0.09	Sandstone
9	0.1 - 2	0.04 - 0.05	135 - 269	8 - 16	0.04 - 0.06	Basement rocks.
10	0	0 - 0.03	0 - 135	0 - 8	<0.04	

A Multidisciplinary Approach to Mapping Potential Urban Development Zones in Sinai Peninsula, Egypt Using Remote Sensing and GIS

35

Table 5. Standardization of the suitability factors: accessibility.

Suitability scale	Factors for accessibility		
	Distance to roads (meters)	Distance to ports (meters)	Distance to airports (meters)
0	0 - 200	0 - 1000	0 - 1000
1	17,420 - 19,355	308,493 - 342,769	84,933 - 94,369
2	15,489 - 17,419	274,217 - 308,492	75,496 - 84,932
3	13,549 - 15,484	239,940 - 274,216	66,060 - 75,495
4	11,614 - 13,548	205,663 - 239,939	56,623 - 66,059
5	9678 - 11,613	171,386 - 205,662	47,186 - 56,622
6	7743 - 9677	137,109 - 171,385	37,749 - 47,185
7	5807 - 7742	102,832 - 137,108	28,312 - 37,748
8	3872 - 5806	68,555 - 102,831	18,875 - 28,311
9	1936 - 3871	34,278 - 68,554	9437 - 18,874
10	201 - 1935	1000 - 34,277	0 - 1000

Table 6. Standardization of the suitability factors: cost of construction.

Suitability scale	Factors for cost of construction			
	Land cover	Distance to power supply (meters)	Distance to water supply (meters)	Slope (degrees)
0	Water bodies, built-up, sabkha	0 - 200	0 - 200	>30
1	Natural vegetation	103,523 - 115,025	142,945 - 159,937	
2	Agriculture	92,021 - 103,522	127,951 - 143,944	24 - 30
3		80,518 - 92,020	111,957 - 127,950	
4	Sand dunes	49,016 - 80,517	95,963 - 111,956	
5		57,513 - 69,015	79,970 - 95,962	
6		46,011 - 57,512	63,976 - 79,969	15 - 23
7		34,508 - 46,010	47,982 - 63,975	
8		23,006 - 34,507	31,988 - 47,981	8 - 15
9		11,503 - 23,005	15,995 - 31,987	
10	Desert lands	200 - 11,502	200 - 15,994	0 - 8

Πc_l is the constraint binary map description of symbols.

5. Results and Discussion

In this section, the resultant factors maps that were created using ESRI Spatial Analyst various functions are presented. Factors constituting each theme are grouped together for each of the five criteria themes as follows:

5.1. Factor Maps and Standardization Tables

Factor maps for Themes are depicted in the following figures; land resources (**Figure 3**), land stability (**Figure 4**), accessibility (**Figure 5**) and construction costs (**Figure 6**) Such maps are the preliminary results which were further processed to produce the sub-models themes index maps:

Figure 3 shows the factors of the land resources theme and the spatial distribution of potential land resources in the North, middle and southern zones. **Figure 4** shows

the spatial distribution of geologic stable and less stable lands. **Figure 5** shows the factors of the accessibility theme and spatial distribution of most and least accessible zones based on the road networks. Factors for the least cost theme are shown in **Figure 6**. Standardization of the criteria maps are also presented in **Tables 1**, **4** and **5**.

5.2. Deriving the Factor Weights

Tables 7(a)-(c) depict the Analytical Hierarchy Process AHP used to calculate the factor weights. **Table 7(d)** depicts the calculation of the consistency ratio (CR).

5.3. Deriving the Sub-Models for Each Theme

The four sub-models themes (**Figures 7(a)-(d)**) were derived by running Multicriteria evaluation model Equation (1) using equal factor weights. Land Protection theme binary sub-model is shown in **Figure 8**.

The model produced a suitability index map for poten-

(a)

(b)

(c)

(d)

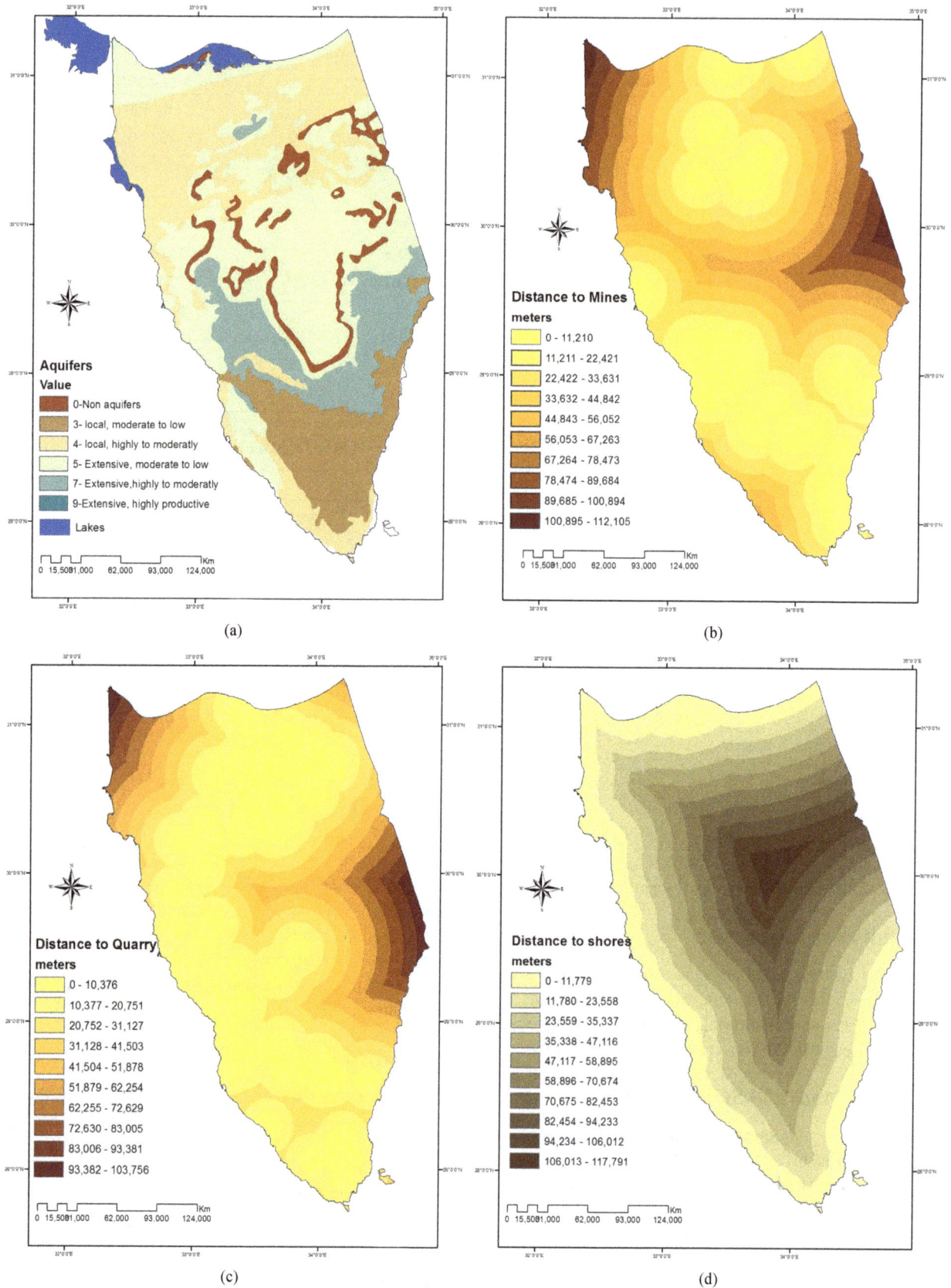

Figure 3. Factors maps of land resources: (a) Aquifers; (b) Distance to mining sites; (c) Distance to quarry sites; (d) Distance to shorelines.

A Multidisciplinary Approach to Mapping Potential Urban Development Zones in Sinai Peninsula,
Egypt Using Remote Sensing and GIS

37

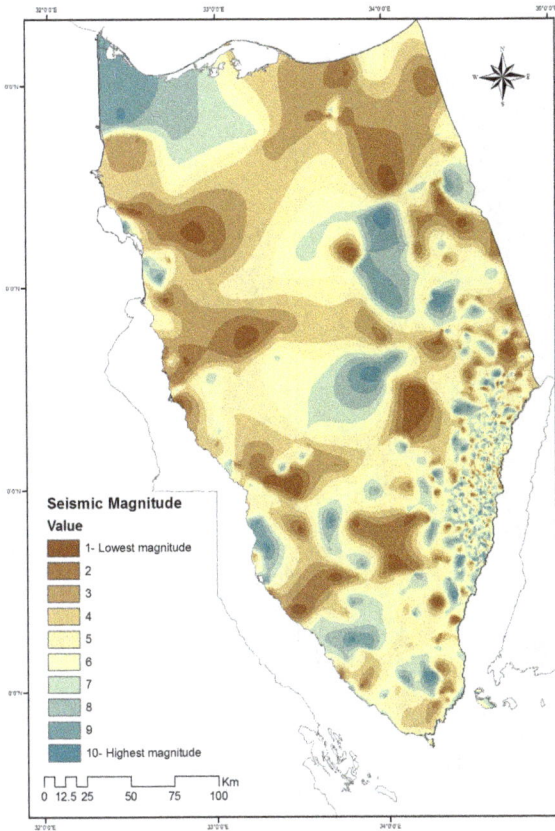

(a)

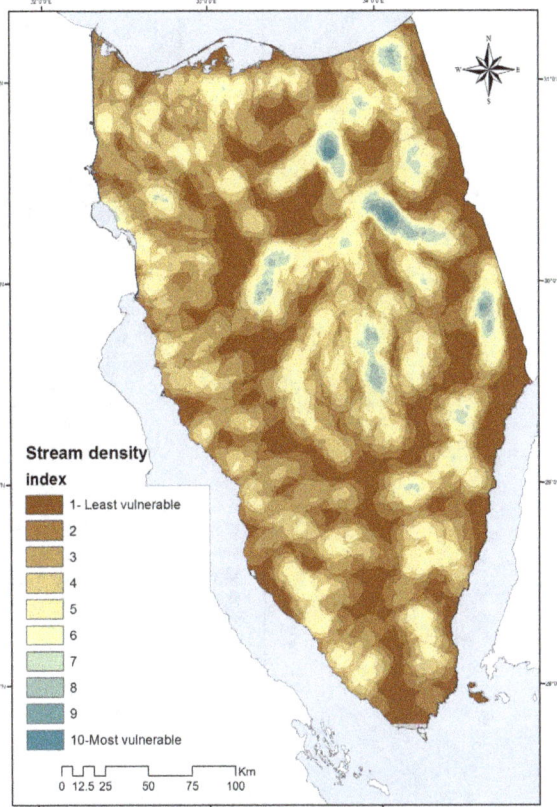

(b)

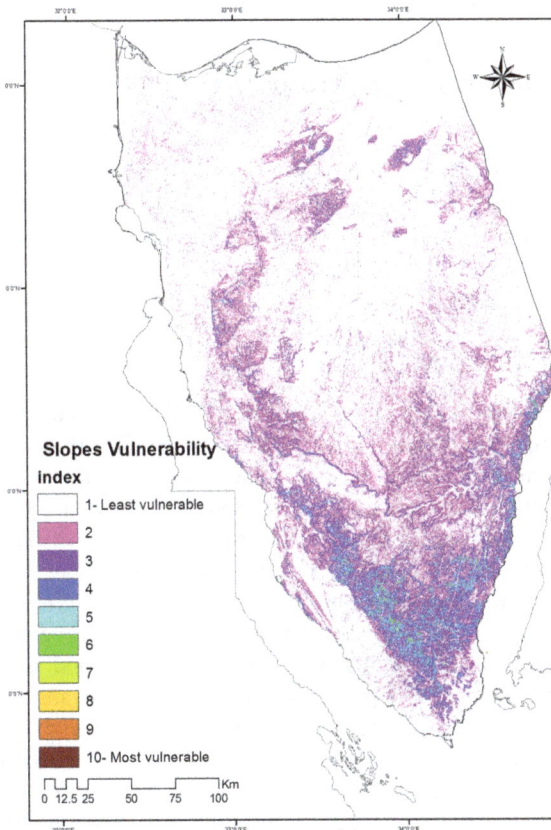

(c)

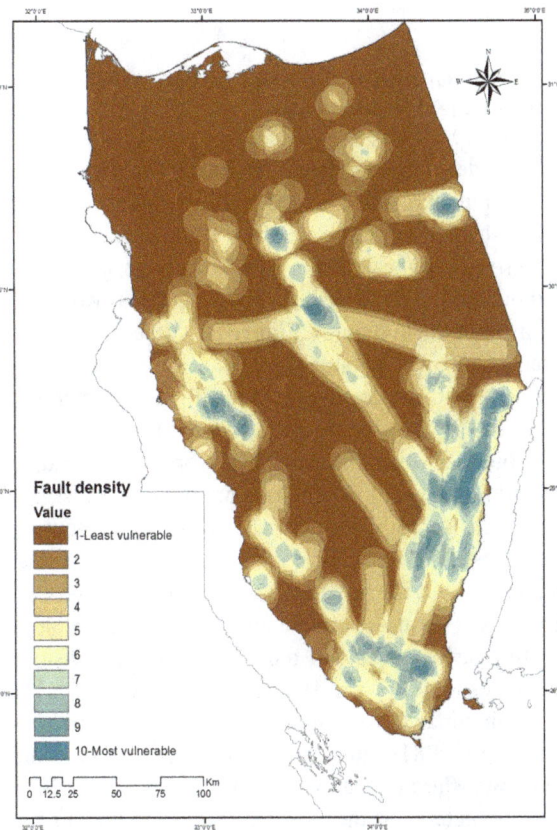

(d)

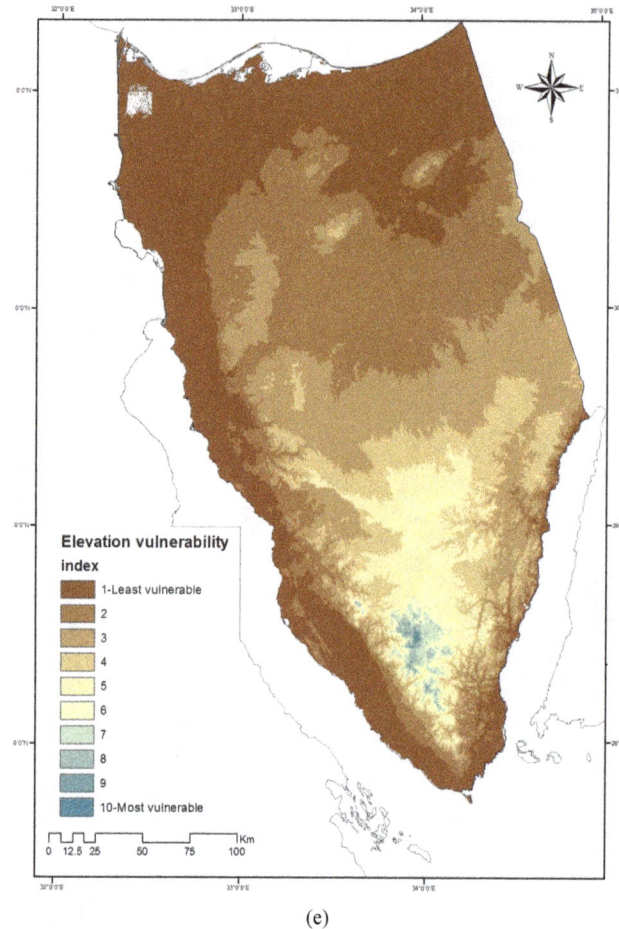

(e)

Figure 4. Standardized factors maps for land stability theme: (a) Seismic magnitudes zones; (b) Stream density; (c) Slope; (d) Fault density; (e) Elevation zones.

tial urban development lands in Sinai Peninsula. The model output explains that most of the lands in Sinai Peninsula is suitable for new urban communities (**Table 8**). Zones with highest suitability values (with grid value 10) amounted to 5326 square kilometers equivalent to 8.9% of the total study area **Figure 9**. These zones were selected as the most suitable sites. They are the most suitable zones for urban development fulfilling the suitability criteria used in the five themes. The spatial distribution shows potential zones distributed in North Sinai, middle zones and South Sinai. Such zones are high in accessibility, with safe geology and have potential land resources and low cost of construction within the studied criteria. Such zones avoid the protected zones of natural and cultural values. The most suitable zones for urban development are distributed among the administrative divisions of the Peninsula. They are described as follows:

In North Sinai Governorate, the most suitable zones amounted to 2081 square kilometers. In South Sinai Governorate, the most suitable zones amounted to 3015 square kilometers. In the portion of land falling in the Suez Governorate, most suitable zones amounted to 163

square kilometers In portion of lands falling in Ismailia Governorate, the most suitable zones amounted to 66 sq. km. For the portion of lands falling in Port Saied Governorate the most suitable sites amounted to 1.0 square kilometers.

6. Conclusion

Spatial multicriteria decision model was used in this study to identify the most suitable sites for urban development in Sinai Peninsula. The model integrates various remote sensing data and geographic information layers in a multidisciplinary approach. Applying such a method, it was possible to create a step by step zoning map for potential urban development sites, which was the main objective of this research paper. The other objective is that some spatial factors were studied. The method is well established and provides a comprehensible and logic procedure. If this technique is adopted by governments, it should involve the participation of several stakeholders, decision makers, scientists and public participation in the land planning process. Such public participation should aim at establishing and maintaining a high degree of

A Multidisciplinary Approach to Mapping Potential Urban Development Zones in Sinai Peninsula, Egypt Using Remote Sensing and GIS

39

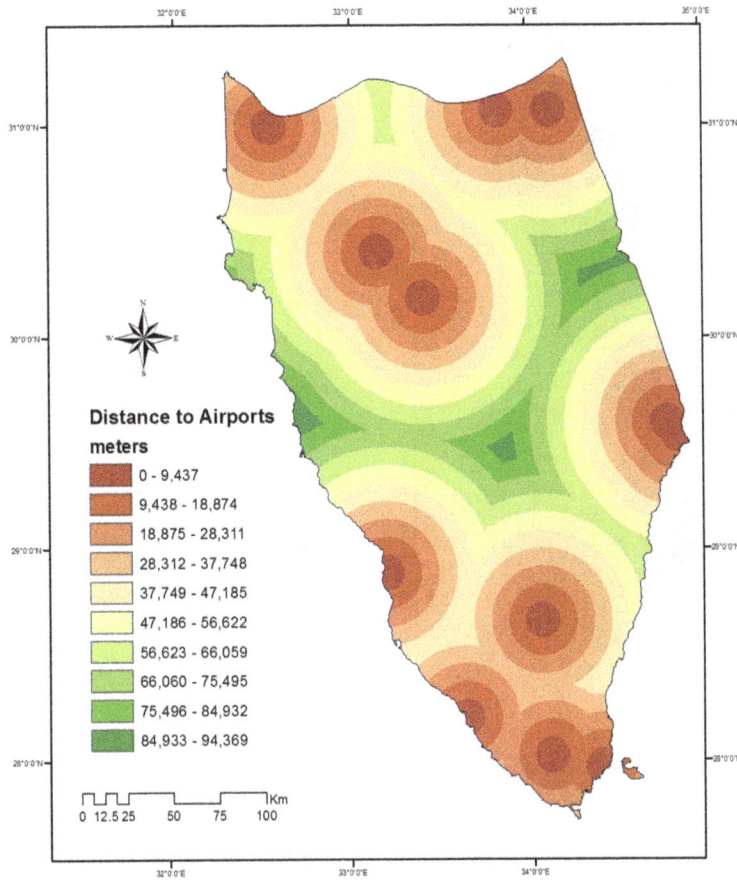

(a)

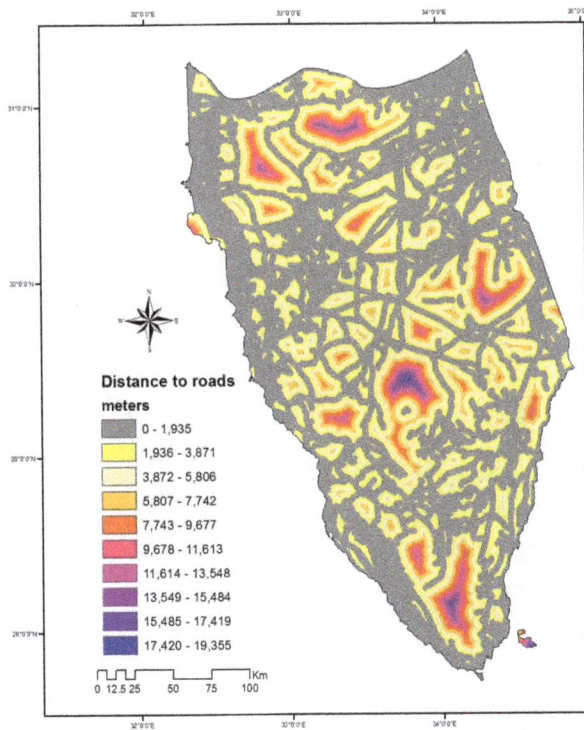

(b)

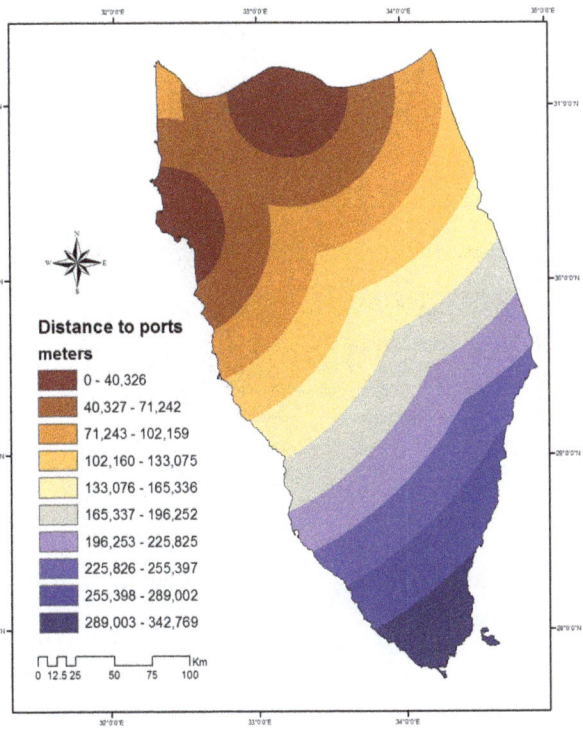

(c)

Figure 5. Factors for accessibility theme: (a) Distance to airports; (b) Distance to roads; (c) Distance to ports.

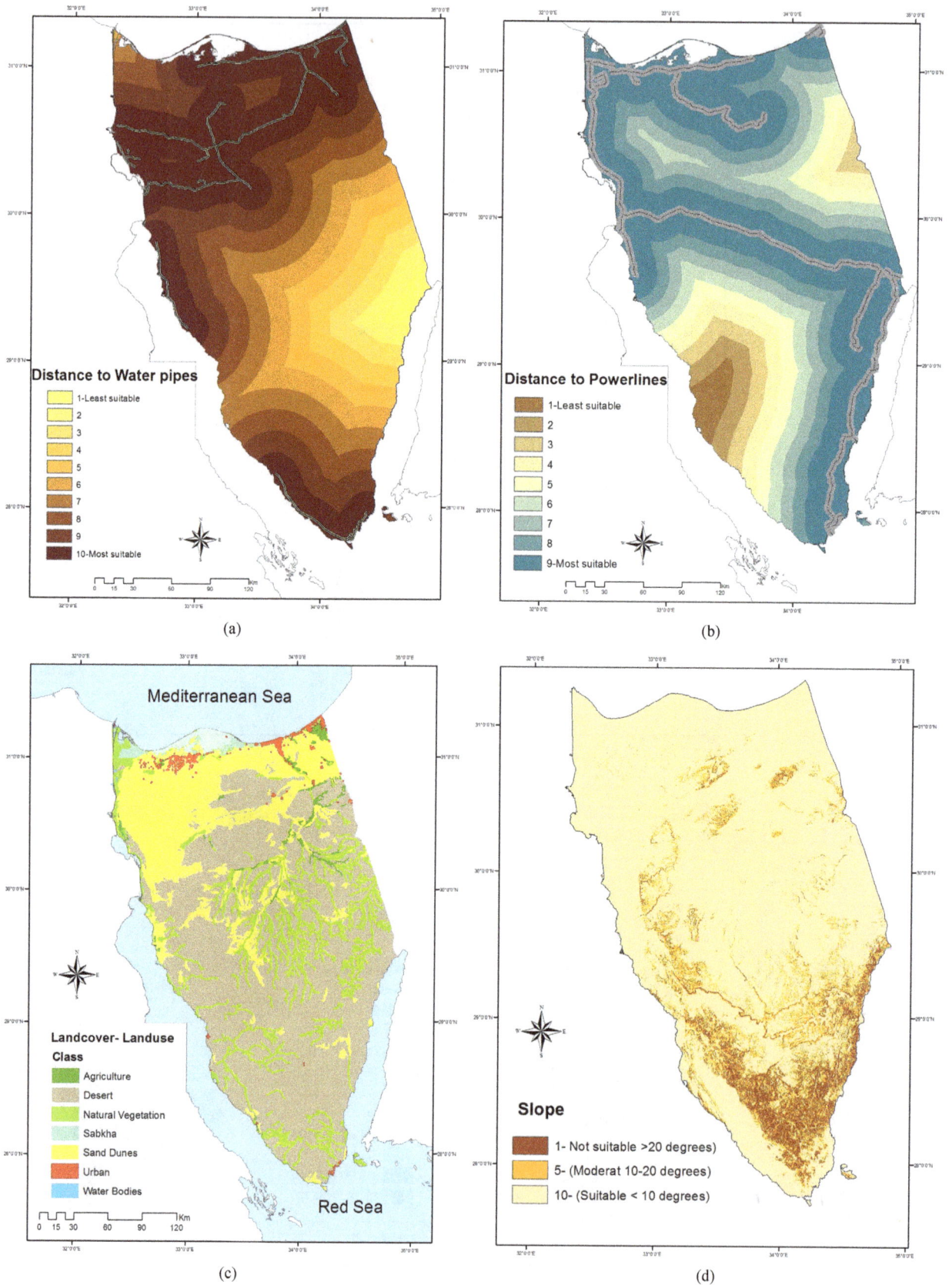

Figure 6. Factors maps for construction costs theme: (a) Distance to water pipes; (b) Distance to power lines; (c) Landcover classes; (d) Slopes.

A Multidisciplinary Approach to Mapping Potential Urban Development Zones in Sinai Peninsula, Egypt Using Remote Sensing and GIS

41

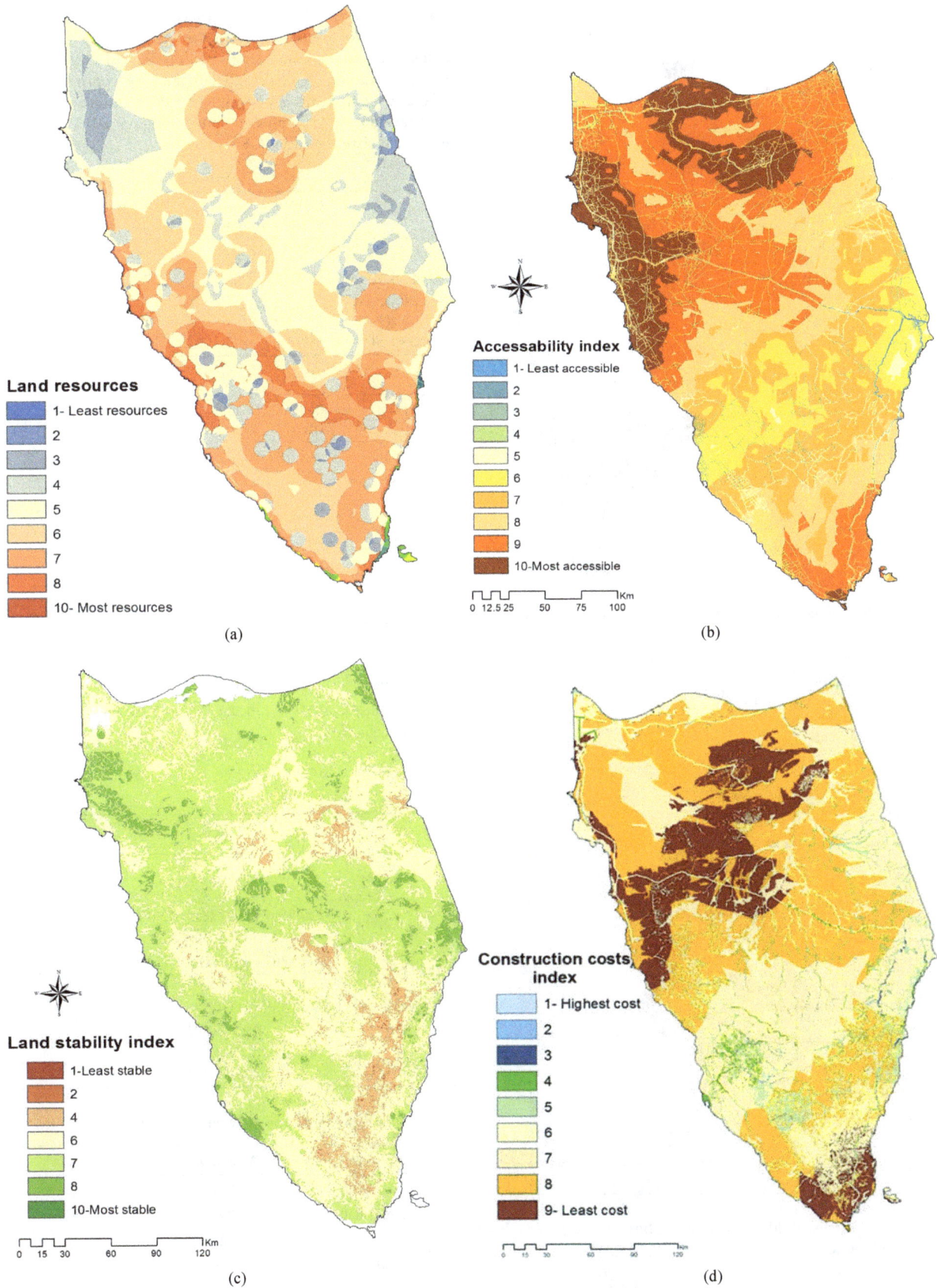

Figure 7. Themes sub-models index maps: (a) Land resources; (b) Accessibility; (c) Land stability; (d) Construction costs.

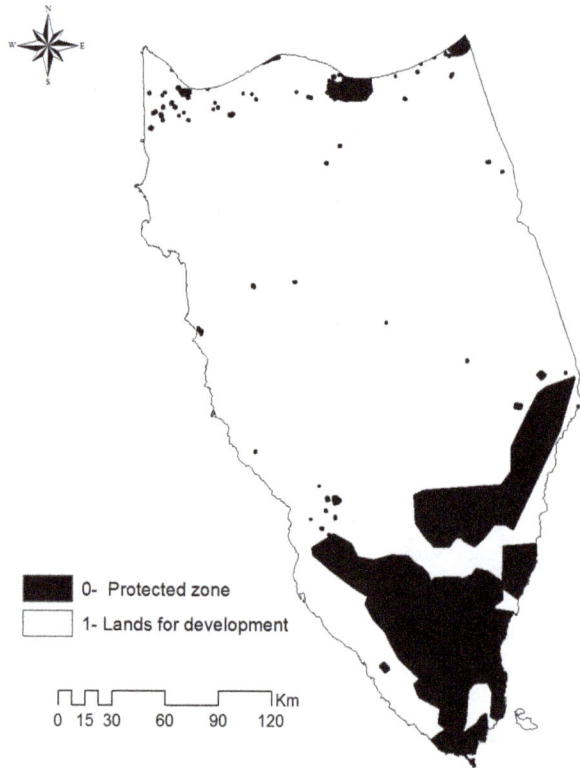

Figure 8. The land protection theme binary sub-model.

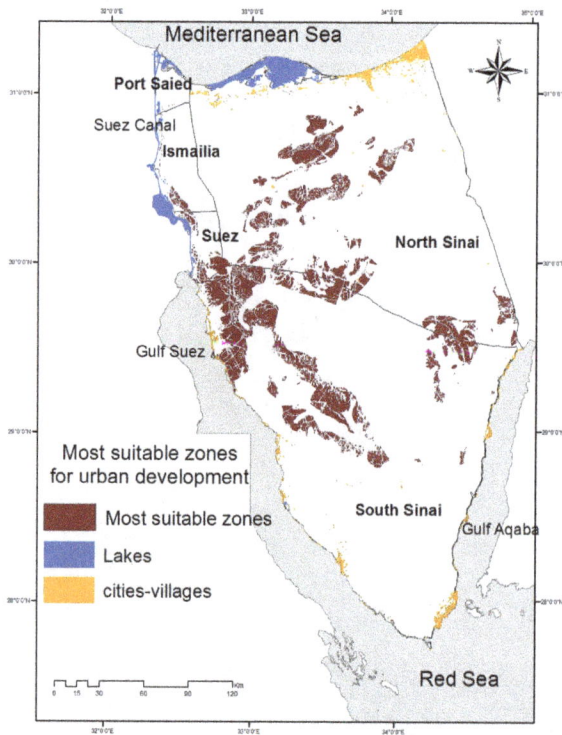

Figure 9. Most suitable zones for urban development in Sinai Peninsula.

transparency, and a sense of shared responsibility for all parties involved in the land-use planning process. This

Table 7. (a)-(c) depicts the analytical hierarchy process AHP used to calculate the factor weights. (d) depicts the calculation of the consistency ratio (CR). (a) Pairwise comparison matrix; (b) Normalized pairwise comparison matrix and computation of weights; (c) Normalized pairwise comparison matrix and calculated weights; (d) Calculation of the consistency ratio (CR).

(a)

	Land safety	Land resources	Construction costs	Accessibility
Land safety	1	9/7	9/5	9/4
Land resources	7/9	1	7/5	7/4
Construction costs	5/9	5/7	1	5/4
Accessibility	4/9	4/7	4/5	1

(b)

	Land safety	Land resources	Construction costs	Accessibility
Land safety	1	1.28	1.8	2.25
Land resources	0.77	1	1.4	1.75
Construction costs	0.55	0.71	1	1.25
Accessibility	0.44	0.57	0.80	1
Total	2.77	3.57	5	6.25

(c)

	Land safety	Land resources	Construction costs	Accessibility
Land safety	0.360	0.359	0.360	0.360
Land resources	0.279	0.280	0.280	0.280
Construction costs	0.199	0.200	0.200	0.200
Accessibility	0.159	0.159	0.159	0.160
Total				1.000

(d)

Criterion	Weighted sum vector	Consistency vector
Land safety	1 (0.36) + 9/7 (0.28) + 9/5 (2) + 9/4 (0.16)	1.44/0.36 = 4
Land resources	7/9 (0.36) + 1 (0.28) + 7/5 (0.20) + 7/4 (0.16)	1.12/0.28 = 4
Construction costs	5/9 (0.36) + 5/7 (0.28) + 1 (0.20) + 5/4 (0.16)	0.80/0.20 = 4
Accessibility	4/9 (0.36) + 4/7 (0.28) + 4/5 (0.20) + 1 (0.16)	0.64/0.16 = 4

Table 8. Area under different suitability categories.

Suitability categories	Area in sq. km	Area in %
Most suitable	5327	8.9
More suitable	29,980	50.0
Suitable	12,630	21.0
Marginally suitable	1590	2.60
Least suitable	76	0.12
Not suitable	10223.9	17.0

A Multidisciplinary Approach to Mapping Potential Urban Development Zones in Sinai Peninsula, Egypt Using Remote Sensing and GIS

43

issue is quite valuable especially in developing countries.

REFERENCES

[1] D. F. Brown, "Back to Basics: The Influence of Sustainable Development on Urban Planning with Special Reference to Montreal," *Canadian Journal of Urban Research*, Vol. 15, No. 1, 2006, pp. 99-117.

[2] I. Bystron, C. MacDonald and J. Stansfield, "Department of Geography," University of Guelph, Canada, 2013. http://www.uoguelph.ca/geography/research/geog4480_w 2013/index.html

[3] L. J. Luo, Z. He and Y. Hu, "Study on Land Use Suitability Assessment of Urban-Rural Planning Based on Remote Sensing—A Case Study of Liangping in Chongqing," *The International Archives of the Photogrammetry, Remote Sensing and Spatial Information Sciences*, Vol. 37, Part B8, Beijing, 2008.

[4] M. Cerreta and P. De Toro, "Urbanization Suitability Maps: A Dynamic Spatial Decision Support System for Sustainable Land Use," *Earth System Dynamics*, Vol. 3, 2012, pp. 157-171.

[5] I. L. McHarg, "Design with Nature," Wiley, New York, 1969.

[6] L. Hopkins, "Methods for Generating Land Suitability Maps: A Comparative Evaluation," *Journal of American Institute*, Vol. 34, 1977, pp. 19-29.

[7] R. K. Brail and R. E. Klosterman, "Planning Support Systems," ESRI Press, Redlands, 2001.

[8] M. G. Collins, F. R. Steiner and M. J. Rushman, "Land-Use Suitability Analysis in the United States: Historical Development and Promising Technological Achievements," *Environmental Management*, Vol. 28, No. 5, 2001, pp. 611-621.

[9] J. M. C. Pereira and L. Duckstein, "A Multiple Criteria Decision-Making Approach to GIS-Based Land Suitability Evaluation," *International Journal of Geographical Information Systems*, Vol. 7, No. 5, 1993, pp. 407-424.

[10] R. Store and J. Kangas, "Integrating Spatial Multi-Criteria Evaluation and Expert Knowledge for GIS-Based Habitat Suitability Modeling," *Landscape and Urban Planning*, Vol. 55, No. 2, 2001, pp. 79-93.

[11] G. F. Bonham-Carter, "Geographic Information Systems for Geoscientists: Modeling with GIS," Pergamon Press, New York, 1994.

[12] J. C. Cambell, J. Radke, J. T. Gless and R. M. Whirtshafter, "An Application of Linear Programming and Geographic Information Systems: Cropland Allocation in Antigue," *Environment and Planning A*, Vol. 24, No. 4, 1992, pp. 535-549.

[13] S. Kalogirou, "Expert Systems and GIS: An Application of Land Suitability Evaluation," *Computers, Environment and Urban Systems*, Vol. 26, No. 2-3, 2002, pp. 89-112.

[14] W. Miller, W. M. G. Collins, F. R. Steiner and E. Cook, "An Approach for Greenway Suitability Analysis," *Landscape and Urban Planning*, Vol. 42, No. 2-4, 1998, pp. 91-105.

[15] R. Gunasekera, "Use of GIS for Environmental Impact Assessment in Interdisciplinary Approach," *Interdisciplinary Science Reviews*, Vol. 29, No. 1, 2004, pp. 37-48.

[16] J. R. Eastman, P. A. K. Kyem, J. Toledano and W. Jin, "GIS and Decision Making," UNITAR, Geneva, 1993.

[17] R. L. Church, "Geographical Information Systems and Location Science," *Computers and Operations Research*, Vol. 29, No. 6, 2002, pp. 541-562.

[18] R. Janssen and P. Rietveld, "Multicriteria Analysis and Geographical Information Systems: An Application to Agricultural Land Use in the Netherlands," Geographical Information Systems for Urban and Regional Planning, Kluwer Academic Publishers, Dordrecht, 1990, pp. 129-139.

[19] Central Agency for Public Mobilization and Statistics, CAPMAS, National Information Center (NIC), Nasr City, Cairo, 2013.

[20] Center of Housing and Building Researches, Ministry of Housing, Utilities and Urban Development, "Egypt, Mines and quarry map of Egypt," 2007.

[21] Shutle Radar Topography Mission, United States Geological Survey (USGS), 2000. http://seamless.usgs.gov

[22] Egyptian Military Survey Department, "The Topographic Map of Egypt," 1995.

[23] Egyptian General Petroleum Corporation, CONOCO-Coral, "The Geological Map of Egypt, Scale 1:500,000," Egypt, Cairo, 1987.

[24] Ministry of State for the Environment, Egyptian Environmental Affairs Agency (EEAA), Department of Nature Conservation, National Biodiversity Unit, Egypt: "National Strategy and Action Plan for Biodiversity Conservation," 1998.

[25] Research Institute for groundwater (RIGW), National Water Research Center (NWRC), The "Hydro-Geological Map of Egypt, Scale 1:200,000," 1999.

[26] ESRI, "Working with the ArcView Spatial Analyst," Environmental Systems Research Institute, Inc., Redlands, 1996.

[27] General Organization for Physical Planning, GOPP, "The Environmental Vision of the Strategy of Urban Development in Egypt," Unpublished Report, GOPP, Cairo, 2007.

[28] J. R. Eastman, W. Jin, P. A. K. Kyem and J. Toledano, "Raster Procedures for Muli-Criteria/Multi-Objective Decisions," *Photogrammetric Engineering & Remote Sensing*, Vol. 61, No. 5, 1995, pp. 539-547.

[29] D. T. Bailey, "Development of an Optimal Spatial Decision-Making Using Approximate Reasoning," Ph.D. Dissertation, Queensland University of Technology, 2005.

[30] T. Saaty, "A Scaling Method for Priorities in Hierarchical Structures," *Journal of Mathematical Psychology*, Vol. 15, No. 3, 1997, pp. 234-281.

[31] T. A. Saaty, "Decision-Making for Leaders," 2nd Edition, RWS Publication, Pittsburgh, 1992.

[32] T. L. Saaty, "The Analytic Hierarchy Process," McGraw-Hill, New York, 1980, pp. 20-25.

5

Cartographic-Environmental Analysis of the Landscape in Natural Protected Parks for His Management Using GIS: Application to the Natural Parks of the "Las Batuecas-Sierra de Francia" and "Quilamas" (Central System, Spain)

Antonio Miguel Martínez-Graña[1], Jose Luis Goy[1], Caridad Zazo[2]

[1]Geology Department, External Geodynamics Area, Sciences Faculty, University of Salamanca, Salamanca, Spain

[2]Section Geology, National Museum of Natural Sciences, Madrid, Spain

ABSTRACT

In this work we report a methodological procedure with an integrated physical-perceptual approach that allows units of landscape in protected natural areas to be differentiated. First, indirect methods were applied by means of a mapping procedure, which identified the physical components of major relevance. We then generated maps of natural units, anlyzing the "printed" landscape of a territory. Secondly, we developed direct methods to identify and describe the reresentative elements of the landscape, analyzing the "perceived" landscape. The identification and delimitation of these landscape units with geographical information systems provide detailed maps facilitate the tasks of planning and management. The procedure was validated by means of its application in two protected natural spaces. The treatment used here considers landscape not only as an aesthetic element but also as something "live" elaborating maps that should be of use in land planning and management of natural areas.

Keywords: Landscape Cartography; Natural Hazard; Territorial Planning; GIS Techniques; Management Protected Parks

1. Introduction

Study of the landscape has developed based on different disciplines—geology, geography, architecture, biology—generating different definitions, including intangible and tangible values of the same one, constituting a multisensory perception of a system of ecological relations that differentiate a perceptible part and an intangible part: functional and causal factors [1]. The landscape is understood as a set of interrelationships derived from the interaction of its natural components: rocks, water, air, plants, animals, and human beings [2] and their disposition and distribution in the area. Currently, increasing importance is being accorded to the landscape since these environmental values are an important factor in human quality of life, and hence landscape studies may offer a solution to problems generated in the planning of urban development and land management [3-5].

The absence of a clear concept of landscape and the difficulties involved in obtaining manageable information in environmental studies have determined the slow development of its analysis in relation to other natural variables. Correct analysis of the landscape is usually complex since it must bear in mind all the components of the physical medium (geology, geomorphology, vegetation, fauna, soils and human activities) and their interacttions [6,7]. This has elicited a multiplicity of approaches, most of them complementary, to the physical medium, although with a common basis: territorial reality, objective procedures being used in the analysis, and subjective evaluations of the natural quality to estimate the way it is perceived or its beauty [8,9].

Currently, the term "landscape" has evolved beyond previous considerations and is held as a resource on the basis of its aesthetic value generated as an entity derived from the interaction between geology, geomorphology, climate, soils, ecology, vegetation, hydrology, fauna and anthropic activity. A correct analysis of the landscape must necessarily involve study of these components, perceived in different ways, both visual and auditory.

In the last decade, land management has become more flexible and selective, both at the conceptual and methodological level, and also at the practical level. Thus such that territorial planning establishes a model that is integrated, on the one hand, by structures and territorial systems that increase its internal cohesion (equipment and public services, human settlements, systems of communications and transport). On the other it is constituted by very natural territorial sectors, with special and prominent constitutive characters that require specific measures of protection and management, since they are protected natural spaces in which the maintenance of geobiodiversity can be guaranteed and the environmental conditions of the spaces can be supported. The proposals for the territorial planning of certain structures (infrastructures and urban areas) must be expressed in spatial terms (locations of actions and zoning), and their analysis and diagnosis influences the processes and factors that determine the evolution of the territory and its associated landscape [10-12].

Landscape analysis is useful and effective in territorial planning since it contributes to the correct location and disposition of elements and use of the territory, uncovering the degree of reception and the impact of the use of the physical space. Accordingly, the landscape constitutes a meeting point between the technical, scientific, social and political aspects, allowing civilian participation in proposals of territorial planning (since the term landscape includes the physical spaces where people carry out their daily activities for reasons of work or residence), establishing aims for the conservation of the landscape quality of the territory, as established by the European Landscape Convention.

2. Background

Strong but unequal urban and industrial development and that of other human activities in recent decades have influenced landscape quality directly. In 1985, Directive 85/337 regulating Environmental Impact Assessment [13] portrays the landscape as a resource that is subject to rapid deterioration and is difficult to replenish. Thus, in territorial planning it is necessary to attend to its preservation and conservation, and it must be analyzed as another, more independent, factor apart from the abiotic, biotic and socioeconomic factors. Later, in 1992, by means of the European Perspective of Territorial Planning the European Union established the need for an interrelationship between the landscape and territorial planning, favoring as from 1999 an approach towards a balanced and sustainable territorial development across the European Territorial Strategy, as indicated in the Declaration of Lisbon. In Spain, the National Plan of Thematic and Environmental Cartography considers the landscape as a strategic variable, whose mapping defines

the landscape as homogeneous units derived from the sum of their components and elements [14].

The European Landscape Convention was convened in Florence in 2008, and the need to determine the effects on the landscape of certain urban development plans and projects, by means of analysis and detailed reports of landscape incidents, was discussed. In Spain the European Landscape Convention was ratified in February 2008 [15]. This promotes measures of protection, management and planning, and the identification and qualification of the territorial landscapes of each region. It also analyzes their characteristics as well as the pressures transforming them, and establishing the aims of landscape quality on the basis of the particular values of each sector, taking into account the participation of the population. In Spain, the drawn-out delay in the implementation of infrastructures during the last decades has led to less development and hence a lower degree of territorial development. Bearing in mind the direct relationship between the degree of use of a territory and its conservation, this implies greater conservation (uncontrolled territories subject to little attention offer a more "natural" image). The low density of the population and its irregular distribution have generated an interrelated mosaic of very diverse landscapes, which determine the territorial perceptual reality of Spain, with a high intrinsic quality unlike the rest of Europe. This is confirmed by the fact that approximately 30 % of the territory is under some degree of national and/or international protection: Natural Protected Spaces, Zones of Special Protection for Birds, and Sites of Community Importance [16,17].

Since 2006, the integration of the landscape in territorial planning has been implemented first by means of strategies and territorial directives that involve treatment of the landscape at general level, defining types and processes that generate them, passing through European, state or autonomous scales. Secondly, it is addressed in plans and planning projects where an analysis of the formal landscape, at regional or local scale, is carried out under the aegis of the procedures of Strategic Environmental Assessment and Evaluation of Environmental Impact. This includes the establishment of maps of landscape units in which these characterize each type of landscape, integrating the components (geological, geomorphologic, botanical, historical and cultural), uses and present activities. In this way our understanding of the landscape and its evolution, especially in natural spaces of interest, is established for application of the European Landscape Convention [18].

At regional level, specific laws have been enacted, such as in Catalonia with Bill 8/2005 concerning the protection, management and planning of the landscape [19].

This regulates studies and reports concerning impact and landscape integration, and establishes the need to elaborate landscape catalogues. These offer analyses of descriptive and prospective documents determining the typology of the landscapes of a Community, identifying their values and state of conservation, and proposing quality objectives to be achieved, including the mapping of landscape units, maps of landscape fragility, maps of landscape evaluation (integrating risks and impacts) and maps of landscape quality. This legislation also establishes the drafting of landscape charters as instruments for the reconciliation of strategies between public and private agents with a view to implementing measures of landscape protection, management and planning. It also underscores the need to perform studies addressing landscape integration, determining the consequences of actions and of civil building projects on the landscape. Currently, the landscape, considered as a resource to be conserved and preserved, has a strong impact on the po-

pulation, and is a component included within the mapping of the natural heritage as a thematic layer determined by the territorial sectors of greatest landscape quality, and hence those to be protected [8]. In the present work our intention is to improve a methodology aimed at obtaining a mapping of landscape units that will allow us, easily and at low cost, to generate a landscape catalogue for all types of natural spaces.

3. Methods

The study of the landscape is linked to the study of natural areas and to that of their evolution and transformation, either as a consequence of natural processes or as a result of human intervention. The landscape, within the context of the perception of the environment, plays an important role in human well-being and quality of life [20]. Accordingly concerns about systematic landscape analyses are steadily increasing. The methodology used here (**Figure 1**, top) allows Landscape Units to be mapped by

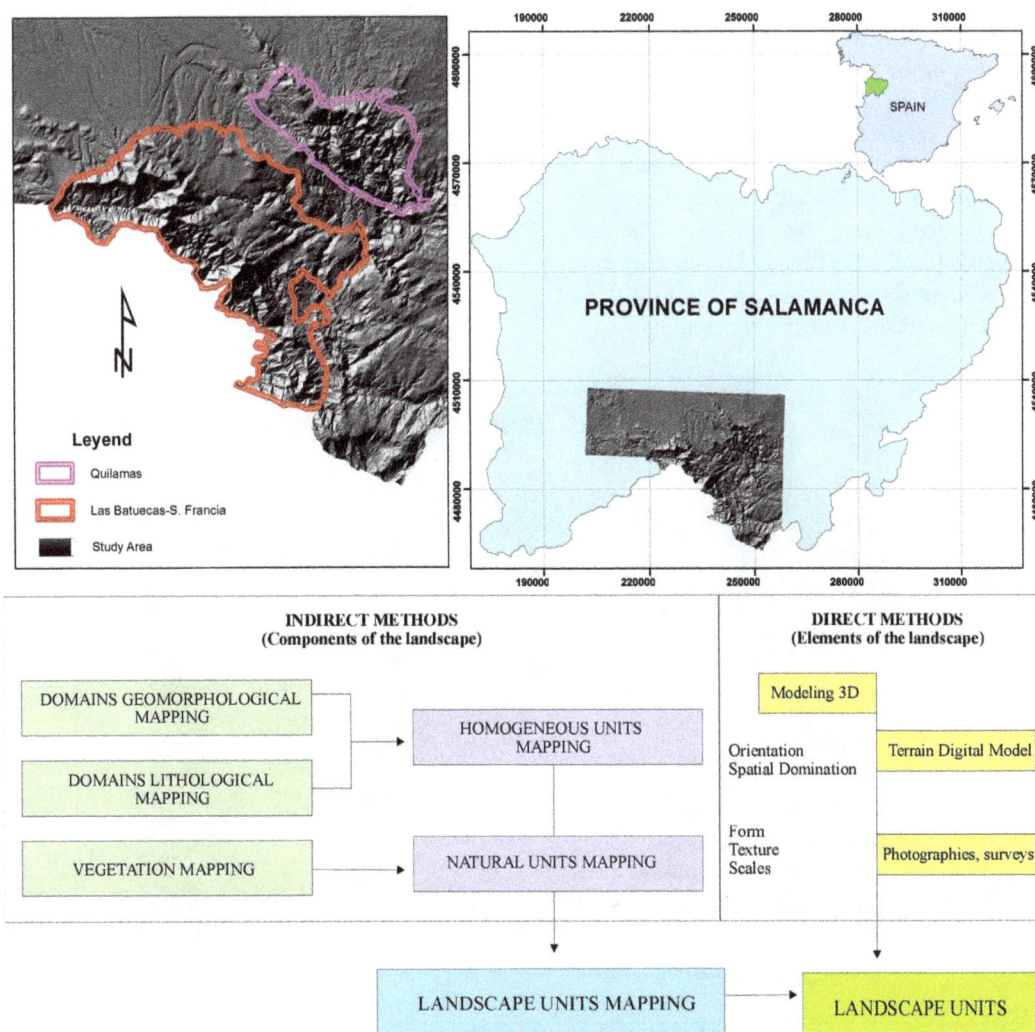

Figure 1. Study area (top) and methodological scheme (down).

integrating indirect methods that analyze the total land-scape, regrouping natural components [21,22], and with direct methods, describing the elements of the "aesthetic" landscape [23,24].

The methodology is confirmed and validated, applying it in two natural spaces protected from the Spanish central mountain system: the Las Batuecas-S. de Francia park and the Quilamas park (**Figure 1**, lower panel). These natural parks were chosen because they show a direct relationship between landscape quality-visual landscape-and the geological-geomorphologic features, and also feature important areas of geological interest. This value of landscape quality may be reduced owing to anthropic activities, such that landscape analysis is mandatory in planning procedures, both for specific projects (the mining industry and infrastructures (e.g., highways, railroad, housing estates…) and regional strategic projects (housing estates, industrial areas and recreational areas). This analysis is performed from the geological standpoint, using cartographic representation techniques of the natural components, together with aspects from direct observation of the terrain as a new approach for the physical-perceptual analysis of the territory.

To perform the landscape analysis, we divided the territory spatially into landscape units, constructing an ordered and coherent aggregation of the elementary components of each sector, which were classified, re-classified, valued numerically, and then mapped. The landscape unit is a structurally, functionally or visually coherent territorial area subject, partly or totally, to different regimes of protection, management and planning.

In the first phase, we applied indirect objective and quantitative methods based on the evaluation of the constituent components of the landscape, determining the natural units, and their value was calculated from a few weighting coefficients. This was the theoretical valuation of the content of the "image" of a territory. The abiotic factors—geomorphology and lithology—are the most representative and characteristic components of the landscape in the Protected Natural Parks of Las Batuecas-Sierra de Francia and Quilamas [25]. This method allowed us to identify the landscapes typical of the study sector on the basis of their characteristics and cartographic distribution through the components and elements of the environment. Thus, we obtained a functional methodology for the elaboration of a mapping of units of landscape on the basis of natural homogeneous units—geomorphology-lithology-vegetation, which should be effective for the territorial planning and assignment of land use. In addition, it allows an inventory and a zoning to be made of the places of major perceptual relevance in order to implement legally approved categories of landscape protection. The geomorphologic component was obtained from the synthesis of the maps of geomorphologic units, obtaining the cartography of geomorphologic domains, which can be re-classified according to their representation in the landscape, becoming simplified in the following domains and with their value perceptual weighted: summits (1), watersheds (1), hillsides (2), fitted fluvial valleys (3), surfaces, glacis and pediments (4), fluvial and tectonic scarps (5) and fluvial terraces and valley floors (6) (**Figure 2**).

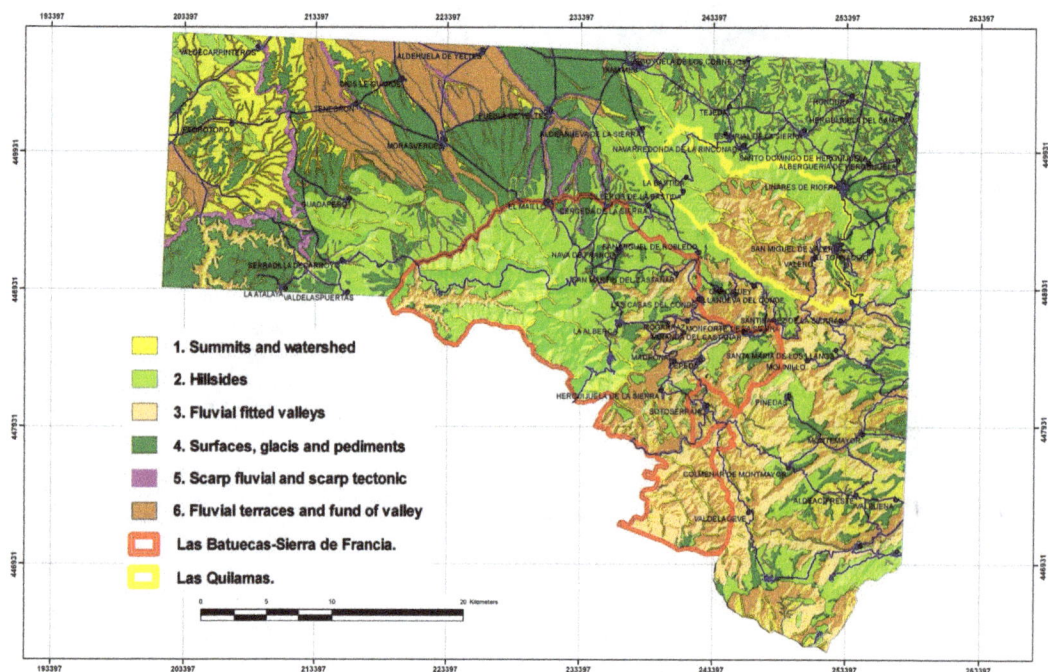

Figure 2. Mapping of geomorphological domains.

The lithologic component can be established from a synthesis of the geological mapping, extracting the lithological zoning and simplifying the lithological units according to their landscape effect, as: granites (A), quartzites (B), slates, schists and greywacke (C), limestones and dolomites (D), reddish sandstones and brown arkose (E), and conglomerates, gravels, sands and silt (F) (**Figure 3**).

Integrating the geomorphologic and lithologic maps by means of Geographic Information System—GIS-(ArcGis v.10) techniques, the different units were obtained and simplified, eliminating and/or grouping some on the basis of the following criteria:

1) Very small and dispersed units, which are not perceptually representative and are included in other larger, well-known units.

2) Units that are similar to others, such as for example the groups "2.B" hillsides developed over quartzites, and "2.C" hillsides on slates, would remain integrated within the dominant group "2.C", since at landscape level they behave equally. Likewise, units "3.A", fluvial valleys.

3) The grouping of some units, which are thought to go together owing to their perceptual impact. These units constitute relatively homogeneous portions of the area as regards environmental conditions and landscape components.

4) Elimination of some "mistaken" units generated

with the layer of the limits of the study zone.

Finally a map with 14 Homogeneous Units was obtained (**Figure 4**). Once the Homogeneous Units had been obtained, the third component, which affects the landscape decisively was integrated: *i.e.*, vegetation (**Figure 5**). To accomplish this, the vegetation map was reclassified for use in the landscape, simplifying the vegetation units in four groups: Arboreal Formations (FA), Sub-shrubby and shrubby Formations (FD), mixed Wooded Formations (FM) and Pastures, Crops and Fallow land (PaCyB). In some sectors there is no vegetation (SV). These vegetation units were grouped as described below: Arboreal Formations (FA): including coniferous (repopulation with pines), leafy evergreen and deciduous (sclerophyllous) forest, dehesa (savannah-like terrain) with holm-oaks, deciduous forest, dehesa of oaks and ash-trees and mixed leafy species (mixed formations of deciduous and sclerophyllous).

Sub-shrubby and shrubby Formations (FD): including all the scrubs, that is both Shrubby Formations (fructicose shrub formation, outcrops with rockrose and heather) and Sub-shrubby Formations (lavender, thymes, brooms, and Genista shrubs) fitted within granite, and "4.C", surfaces on slates and granites, were integrated since they are intermingled over the landscape, thus defining specific singular units. Finally, units "5.C": scarp on slates and "5.E", scarp on conglomerates and sandstones, were

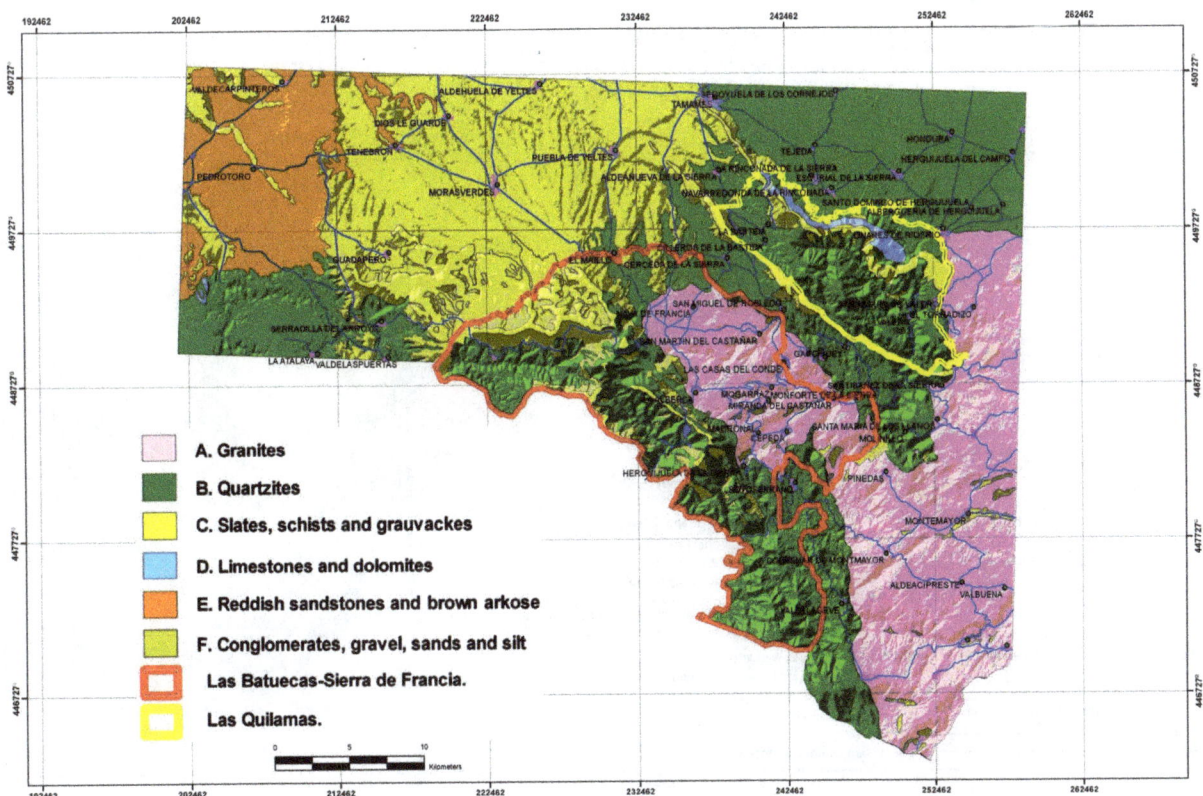

Figure 3. Mapping of lithological domains.

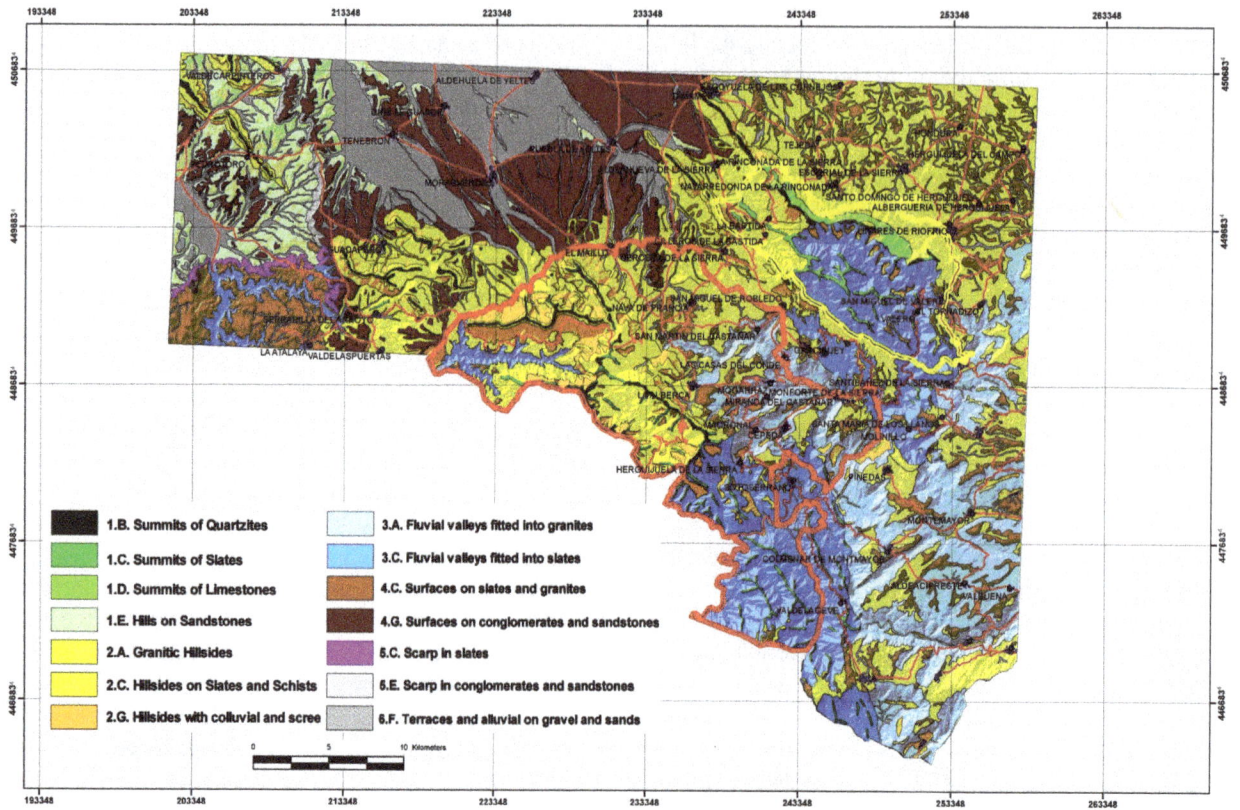

Figure 4. Mapping of homogeneous units.

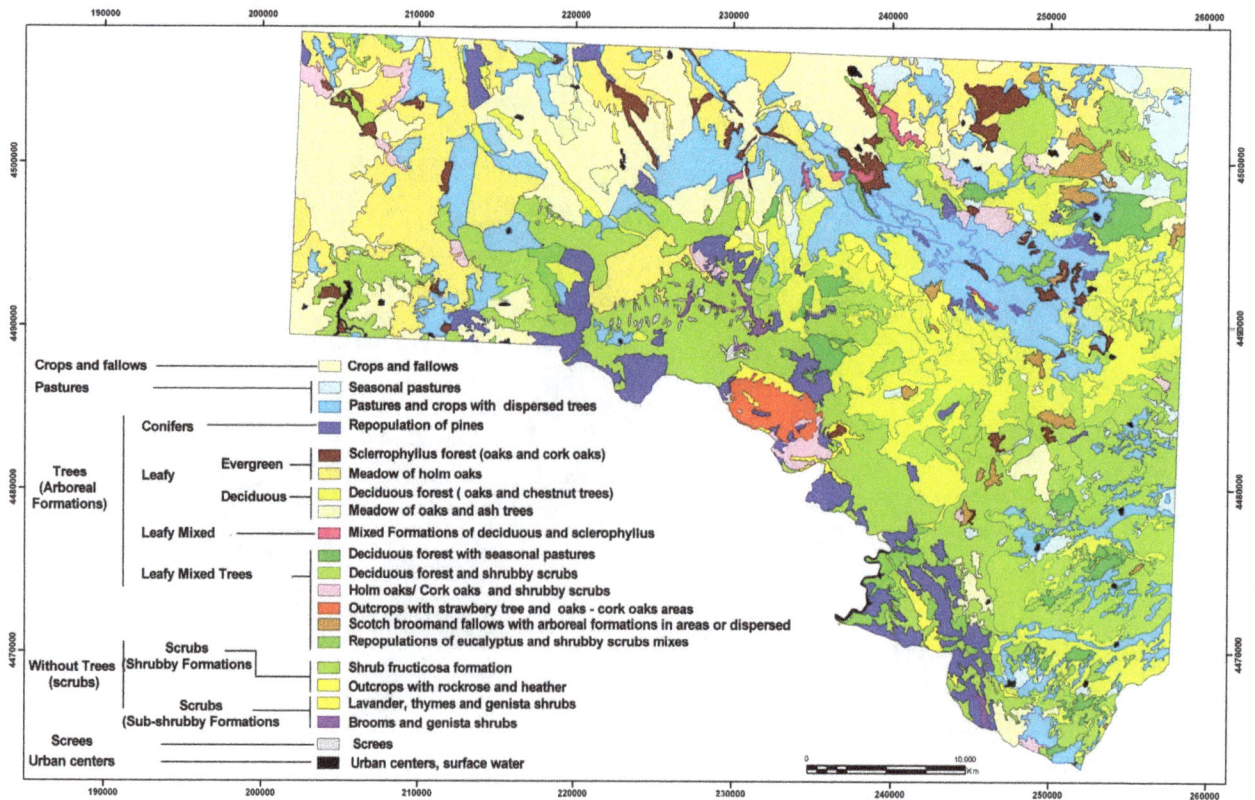

Figure 5. Mapping of vegetation.

grouped because they affect the landscape, especially the geomorphologic domain, but not the lithological differences.

Mixed Wooded Formations (FM): including a mosaic of deciduous forest with seasonal pastures and shrubby scrubs, deciduous forest with scrubby (sub-shrubby) formations, oak-cork and oak areas, scotch broom and genista shrubs with arboreal formations, either dispersed or in clumps, repopulations of eucalyptus and shrubby scrub mixes.

Pastures, Crops and Fallow Land (PaCyB): including vivacious seasonal pastures and pastures with dispersed arboreal elements. In this unit, we integrated the crops and fallow areas.

Integration of the mapping of homogeneous units and the mapping of the vegetation simplified for the landscape generated cartography of natural units with 21 units, some of them lacking perceptual importance, and others very delimited within each unit. Accordingly, we were analyzing the possible integration of the type and spatial position of the vegetation units in the homogeneous units, obtaining a mapping of natural or environmental units (**Figure 6**).

The methodological procedure developed prioritizes the groups of landscape units on the basis of geomorphological domains, since they are units that print the spatial configuration of the relief. We also took into ac-

count the lithological units in sectors where they had added landscape value, such as high and slope sectors, coinciding with shrubby and sub-shrubby or mixed formations, understanding that in zones with wooded or mixed formations the density and/or wooded stands eliminate the differential colour of the lithological substrate.

Accordingly, and in light of the peculiarities of the study zone, on the slopes we differentiated the lithological unit of the slates, which impose a green-to-dark-grey colour, and the light colour of quarzites. In the summit and ridge domain, Armorican quartzite predominates with respect to the other lithologies (granites, slates...), which is of special relevance, since with their original white crests they stand out on the terrain.

We next addressed the type and disposition of the plant masses, grouping them according to their singularity, the degree from which they stand out from the surroundings according to their height and nature, in different geomorphological domains and in four groups: Wooded formations (FA), Shrubby and sub-shrubby formations (FD), Mixed formations (FM) and in more anthropic units (PaCyB: pastures and crops, fallow land). Once we had reclassified the natural or environmental units, we obtained 11 landscape units that provided the cartography of landscape units (**Figure 7**).

In the mapping of the landscape units, we observed

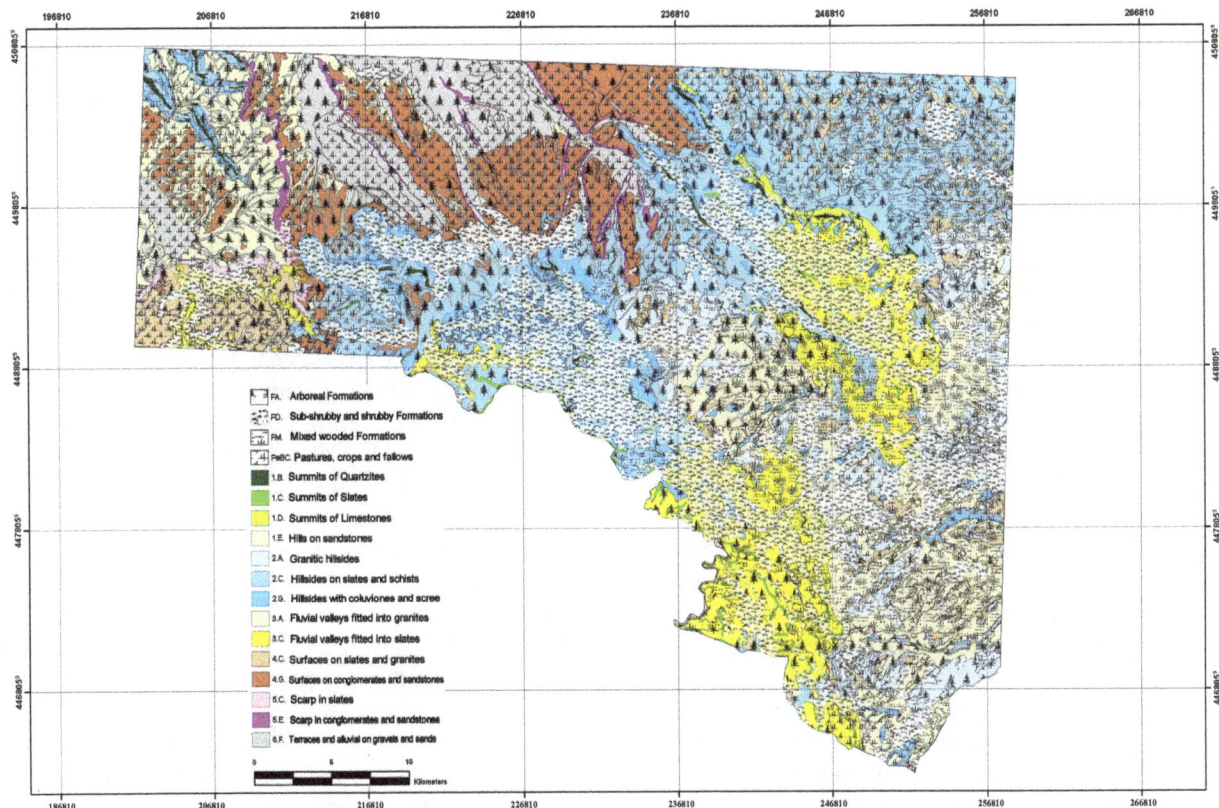

Figure 6. Mapping of natural or environmental units.

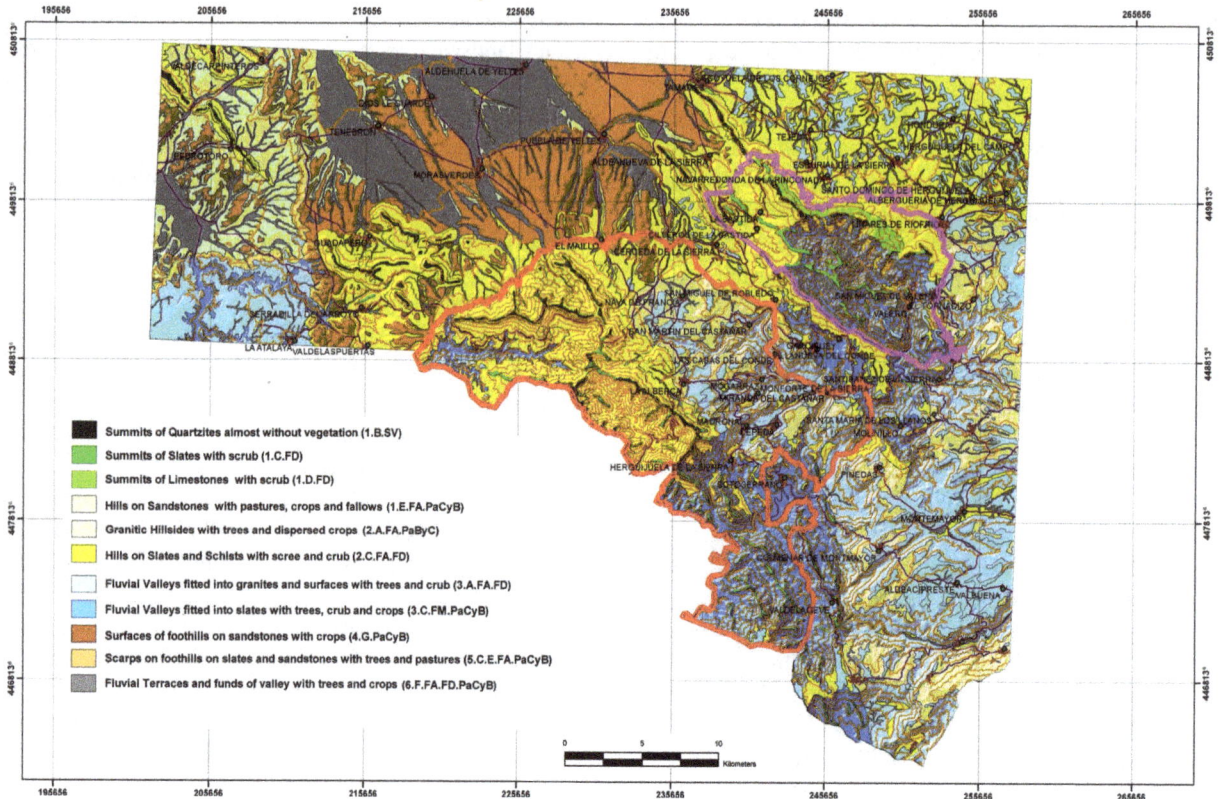

Figure 7. Mapping of landscape units.

that the highest sectors (Unit 1: 1.B.SV, 1.C.FD, 1.D.FD, 1.E.FA.PaCyB, green colours on the map) show the following units: summit (Ridge) sectors without vegetation (SV) covered by colluvials, especially screes, which owing to their disposition and colour stand out perceptually (visually). It is also necessary to mention the hill and hillside sectors with wooded formations (FA) and, to a lesser extent, pastures and croplands-fallow lands (PaCyB) in the western sector. These are not very abundant units although they are representative of the sub-shrubby and shrubby formations of the summits and scarps (El Maillo-Puebla de Yeltes).

The hillsides (Unit 2: 2.A.FA.PaCB, 2.C.FA.FD, yellow on the map) are highly representative domains in the centre of the Natural Area of Las Batuecas-Sierra de Francia and in the NE sector of the study zone, wooded formations being outstanding in the SW sector, with large patches of shrubby and sub-shrubby formations predominating in the rest of the natural park on the hillside of the S sector. In the NE sector there is a predominance of pastures and croplands-fallow land.

The sectors of fitted valleys Unit 3: 3.A.FA.FD, 3.C.FM.PaCyB, in blue), are distributed within the natural parks of the Quilamas and the S and SE parts of Las Batuecas, with a predominance of shrubby and sub-shrubby vegetation (FD), in the between-sierra sectors (Garcibuey, Sequeros), followed by mixed formations (FM)

in zones adjacent to the previous ones but of lesser extension in the eastern zone (Fuente de San Esteban, Linares...) and SE (Pinedas, Montemayor del Río...). The units with wooded formations are less well represented and are distributed in the S and SE sector surrounding the river valleys.

The mainly flat surfaces (Unit 4: 4.G.PaCyB, in brown) are found at the periphery of the natural parks, dispersed among the granitic valleys, or are seen in the N and E sectors as discontinuous below pasture formations over sediment-filled surfaces (N sector) and over erosive surfaces (E sector). The pediment has mixed formations, showing highly dispersed remnants of shrubby and sub-shrubby formations (FD). Fluvial and tectonic escarpments (Unit 5: 5.C.E.FA.PaCyB, in orange) are important on the right bank of the Yeltes river, the fluvial escarpment that goes from El Maillo to Aldehuela de Yeltes being displaced and retracted, as is the tectonic escarpment, to the NE of the study zone.

Finally, the terraces and valley floors (Unit 6: 6.F. FA.FD.PaCyB, in grey), mainly associated with the river courses of the northern zone of the study area (Tenebrón, Dios le Guarde, Morasverdes...), have wooded formations (FA) in the northern and eastern sectors, and on the valley floors there is a preponderance of riverbank wooded formations and/or repopulated forest.

This systematic identification of the landscape by

means of indirect methods, from the inventory of the different components forming it, allows easy integration and characterization of the landscape mapping, which is of great use for environmental planning.

In a second phase the landscape analysis was complemented with direct methods [26], describing the different landscapes on the basis of their visual characteristics by direct observation of the terrain and photos taken during the different seasons of the year and then performing a subjective evaluation. We performed an analysis of the visual perception and aesthetics of each sector, without separating their components, such that the landscape would offer a perceived and printed "image" of a territory.

To accomplish this, different direct methods of controlled subjectivity were used, with evaluations by categories. We noted that this method was strongly influenced by the observer, who shows a descriptive attitude conditioned by his/her individual expectations (educational level, age, social status…). With the mapping of homogeneous units we carried out an analysis of the most representative units of the study zone, assessing the areas of singular landscape-touristic interest according to questionnaires concerning the best evaluated components and elements present in the zones, and identified the different landscape components with 3D modeling (**Figures 8(a)** and **(b)**) from the Digital Model Terrain on which

we implemented different thematic layers (geomorphology, lithology and vegetation). We also used direct methods of representative subjectivity, in which the values were obtained from questionnaires given to population groups, and also from fragments of the terrain by photos, which always give only a limited view of landscapes, then making a description from different observational points of the landscape elements (colour, scale, textures, shapes, lines and spatial distribution) in each perceptual scenario, visual organization of the park (existing visual contrast: between colours and shapes, spatial dominance [23], visual vulnerability and visual absorption capacity (**Figures 8(C)** and **(D)**).

Both methods were used in the territorial analyses. Initially, the direct method predominated owing to its ease of application. Currently we are starting to develop indirect methods, in parallel with the advance in IT tools for cartographic treatment, such as geographic information systems (GIS). However, the assessment of a landscape via direct methods (subjective assessment by the observer) does not necessarily have to coincide with the indirect assessment of the quality of that landscape. For example, an arid landscape with certain endemisms may be of high value as regards its specific nature (e.g., biological/geological interest) but, in contrast, may not be particularly attractive to the aesthetic preferences of most people.

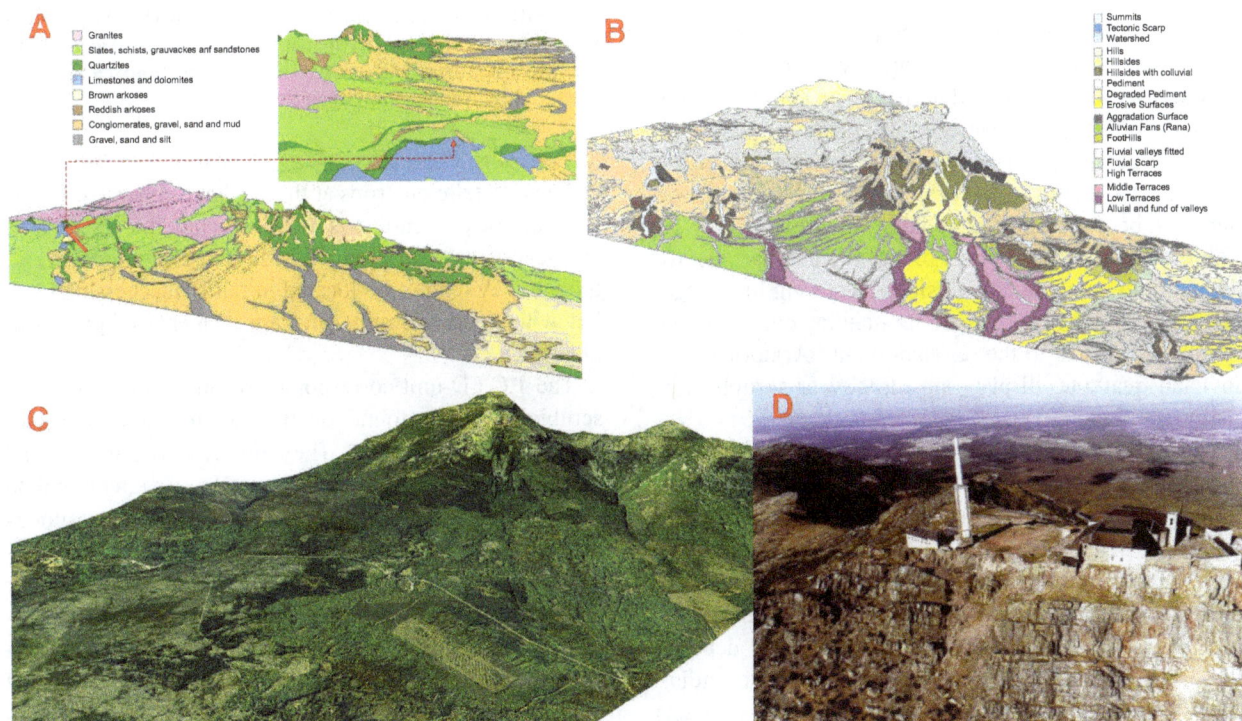

Figure 8. 3D Models showing the spatial distribution of the lithology (A), and of the geomorphological domains (B). Digital Terrain Model of the "Peña de Francia" (C) and dominant lithology in the landscape of the summit of the "Peña de Francia" (D).

4. Results

Analysis of the landscape of the natural parks of Las Batuecas-Sierra de Francia and the Quilamas, the attitudinal and topographic variation, the geobotanical differentiation and the different forms of human activities and life-styles govern the occupation of the territory and allow a landscape differentiation to be made according to the large geomorphological domains that form part of the landscape units, characterized by their natural components and elements, as detailed below.

Saw Landscapes (Units of Landscape: 1.B.SV, 1.C.FD, 1.D.FD, 1.E.FA.PaCyB). The units included in the saw sectors are open, or panoramic, landscapes in which the limitation to sight goes far beyond the point at which elements can be properly recognized. It is possible to note the disposition of the regional relief, with a predominance of horizontal lines, and the sky occupies a large part of the scenic background. These panoramic landscapes are seen from the hills and summits. The quartzite outcrops, which form abrupt walls and steep watersheds, stand out over non-arboreal formations in broad sectors of the upper slopes of the Quilamas and Francia sierras and the surroundings of Guadaperro.

In this sector the landscapes are ordered as a function of their altitude, which in turn determines the climatic conditions and hence the geodynamic processes (screes and slides), the installation of the vegetation and human settlements.

The 1.B.SV landscape unit corresponds to quartzite summits almost completely lacking in vegetation. In these high sectors, there is a predominance of abiotic components since the forest vegetation disappears and is replaced by shrubland and pastures, which are elements of lesser landscape interest than forest. By contrast, the abiotic elements become more abundant and stand out more, with the presence of summits of Armorican quartzite, rocky summits, screes and high slopes featuring little vegetation. These outcrops of quartzite standing out at the top, many of them due to the resistant nature Armorican Ordovician quartzite, display sub-shrubby or shrubby formations.

The above sharp outcrops of quartzite, periglacial morphologies such as detritus left by the snow cover and continuous structures on the summits draw attention to the landscapes featuring such elements in environments relatively close to the observation points (**Figure 9(A)**). At perceptual level, these sectors offer a panoramic space, with lines of diffuse borders due to periglacial deposits with a medium-sized grain texture in groups. Depending on the vegetation they display important colour contrasts (intense flowering of broom and grasses), especially in spring and autumn. They form zones of singular scientific interest as well as being sectors of considerable co-

Figure 9. Saw Landscapes: Summits of quartzites with structures of periglacial (snow cover) in high areas of the hillside (A). Outstanding morphologic intensity of crest quartzites whose structural lithological alignments attract visual attention (B). Summits and crests in Las Quila mas, with arboreal and sub-shrubby formations, where the incisions of the water courses are striking (C). Summits where hills and crests alternate, defining lines on the horizon. Wooded and different zones of shrubs are differentiated (D). Limestones hills with alignments due to karstification (E). Sandstones hills conferring sinuosity to the area, with natural and anthropic (pastures and crops) forms. The landscape is reticulated by human activity (F).

lorfulness and beauty.

The quartzite outcrops at the top have generated highly vertical rocky headlands of landscape interest. These landscapes have a strong lithostructural component, although they tend to be located in rather inaccessible sites but with good visibility, and are elements of great singularity.

The 1.C.FD unit corresponds to summits of slate with scrubland and is found on the summits and ridges of Tamames (**Figure 9(B)**). They are present as gentler and more rounded morphologies, attracting less attention than the previous unit (**Figure 9(B)**) Unit 1.D.FD corresponds to limestone ridges and summits with scrub, also representative of some sectors of the Quilamas natural park (**Figures 9(D)** and **(E)**). The 1.E.FA. PaCyD unit is located in zones of hills and hills on sandstones, of low altitude or close to escarpments, where there are very well developed pastures and/or crops, as occurs with the vicinities of Serradilla del Arroyo, Serradilla del Llano (**Figure 9(F)**) and Tenebrón.

Hillside landscapes (2.A.FA.Pa.CB/2.C.FA.FD land-

Cartographic-Environmental Analysis of the Landscape in Natural Protected Parks for His Management Using GIS: Application to the Natural Parks of the "Las Batuecas-Sierra de Francia" and "Quilamas" (Central System, Spain)

55

scape units). These form transition zones between the highest sectors—summits, ridges and hills—and the lowest sectors: river valleys that are more or less steep-sided and alluvial plains. These sectors exhibit considerable erosive and depositional dynamics.

They feature singular granitic morphologies in which there are important megaforms and microforms, together with periglacial structures (striated surfaces) and arranged in angular blocks. There are also fluvial-torrent shapes. In the granitic zone there are complex forms with diffuse borders and a coarse texture with little contrast.

The very steep-sided slopes features scree and a texture ordered by the disposition of arboreal groups, with defined edges and they are three-dimensional in shape with a distance effect that is marked by the extent of the hillside. The gentler slopes display three-dimensional forms, lines with defined edges, an ordered texture and offer a panoramic space.

Unit 2.A.FA.Pa.CyB corresponds to landscapes of granitic hillsides with arboreal formations and dispersed pastures and croplands, as may be seen at San Miguel de Valero, San Esteban de la Sierra… In some granitic sectors, on the hillsides it is possible to observe characteristic forms of this type of lithology, such as granite boulders, hanging boulders, indentations, mounds and tafoni, mixed with mixed forms, although these cannot be differentiated since on the terrain they are not well visible since they are integrated in mosaics of plant formations in groups or are not contrasted because they are close to water courses in fitted valleys and outstandingly steep fluvial-torrent landscapes, as is the case of the River Alagón during its passage through San Esteban de la Sierra and San Miguel de Valero (**Figure 10(A)**).

The 2.C.FA.FD unit includes hillsides on slates and schists with trees at the bottom and colluvials and shrub on the medium-high part. This unit is seen to the N of Tamames, in the Honfría sector at Linares de Riofrío, and in the valleys of Las Batuecas (**Figure 10(B)**). In some sectors these hillsides show shrubby and/or sub-shrubby formations, such as in the Quilamas, with steeply sloped sides. The hillsides exposed to shade show greater diversity, with shrubs on the sides of paths and covers of broom, thyme, lavender and an abundant array of herbaceous species. Where there are mixtures of arboreal and shrubby-sub-shrubby formations, these units stand out from the substrate, as occurs in the neighborhood of Guadapero and some sectors within La Bastida district. On the lower parts of the hillsides, there are certain plant formations that are dominant in the perceptual environment, either due to their variety, structure or their plant density. Examples are the well conserved forests of oak, chestnut and pine, together with the sectors featuring heather on the upper parts of the hills, clearly differentiating the altitudinal structuring of the

Figure 10. Hillside Landscapes: Hillsides with a granitic morphology, developed due to the alignment of fractures and cracks, with dispersed arboreal vegetation (A). Landscape showing quartzite crests without vegetation, with steeply sloped screes on the high part of the hillside, and with less dense arboreal forest owing to the lack of soil at the (B). Hillsides in Las Quilamas with colour contrasts in stands of chestnut-trees and oaks, with holm-oaks, oaks and dispersed pines. Highly contrasted arboreal formations and two-dimensional forms (C). Sinuosity of the traces followed by the fluvial network; reliefs fitted in the Las Batuecas valley(D). Symmetric valley of the river Francia, fitted and two-dimensional forms. Miranda del Castañar (E). Valley fitted in a substratum of gray slates and light coloured pastures of The Bastida.

arboreal and shrubby-subshrubby formations, as is the case of the slopes of the Quilamas (**Figure 10(C)**).

Landscapes of fitted valleys (3.A.FA.FD, 3.C.FM. PaCyB landscape units). The fitted valleys are sectors where the relief is of special importance, creating particular landscapes of great scenic value and a high degree of natural preservation. They are enclosed landscapes, where the visual limits are close to the point of observation owing to the existence of steep walls that act as visual barriers, closing off the scenic view. These landscapes are seen in the interior valleys of the Batuecas and the Quilamas, which have undergone frequent gravitational events such as rocks slides and falls. There are two landscape units represented in the fitted valleys, although they are the most spatially extensive ones and hence important, above all in the S and SE part of the study zone.

The first is the 3. A.FA.FD unit, which is formed by river valleys fitted in granites with erosive surfaces and featuring dispersed trees and bushes (**Figure 10(D)**). On the hillsides, there are rocky outcrops; these depend on the depth of the soil, above all in sectors with resistant granitic bedrock, allowing arboreal formations of oaks and holm-oaks, even in the surrounding of areas with fruit-trees, such as in Valero and San Esteban de la Sierra. The 3.C.FM.PaCyB unit represents valleys fitted in slates, with groups of trees, bushes and disperse cropfields. These are river valley sectors with an important tree density, associated with riparian forest. Important examples are the riverside forests found in the valley of Las Batuecas (**Figure 10(E)**). In general, the valleys associated with the saws of Las Batuecas-Sierra de Francia and Quilamas show a considerable visual barrier, with V-shaped morphologies and narrow vistas that increase the perception of the surrounding intrinsic landscape. The basin vistas are lengthened owing to the space allowed by the course of the rivers (**Figure 10(F)**). Also, differences in level indicate their young age, which is highlighted by specific geomorphological characteristics: rapids, scoured bottoms with pits and cascades, such as that found in the valley of Las Batuecas.

Scarp landscapes (5.C.E.FA.PaCyB Landscape Units). These correspond to sectors with scarps over slates and sandstones and conglomerates, with disperse trees and pastures. This type is geometric in shape, with defined edges, a coarse texture and considerable contrast. The scale shows a noteworthy effect of site and is highly attention-drawing. This unit is outstanding in the case of the river escarpments, such as the linear trace of the escarpment that runs from El Maillo to Aldehuela de Yeltes (**Figure 11(A)**), which has generated an asymmetric valley through retraction of the southern margin coinciding with the scarp. Additionally, in the western zone of the sector there are scarps on sandstones that have govern the morphology of the landscape, as is the case of the tectonic escarpment of the W sector.

Surface landscapes (4.G.PaCyB Landscape Unit). These include the sectors with more or less extended terrains that form the deposit-filled, erosion, pediment and piedmont surfaces. These surfaces are very broad and are important in the N sector of the study area. They are also are somewhat more disperse between both sierras (Francia and Quilama), where the morphology is sometimes governed by geomorphological factors, such as at Nava de Francia. They display two-dimensional forms, with silhouetted lines, a fine texture and high contrast. The scale shows a distance effect owing to the broadness of the unit and a panoramic space. The unit found in this type of landscape is 4.G.PaCyB, which forms piedmont surfaces over outcrops of conglomerates and sandstones with crops. Sometimes these surfaces are associated with holm-oak and oak forests, located between villages, and on eroded or sediment-filled surfaces. This unit is distributed along the zone of confluence between the lower parts of hillsides and surfaces,

Figure 11. Landscapes of the fluvial escarpment between El Maillo-Aldehuela de Yeltes. The lithology is well visible and there is a tract of associated vegetation of riparian galleries (A). Surface landscapes: flat surfaces of foot hills with pastures (B). Landscapes of terraces and valley floors: alluvial of the river Batuecas inside the Las Batuecas valley with dense bank vegetation (C), alluvial riverbed with a coarse texture of the river Yeltes, with defined edges and erosive forms of fluvial margin. Aldehuela de Yeltes (D), riverbed, with human activities intercalated with the vegetation of black poplars at Morasverdes; and fluvial terraces with arboreal formations of holm-oaks (left) and oaks (right).

showing diffuse borders owing to the presence of shrubby-subshrubby plant formations. These formations are found in the study zone both at the edges of anthropic surfaces in the W sector and close to the mid-stretches of rivers in the central sector. In some sectors, the presence of dehesa environments for traditional sustained use is very common, where the sclerophyllous forest has been cleaned of bushes, the mature elements of the forest coexisting with semi-natural communities (pastures. Crops and livestock) in a sustainable relationship (**Figure 11(B)**).

Terrace and valley floor landscapes (6.F.FA.FD. PaCyB Landscape Unit), generally with disperse groups of trees and crops. They integrate river course and/or valley sectors where there is a mixture of arboreal formations with shrubby and subshrubby mosaics. These units are associated with braided river channels. This landscape is organized around the fluvial axes harboring most villages are located Morasverdes, Tenebron and Dios le Guarde and the surfaces have been converted into irrigated lands. These landscapes are linked to the presence of surface water, which allows the development of river bank vegetation in its surroundings.

There is a predominance of biotic elements (riverside vegetation) and abiotic ones (gravels from river beds) (**Figure 11(C)**; **Figure 11(D)**) Anthropic elements are also seen, and attract much visual attention, above all in the cropland surfaces and, to a lesser extent, in the villages and highways. The relief here becomes less important since its forms are not very prominent, where some of the biotic elements are relegated to a lesser position due to transformation through human action. It is sometimes possible to fairly large observe sectors with open tree formations and, noting from inside the adjacent scenic components, as is the case of the felled forests of oak, chestnut and pines present on the lower slopes of the Quilama and Francia saws, extending to the river terraces. At the Quilamas stream there is a series of stepped surfaces on the summit zones, where the relief reflects strong energy in the encasement of the river bed. These slopes support a rich and varied vegetation of arboreal formations on the low and medium parts and shrubby formations higher up. Overall it is a zone of great of great natural attraction where biotic and abiotic elements predominate and the anthropic elements have been relegated to second place. This unit has sectors with narrow vistas, which increase their perception, with deep visual extensions, medium textures, showing sharp visual changes when they are crossed by roads, agricultural tracks, etc. The agricultural activities have generated territorial plots with structures (walls, fences, banks, orchards...) that have afforded a reticulated landscape, n which the anthropic component predominates over the natural ones, the remnants of forests, testimony to period of greater development, are very singular. In some sectors there are forest repopulations that draw attention to mixed formations containing species of great singularity. There are also sectors with terraces in which human activity has been so intense, that the initial natural assets have been replaced by features deriving from anthropic activity (**Figure 11(E)**).

An example of these landscapes are all the sectors of pastures and croplands in the neighborhoods of villages, where traditional sustainable use of the land has left an imprint of human activities, the remains of the traditional ecosystem (deciduous and sclerophyllous forests) coexisting with seminatural communities (meadows, croplands...) (**Figure 11(F)**). In spring, these meadows of vivacious species and pasturelands are highly coloured and together with the presence of the seasonal livestock they generate sectors with an important visual effect. The same is the case for orchards and croplands owing to the variety in the ordering of their distribution, as well as the presence of people and machinery exploiting such areas. These sectors have traditional and cultural value. Finally, it should be noted that the large number of ponds distributed throughout these sectors, many of them artificial,

have generated added visual value since they are associated with vegetation and the fauna.

5. Results

In the landscape analysis reported here we have established the landscape units of the protected natural parks by indirect methods, performing the description and assessment of the mapping of homogeneous and natural or environmental units, bearing in mind their natural components. Using direct methods we determined the visual characteristics and perceptual factors of the most representative units of the study zone in field studies, using three-dimensional models and photos taken from and air and in the field to appreciate the spatial distribution and relevance of the different parameters. This analysis defines the landscape units, which divide the territory into homogeneous sectors or extensive irregular units, depending on the structure, functionality and visual characteristics of the natural components. The geomorphological component has greater weight than the lithological one and the vegetation, since the spatial configuration of the relief predominates in the landscape. Based on this study some important conclusions are observed:

1) The landscape units in the summit sectors show the following relevant features: summit sector with no vegetation or with shrubs covered by colluvials, especially screes, which stand out in the landscape owing to their disposition and colour. Likewise, note should be taken of the hill and hillock sectors with arboreal formations and, to a lesser extent, pasturelands, croplands and fallow land, in the W sector.

2) The hillsides afford highly representative landscape units in the centre of the natural parks of Las Batuecas-Sierra de Francia and in the NE sector of the study zone, with a predominance in the rest of the natural park of large patches of shrubby-subshrubby formations.

3) The fitted valley sectors are distributed within the Quilamas natural park and the central and southern parts of the Batuecas-Sierra de Francia, with a predominance of shrubby and subshrubby formations in the inter-range sectors, followed by mixed formations in zones adjacent to the previous ones but of less extent in the E and SE zone. Less spatial representation is seen for the units with arboreal formations, which are distributed in the S sector surrounding river valleys.

4) The unit of surfaces with crops is distributed across the periphery of the natural parks being limited to the N and W sectors, with a predominance of croplands and pasturelands on the sediment-filled surfaces, N sector, and on erosive surfaces: W sector.

5) The escarpment unit over slates and sandstones with trees and pasturelands stands out on the right bank of the river Yeltes, displacing and retracting the fluvial escarpment that runs from El Maillo to Aldehuela de Yeltes,

and the tectonic escarpment to the NW of the study zone.

6) Finally, the terrace and valley floor unit, mainly associated with the river courses in the N of the study zone, features arboreal formations in the N and E sectors, and on the valley floors there is a predominance of riverside arboreal formations and/or forest repopulations.

7) The treatment used here considers landscape not only as an aesthetic element but also as something "live" that evolves over time due to the presence of natural phenomena, a dynamic landscape, and the anthropic involvement that has increased or decreased this rate of evolution. Regarding environmental planning, study of the visible landscape should be based on maps that should be of use in land planning and management. Accordingly, the landscape assessment made here has a dual aim: on one hand, the presence and distribution of the landscape components (forms and slopes of the terrain, the presence of vegetation, land use...) and, on the other, the absence of impacts that will degrade the natural environment (residues, infrastructures, noise...); that is, on one hand we analyze the "total landscape" that identifies the landscape with the environment on the basis of the thematic components of the territory. We emphasize the geomorphological factor since this imprints a spatial dominance of forms and processes through the disposition of the relief and its active processes, the lithological coloring and the distribution of the vegetation according to height. On the other, we address the "visual landscape", where we assess the natural environment on the basis of aesthetic or perceptual criteria by modeling the terrain in 3D, thus allowing a more efficient spatial analysis. The first approach provides systematic information about the territory, whereas the second one identifies what the observer is able to perceive in that territory.

8) These landscape units offer a causal interpretation of the forms of the terrain and of the changing elements (a natural dynamic landscape) and their structure and spatial relationships.

9) The identification and characterization of the different landscape units has allowed us to establish the compatible and incompatible uses, attending to criteria of assessment, protection, management and ordering, always from the perspective of the sustainable development of each unit. The maintenance of the functionality of these natural parks through conservation, recovery and landscape integration of the underlying geological, biotic and socioeconomic system is a guarantee of landscape viability.

6. Acknowledgements

Projects CGL2012-33430/BTE and CGL2012-37581-CO2-01.

REFERENCES

[1] F. D. Pineda, "Terrestrial Ecosystems Adyacent to Larg Reservoirs," International Committee on Large Dams, XI Congress, Paris, 1973.

[2] M. Dunn, "Landscape Evaluation Techniques: An Appraisal and Review of the Literature," Centre of Urban and Regional Studies, Birminghan, 1974.

[3] W. Nohl, "Sustainable Landscape Use and Aesthetic Perception-Preliminary Reflections on Future Landscape Aesthetics," Landscape and Urban Planning, Vol. 54, No. 1-4, 2001, pp. 223-237.

[4] R. De Groot, "Function-Analysis and Valuation as a Tool to Assess Land Use Conflicts in Planning for Sustainable, Multi-Functional Landscapes," Landscape and Urban Planning, Vol. 75, No. 3-4, 2006, pp. 175-186.

[5] K. Soinia, H. Vaaralab and E. Poutaa, "Residents' Sense of Place and Landscape Perceptions at the Rural-Urban Interface," Landscape and Urban Planning, Vol. 104, No. 1, 2012, pp. 124-134.

[6] A. García-Quintana, J. F. Martín-Duque, J. A. González-Martín, J. F. García-Hidalgo, J. Pedraza, P. Herranz, R. Rincón and H. Estévez, "Geology and Rural Landscapes in Central Spain (Guadalajara, Castilla-la Mancha)," Environmental Geology, Vol. 47, No. 6, 2005, pp. 782-794.

[7] J. A. Soria and F. G. Quiroga, "Análisis y Valoración del Paisaje en las Sierras de la Paramera y la Serrota (Ávila)," M + A Revista Electrónica de Medio Ambiente, No. 1, 2006, pp. 97-119.

[8] A. M. Martínez-Graña, J. L.Goy and C. Zazo, "Natural Heritage Mapping of the Batuecas-Sierra de Francia and Quilamas Nature Parks (SW Salamanca, Spain)," Journal of Map, Vol. 7, No. 1, 2011, pp. 600-613.

[9] R. Beunen and P. Opdam, "When Landscape Planning Becomes Landscape Governance, What Happens to the Science?" Landscape and Urban Planning, Vol. 100, No. 4, 2011, pp. 324-326.

[10] B. A. Bryan, "Physical Environmental Modeling, Visualization and Query for Supporting Landscape Planning Decisions," Landscape and Urban Planning, Vol. 65, No. 4, 2003, pp. 237-259.

[11] C. He, J. Tianc, P. Shia and D. Hud, "Simulation of the Spatial Stress Due to Urban Expansion on the Wetlands in Beijing, China Using a GIS-Based Assessment Model," Landscape and Urban Planning, Vol. 101, No. 3, 2011, pp. 269-277.

[12] F. Gobattonia, R. Pelorossoa, G. Laurob, A. Leonea and R. A. Monacoc, "Procedure for Mathematical Analysis of Landscape Evolution and Equilibrium Scenarios Assessment," Landscape and Urban Planning, Vol. 103, No. 3-4, 2011, pp. 289-302.

[13] European Union, "Relativa a la Evaluación de las Reper-

cusiones de Determinados Proyectos Públicos y Privados Sobre el Medio Ambiente," European Union, Luxembourg, 1985.

[14] Pacific Northwest Christmas Tree Association, "Plan Nacional de Cartografía Temática Ambiental. Análisis y Desarrollo Ministerio de Medio Ambiente," PNCTA Documento Técnico, Madrid, 1996.

[15] Bank of England, "Instrumento de Ratificación del Convenio Europeo del Paisaje," 2008.
http://www.juntadeandalucia.es/obraspublicasyvivieda/estatcas/sites/consejeria/areas/ordenacion/documentos/ConvencionEuropeadelPaisaje.pdf

[16] I. O. Pastor, S. M. Quintana and E. O. Pérez, "El Paisaje como Elemento en la Evaluación Ambiental Estratégica de Planes de Infraestructuras. Cartografía de la Calidad del Paisaje de España," I Congreso Paisaje e Infraestructuras, Sevilla, 2006.

[17] I. O. Pastor, M. A. C. Martínez, A. E. Canalejo and P. E. Mariño, "Landscape Evaluation: Comparison of Evaluation Methods in a Region of Spain," *Journal of Environmental Management* Vol. 85, No. 1, 2007, pp. 204-214.

[18] Convenio Europeo, "Recomendación del Comité de Ministros a los Estados Miembro Sobre las Orientaciones para la Aplicación del Convenio Europeo del Paisaje," Convenio Europeo, London, 2008.

[19] Diari Oficial de la Generalitat de Catalunya, "De Protección, Gestión y Ordenación dl Paisaje, y Se Regulan los Estudios e Informes de Impacto e Integración Paisajística," Diario Oficial de la Generalit de Catanlunya, Sevilla, 2006.

[20] F. G. Bernáldez, "Ecología y Paisaje," Blume, Madrid. 1981.

[21] Y. M. Ayad, "Remote Sensing and GIS in Modeling Visual Landscape Change: A Case Study of the Northwestern Arid Coast of Egypt," *Landscape and Urban Planning*, Vol. 73, No. 4, 2005, pp. 307-325.

[22] Y. Liu, X. Lv, X. Qin, H. Guo, Y. Yu, J. Wang and G. Mao, "An Integrated GIS-Based Analysis System for Land-Use Management of Lake Areas in Urban Fringe," *Landscape and Urban Planning*, Vol. 82, No. 4, 2007, pp. 233-246.

[23] M. G. Aguiló, "Metodológica para la Elaboración de Estudios del Medio Físico," Secretaria General del Ministerio de Obras Públicas y Transportes, Madrid, 2000.

[24] P. Ghadirian and I. D. Bishop, "Integration of Augmented Reality and GIS: A New Approach to Realistic Landscape Visualisation," *Landscape and Urban Planning*, Vol. 86, No. 3-4, 2008, pp. 226-232.

[25] R. G. Echevarría, "La Fotografía Elemento para el Análisis y la Simulación del Paisaje Forestal," Tesis Doctoral, Universidad Autónoma de Madrid, Madrid, 2000,

[26] J. F. Palmer and L. J. Roos-Klein, "Evaluating Visible Spatial Diversity in the Landscape," *Landscape and Urban Planning*, Vol. 43, No. 1-3, 1998, pp. 65-78.

6

Extraction of Urban Vegetation in Highly Dense Urban Environment with Application to Measure Inhabitants' Satisfaction of Urban Green Space

Fatwa Ramdani

Institute of Geography, Department of Earth Science, Graduate School of Science, Tohoku University, Sendai, Japan

ABSTRACT

Urban environment has functioned not only for ecological reason but also for socioeconomic function, due to this reason extraction of urban vegetation in highly dense urban environment becomes more important to understand the inhabitants' satisfaction of urban green space. With a medium resolution of satellite imagery, the precision is very low. We used high resolution of WorldView-2 satellite to raise the accuracy. We chose Depok City in West Java as a case study area, analyse four multispectral bands, and apply TCT algorithm for getting vegetation density. The relationship between vegetation density and inhabitants' satisfaction was calculated by Geo-statistical technique based on administrative boundary. We extracted three types of urban vegetation density: good, mid and low. The final result shows that the inhabitants are mostly satisfied with good density of urban vegetation in the city forest inside Campus University of Indonesia.

Keywords: Urban Vegetation; Remote Sensing; Vegetation Extraction; Indonesia

1. Introduction

Vegetation is of particular interest as it presents a versatile resource for effectively managing and moderating a variety of problems associated with urbanization. The spatial distribution and abundance of urban vegetation, for example, is recognized as a key factor influencing numerous biophysical processes of the urban environment, including air and water quality, temperature, moisture, and precipitation regimes [1-3].

During the last ten years, Depok City has been growing fast as a satellite city in south of Jakarta. The growing of the city is also followed by the socioeconomic development, urban green space becomes more important for urban environment and urban ecosystem. Trees can reduce pollutant from transportation mod and can decline the effect of urban heat island. In order to tackle the increasing of flood event and bad air condition in the high density urban environment it is necessary to monitor the actual condition of urban vegetation distribution and evaluate it, to measure the quality of life in the highly dense urban environment.

In addition, vegetated areas such as gardens, parks, and forests have been related to positive social outcomes including reductions in crime [4], health benefits [5], and advanced childhood development [6]. Given the associations between vegetated land cover and the biophysical and social processes of urban systems there exists an ongoing demand for effective urban vegetation mapping and classification techniques.

High resolution of satellite imagery such as World-View-2 provides us much better information in every single pixel with 0.5 cm resolution and 6.5 m accuracy specification. The imagery can highly define the urban vegetation in highly dense urban environment, and it has been accurate for monitoring the green space distribution. Favela and Torres [7] developed image-understanding techniques characterized by artificial intelligence, and other researchers had developed diagnostic expert systems [8,9].

In order to observe the inhabitants' satisfaction, many studies designed by researchers such as ask people, and interview face-to-face about their opinions [10,11]. However there are no studies that have been conducted on urban growth and the correlation between urban vegetation and urban inhabitants' satisfaction level in Depok

Extraction of Urban Vegetation in Highly Dense Urban Environment with Application to Measure Inhabitants' Satisfaction of Urban Green Space

61

City.

The objectives of this study are: 1) to provide an effective method of extraction of urban vegetation in highly dense urban environment; 2) to show the spatial distribution of urban green space using WorldView-2 imagery; and 3) to calculate the relationship between the spatial distribution of urban vegetation and inhabitants' satisfaction.

2. Methods

2.1. Study Area

Depok City is geographically located at 6°19'00" - 6°28'00" south latitude and 106°43'00" - 106°55'30" east longitude. Depok City directly is adjacent to the greater Jakarta area. Depok City as the youngest province in West Java has an area of about 20,029 ha [12].

Depok City Land Resources is under pressure with rapid developments in inner city. Based on data analysis from the Spatial Planning Policy Revision of Depok City (2000-2010) in the land use of urban space, residential areas in 2005 reached 891,509 ha (44.31%) of the total land use of Depok City [12].

In year 2005 urban green space was 10,106.14 ha (50.23%) of the area of Depok City, or decreased 0.93% from the year 2000. Increasing in built-up area caused decline in the natural condition of Depok City, mainly due to pressure from the land use for residential development that reached more than 44.31% of the total area of the city. Built-up area in 2005 reached 10,013.86 ha (49.77%) of the total area of Depok City, it was an increase of 3.59% from year 2000 [12].

The population of Depok City in 2005 reached 1,374,522 people, consist of 696,329 males (50.66%) and 678,193 females (49.34%), with an area of only 200.29 km^2, the population density in Depok City is 6863 people/km^2. Population density was classified as "dense", especially if associated with uneven population distribution [13].

Within 5 years (2000-2005) population of Depok City increased by 447,993 inhabitants. In 1999 the population was still under 1 million and by 2005 had reached 1,374,522 people, making progress on average 4.23% per year. The increase was due to high rates of migration each year [13].

High migration into Depok City as a result of rapid urban development can be seen from the increasing development of residential areas. Depok City immigration numbers in the year 2004, showed a fluctuating pattern, where the immigration of 11,899 people and the emigration is 4503 people, or the average number of arrivals per year reaching 7396 migrants. Based on these developments, estimated number of people who come to the Depok City will increase in the future, as more and more

operational activities and commercial services are growing rapidly [13].

2.2. Data

WorldView-2 is a commercial satellite operated by Digital Globe, Inc. This sensor provides two different captures at the same time of observation, a multispectral image and panchromatic image. The multispectral image provides lower resolution (2 m) of every single bands image, which includes coastal, yellow, red-edge, and near infra-red-2. But the panchromatic image provides higher resolution (50 cm) of one band image. The multispectral data has eight bands, which are coastal (400 - 450 mm), blue (450 - 510 mm), green (510 - 580 mm), yellow (585 - 625 mm), red (630 - 690 mm), red-edge (705 - 745 mm), near infra-red-1 (770 - 895 mm), and near infra-red-2 (860 - 1040 mm). The panchromatic image is black and white (monochrome) with spectral characteristics 450 - 800 mm.

From the image of the study area (**Figure 1**), it can be seen that the urban green vegetation is concentrated in the city forest (Campus of University of Indonesia). The geographical distribution of urban vegetation is various, from side-street trees, trees in the parks, bushes along rivers, to trees inside residential area.

Field surveys were conducted for the period between March 5 and April 23, 2012. It was conducted face-to-face interview with university member, residential member, workers, and commuters. They were asked about their satisfaction on urban environmental condition.

The general framework of the study is separated into two steps (**Figure 2**): The first one is imagery processing. At this step, the satellite image was classified in order to extract urban vegetation in highly dense urban environment for monitor the spatial distribution of urban green space. The second step, we measured the relationship between inhabitants' satisfaction and spatial distribution

Figure 1. WorldView -2 image of the study area, shown in false color.

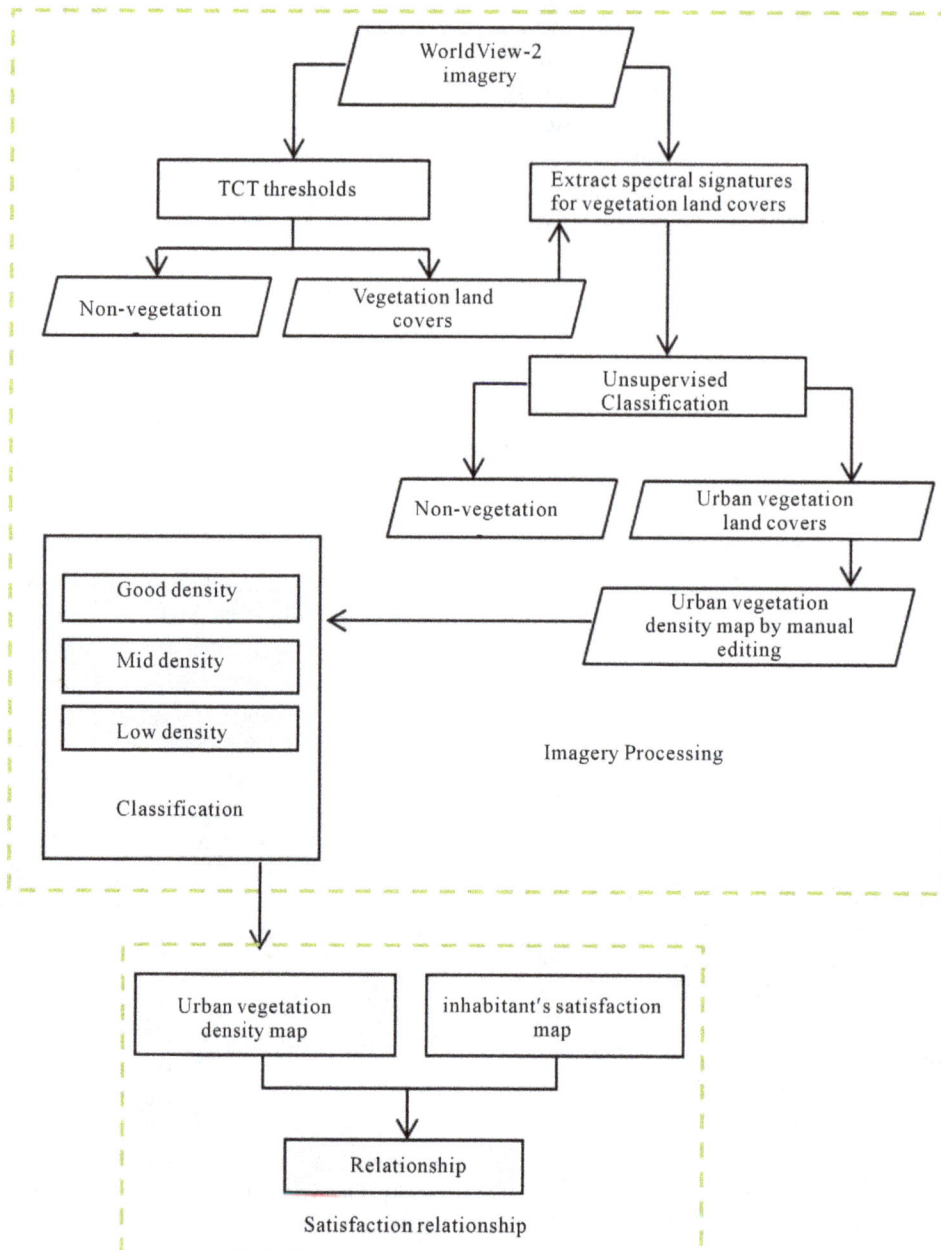

Figure 2. Framework

of urban green space. We applied Tasselled Cap Transformation—TCT algorithm to get vegetation density and the spatial distribution of urban green space. TCT (Tasselled Cap Transformation) is a mathematic formula to calculate brightness level, greenness level, and wetness level from digital number in every band (band1 to band4) of WorldView-2 imagery. TCT was introduced first time by Kauth and Thomas [14] used Landsat MSS. Furthermore, TCT was completed by Crist and Cicone [15] used Landsat TM. In this study we used TCT coefficient from Yrabrough and Easson [16].

The algorithms are shown below;

Wetness level:

band1*(0.319) + band2*(0.542) + band3*(0.490) + band4*(0.604)

Greenness level:

band1*(−0.121) + band2*(−0.331) + band3*(−0.517) + band4*(0.780)

Brightness level:

band1*(0.652) + band2*(0.375) + band3*(−0.639) + band4*(−0.163)

Note: band1 for blue, band2 for green, band3 for red, band4 for near infra-red.

From the TCT image result, we employed isodata unsupervised classification method using greenness image, to extract only the urban vegetation. Greenness image used due to greenness level describe the density of vegetation, higher greenness level is correlated with higher vegetation density [17]. To measure the accuracy level of classification result, 100 points selected randomly from classification result. Points then superimposed and visually assessment was done. The extensive field work survey was also conducted to verify points using global positioning system (GPS). Then we generated confusion matrices to calculate Kappa coefficients (k) and to derive overall accuracy.

3. Analysis

According to isodata unsupervised interpretation, we have extracted three classes of the urban green space from the WorldView-2 image using tree classification method:

- Good density (TCT-greenness greater than 0.8),
- Mid density (TCT-greenness greater than 0.3 but less than 0.79), and
- Low density (TCT-greenness greater than 0.1 but less than 0.29).

These three types can be derived from distinguished their shape, texture, color, size and associations, for example good density is city forest area, which is part of the urban green space in the Campus of University of Indonesia. Mid density of urban vegetation are local guava, banana, and papayas plantation. While low density of urban vegetation situated in street-side and residential trees, situated along the main road.

To highlight the vegetation inside highly dense urban environment, we show the vegetation in red color (false color), we employed an RGB combination using band near infra-red-1, red, and yellow. This combination makes us easier to interpret the urban vegetation and extract it using isodata unsupervised method classification.

To obtain these three different classes, we employed open source software for satellite imagery processing, ILWIS 3.7.1. This open source software allows us to apply the isodata unsupervised method, to calculate the means per band for each class of training pixels as defined in the sample set. The sample was the training phase, where classes of pixels with similar spectral values are defined. Classify is the decision phase, where each output pixel is assigned a class name if the spectral values of that pixel are similar enough to a training class. Relevant information on the classes for which training pixels have been selected in the sample set, can be viewed in the sample statistics. In general, to each output pixel, the class will be assigned of which the spectral values are most similar to (or "nearest") to the spectral values of an input pixel.

The first result of urban vegetation classes is in pixel format, for easier analysis we convert pixel format to vector format. We calculate the urban vegetation density (%) compared the total land area (ha) of every sub district divide by urban vegetation area (ha) and multiply by 100. We only used good density of urban vegetation in calculation, mid and low density of urban vegetation is excluded due to insufficient appearance in the field work activity. These types of vegetation are only small trees or bushes and grass in small part.

We verified the relationship between vegetation density and inhabitants' satisfaction by Geo-statistical technique based on an administrative boundary shown in the map (**Figure 3**). In this study, inhabitants' satisfaction information collected from field surveys was ranked from 1 to 3 levels opinions based on questionnaire results. Rank 1, 2 and 3 indicated low, mid, and high satisfaction, respectively. From **Figure 3**, we can understand that the high density of urban vegetation situated in the northern side of the study area, which is city forest inside University of Indonesia.

High satisfaction level also acquired in the two sub districts located on the northern side of Depok City, associated with high density of urban green space. Mid satisfaction level of urban vegetation acquired in Tanah Baru sub district, in this area there are many local guava, banana, and papayas plantation, local inhabitant feels comfortable and satisfied with the appearances of these plantation, however development of many residential clusters at present become a real threat to these plantations. Low satisfaction level of urban vegetation level acquired in the two sub district of the south side of the study area, in this area vegetation only appear in a small part in street-side and along the residential road network.

4. Result

We extracted the urban vegetation and non-vegetation area using the spectral information provided by the multispectral image using isodata unsupervised classification method. From the confusion matrix analysis we produce high accuracy with 96.0429% overall accuracy and Kappa coefficient is 0.9042 (**Table 1**).

Table 2 showed the R^2 coefficient representing the correlation between the spatial distribution of urban vegetation density and inhabitants' satisfaction. Results show that the value of R^2 coefficient was near unity, indicating that vegetation density and inhabitants' satisfaction are well correlated. Thus, it can be concluded that the inhabitants are mostly satisfied with high density of urban vegetation in the city forest inside Campus of University of Indonesia.

However, due to the availability of the data, the urban vegetation density image does not cover all five sub-district in the study area. The WorldView-2 image only covers 67% of all study areas. The image size is 4 km × 2.7 km equal to 1080 ha, and total area of the five sub-

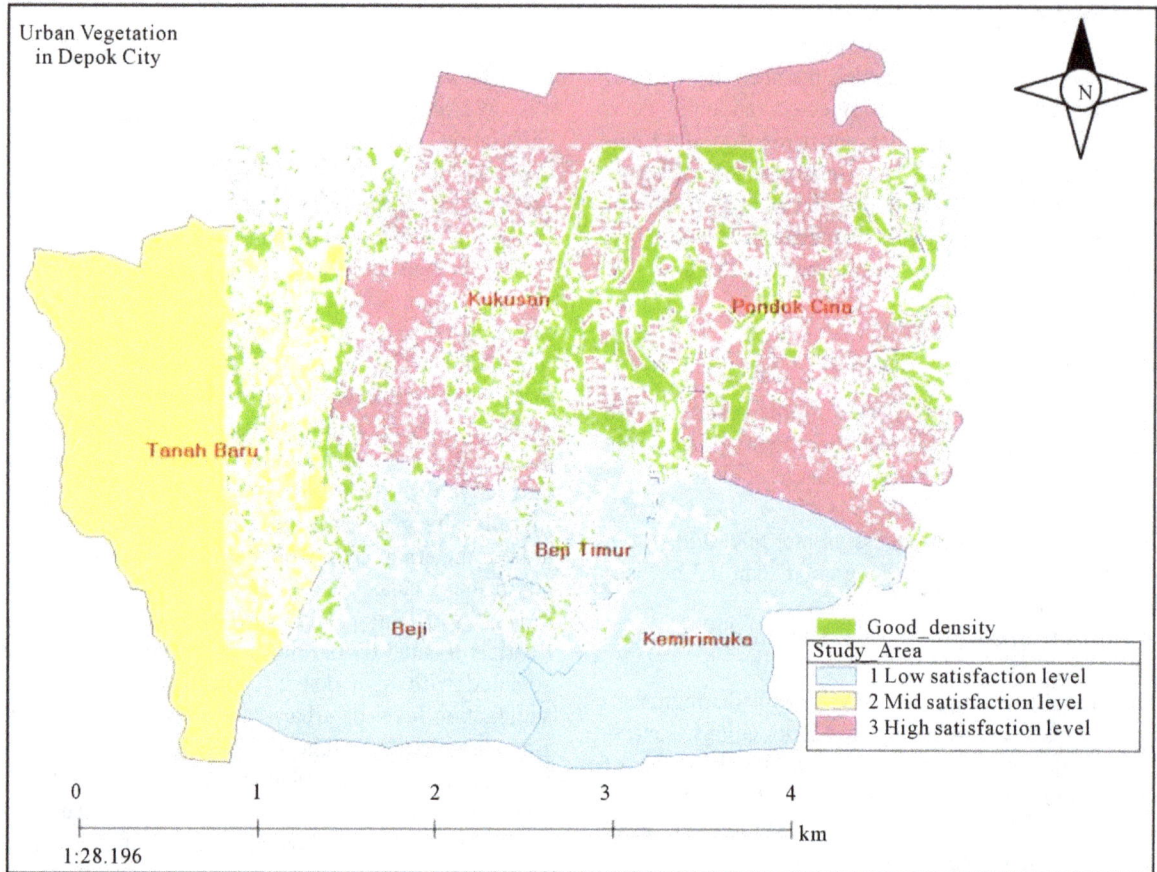

Figure 3. Spatial distribution of urban vegetation density superimposed by inhabitants' satisfaction level.

Table 1. Confusion matrix result.

Ground truth (pixel)			
Class	Non-vegetation	Urban vegetation	Total
Non-vegetation	7776332	167732	7944064
Urban vegetation	278968	3065544	3344512
Total	8055300	3233276	11288576

Ground truth (%)			
Class	Non-vegetation	Urban vegetation	Total
Non-vegetation	96.54	5.19	70.37
Urban vegetation	3.46	94.81	29.63
Total	100.00	100.00	100.00

Class	Prod Acc. (%)	User Acc. (%)	Prod Acc. (pixel)	User Acc. (pixel)
Non-vegetation	96.54	97.89	7776332/8055300	7776332/7944064
Urban vegetation	94.81	91.66	3065544/3233276	3065544/3344512

Table 2. Regression results.

Regression statistics	
Multiple R	0.929
R square	0.863
Adjusted R square	0.817
Standard error	0.328
Observations	156

districts is 1621 ha.

5. Conclusions

This study shows that WorldView-2 imagery could be used for monitoring the urban vegetation extraction in the highly dense urban environment, and to understand the inhabitants' satisfaction of spatial distribution of urban green space density.

High resolution imagery data and GIS technique are very useful for the extraction of information like urban vegetation which is an important attribute for assessing the urban environment. Technique presented in this study produces high accuracy with $R^2 = 0.86$.

Extraction of Urban Vegetation in Highly Dense Urban Environment with Application to Measure Inhabitants'
Satisfaction of Urban Green Space

65

Furthermore, to increase the quality of life in dense population in a growing city such as Depok City could be measured from the point of view of high resolution satellite imagery.

REFERENCES

[1] R. Avissar, "Potential Effects of Vegetation on the Urban Thermal Environment," *Atmospheric Environment*, Vol. 30, No. 3, 1996, pp. 437-448.

[2] C. S. B. Grimmond, C. Souch and M. D. Hubble, "The Influence of Tree Cover on Summertime Energy Balance Fluxes, San Gabriel Valley, Los Angeles," *Climate Research*, Vol. 6, 1996, pp. 45-57.

[3] D. J. Nowak and J. F. Dwyer, "The Urban Forest Effects (UFORE) Model: Quantifying Urban Forest Structure and Functions," In: M. Hansen and T. Burk, Eds., *Proceedings: Integrated Tools for Natural Resources Inventories in the 21st Century. IUFRO Conference*, US Department of Agriculture, Forest Service, North Central Research Station, St. Paul, pp. 714-720.

[4] F. E. Kuo and W. C. Sullivan, "Environment and Crime in the Inner City," *Environment and Behavior*, Vol. 33, No. 3, 2001, pp. 343-365.

[5] S. E. Coen and N. A. Ross, "Exploring the Material Basis for Health: Characteristics of Parks in Montreal Neighbourhoods with Contrasting Health Outcomes," *Health and Place*, Vol. 12, No. 4, 2006, pp. 361-371.

[6] A. F. Taylor, A. Wiley, F. E. Kuo and W. C. Sullivan, "Green Spaces as Places to Grow," *Environment and Behavior*, Vol. 30, No. 1, 1998, pp. 3-28.

[7] J. Favela and J. Torres, "A Two-Step Approach to Satellite Image Classification Using Fuzzy Neural Networks and the ID3 Learning Algorithm," *Expert Systems with Applications*, Vol. 14, No. 1-2, 1998, pp. 211-218.

[8] R. Gamanya, P. De Maeyer and M. De Dapper, "An Automated Satellite Image Classification Design Using Object-Oriented Segmentation Algorithms: A Move towards Standardization," *Expert Systems with Applications*, Vol. 32, No. 2, 2007, pp. 616-624.

[9] X. Huang, J. R. Jensen and H. E. Mackey, "Machine Learning Approach to Automated Construction of Knowledge Bases for Expert Systems for Remote Sensing Image Analysis with GIS Data," *Expert Systems with Applications*, Vol. 11, No. 4, 1996, pp. 8-9.

[10] E. Lange, S. Hehl-Lange and M. J. Brewer, "Scenarios Visualization for the Assessment of Perceived Green Space Qualities at the Urban-Rural Fringe," *Journal of Environmental Management*, Vol. 89, No. 3, 2008, pp. 245-256.

[11] T. T. H. Pham and D.-C. He, "How Do People Perceive the City's Green Space? A View from Satellite Imagery (in Hanoi, Vietnam)," *Proceedings of the IEEE International Geoscience & Remote Sensing Symposium*, Boston, 6-11 July 2008, pp. 1228-1231.

[12] Government of Depok City, 2001.
http://www.depok.go.id/profil-kota/geografi

[13] Government of Depok City, 2011.
http://www.depok.go.id/profil-kota/demografi

[14] R. J. Kauth and G. S. Thomas, "The Tasseled Cap—A Graphic Description of the Spectral-Temporal Development of Agricultural Crops as Seen by LANDSAT," *Proceedings of the Symposium on Machine Processing of Remotely Sensed Data*, University of West Lafayette, Indiana, 1976.

[15] E. P. Crist and R.C. Ciccone, "Application of the Tasseled Cap Concept to Simulate Thematic Mapper Data," *Photogrammetric Engineering and Remote Sensing*, Vol. 50, 1984, pp. 343-352.

[16] L. D. Yarbrough and G. Easson, "Quickbird 2 Tasseled Cap Transform Coefficients: A Comparison of Derivation Method," *Pecora* 16 *"Global Priorities of Land Remote Sensing"*, South Dacota, 2005.

[17] F. Ramdani, "Sub-Urban Growth Impact in the Wet-Land Environment. Case Study: Rawa Danau Nature Re-Serve," *Proceedings of the 10th International Geoconference SGEM*, Varna, 2010.

7

Simulation Models and GIS Technology in Environmental Planning and Landscape Management

Giuliana Lauro

Department of Industrial and Information Engineering, Second University of Naples, Aversa, Italy

ABSTRACT

Landscape protection that, in the past, has been mainly concerned with its historical, artistic and cultural heritage, follows, nowadays, a systemic methodology that looks at landscape as a high level aggregate of spatial, ecologically different units that interact each other by exchanging energy and materials. *Strategic environmental assessment*, nowadays, has been adopted in Europe in landscape planning, whose task is to verify the compatibility of territory transformations with respect to their levels of criticality and vulnerability, to evaluate possible future scenarios as consequence of interventions by checking if they are in line with preservation and valorization of environmental. To this aim, we make here a short survey of three different simulation models that can be used as Decision Support System in landscape planning and management. They adopt tools of the Landscape Ecology and are based on GIS (Geographic Information System) technology. The first one consists of a **planar graph**, the so called *ecological graph*, whose construction needs the computation of suitable indices of environmental control, proper of Landscape Ecology, such as *biodiversity, biological territorial capacity, connectivity*. The planar graph, for the considered environmental system, returns a picture of its actual ecological health condition and provides very detailed indications and operational assistance for choosing among possible ecological sustainable interventions. The second one, based on the data used to construct the ecological graph, uses the *least-cost path* algorithm from GIS technology in order to build an **ecological network** to prevent and to reduce territorial *fragmentation* caused by intense processes of urbanisation and industrialisation. At last, an integrated GIS-based approach is developed combining an ecological graph model and a mathematical model based on a nonlinear differential equation of **logistic-type with harvesting** to perform qualitative predictions on the sustainability of a given territorial plan.

Keywords: Landscape Ecology; Ecological Network; Dynamical System; GIS Technology

1. Introduction

In recent years, the landscape planning has understood the importance of an "ecological-oriented" analysis of environmental systems more or less affected by degradation and ecological phenomena such as fragmentation and reduction of biodiversity.

To be changed is the very idea of "landscape" which, traditionally identified with that of scenic beauty, has now expanded to indicate all the ways of interactions among nature, environment, land, cultural heritage. The landscape as a "field of knowledge" is the basis for the formulation of the European Landscape Convention, CEP, adopted by the Committee of Ministers of Culture and Environment of the Council of Europe July 19, 2000 and since September 1, 2006, law operating in Italy (Law No. 14 of January 9, 2006). According to this convention

"Landscape means a certain portion of territory, as perceived by people, whose character derives from the natural and/or human interrelationships". In addition to defining the term landscape, Convention determines all the rules for the recognition, protection, preservation, management of the landscape. It is the first international treaty with the objective of promoting the protection, management and planning of European landscapes by promoting European cooperation. The Convention is potentially a real conceptual revolution that brings the community to become the primary stakeholders of the evolution of landscapes in which they live, seen as a cultural strategy for the quality of habitat and also relevant from a social and economic point of view. It encourages citizens to take an active part in decision-making processes that affect the landscape at the local and regional levels. The content of the Convention takes into account the whole of the States Parties and covers natural, rural, urban and suburban areas, including land, inland waters and marine waters. Concerns landscapes that might be considered outstanding, and the landscapes of everyday life, even degraded landscapes (Article 2). The methodologies for successful landscape and environmental planning have been severely challenged when concepts like *ecosystems preservation* and *sustainability* have been questioned. The actual challenge is to build transparent and flexible *decision-making tools* to be used in environmental planning and to embrace a broad range of stakeholder needs together with landscape management requirements. Decision-making process needs the use of **interdisciplinary** models. A modern approach to the study of landscape shows that it should not be viewed as a mere sum of parts but as a system of relations between the different constituent ecosystems and processes that determine its evolution in time. In the language of Landscape Ecology, this means considering the landscape as a system of ecosystems, ecomosaic, organized in a hierarchical structure and interacting with each other through exchanges of energy and matter, in a fragile equilibrium under dynamic disturbances of both natural and anthrop origin ([1-3]). This research, therefore, uses principles and models proposed by the Landscape Ecology to analyze and assess the environmental quality of a landscape through the identification of appropriate indices of control such as *biodiversity*, "*bioenergy*", *connectivity*. We refer to the term "*bioenergy*" as the energy available in the environment, present in different forms like animals, seeds, plants, ruled by suitable metabolic processes. Energy and material "*fluxes*", *i.e.*, bioenergy fluxes, through the territory are therefore necessary fundamental processes for biodiversity conservation and capacity of the system to resist at the both natural and anthrop perturbations (resilience). Barriers and surfaces with low permeability to such fluxes hamper the movement of animals, seeds spreading and in general the likelihood of species survival, as they act as external constraints on the natural ecosystem. Modeling these fluxes is therefore necessary to assess the most suitable plan strategies for natural resources conservation management and landscape functionality preservation. A tool in this direction can be represented by the construction of the so called Ecological Graph ([4,5]), as we'll see in the next Section, that furnishes a picture of landscape ecological health condition and provides very detailed indications and operational assistance to guide toward sustainable interventions. Moreover, this model also allows to determine the status of territorial fragmentation and, hence, to verify the need for *ecological network*, functional to dispersion of animals and plants. As said, the flows of gens and individuals between populations is essential for the survival of those species that are sensitive to the fragmentation of

their habitats, therefore, the loss of ecological connectivity, *i.e.*, the difficulty met by organisms in their movement between resource patches, constitutes a challenge for biodiversity conservation. In the 1980s, the idea of developing national ecological networks surfaced more or less simultaneously in several European countries. The ecological network concept not just prioritizes the conservation of core areas as natural or semi-natural values, but also prioritizes the importance of buffering, maintaining and re-establishing ecological connectivity and nature restoration. Depending on the species and spatial scale of interest the characteristics of an ecological network may differ widely, therefore ecological networks can be identified at continental, regional landscape and local scales. To this aim, in Section 3, we'll show a procedure ([6,7]), based on the least-cost path algorithm from GIS technology, to build a local ecological network for the National Park of Cilento. Finally, in Section 4, behind the static frame of these two simulation models, we shall propose a mathematical dynamical model in order to investigate the time evolution of the health condition of a territory, namely, of its bioenergy value.

In fact, changes in bioenergy, due to changes in environmental conditions, may produce territorial modifications toward which individual landscapes will tend to move smoothly (attractors) or may produce, instead, critical thresholds that result in radical changes in the state of the ecological system. In this sense, ecological systems are, in fact, said metastable. The investigation on mestability can be performed by means of the study of the equilibrium solutions of suitable differential equations ([8,9]) that model dynamics of the territory's evolution. The primary objective of these models is to perform qualitative predictions on the sustainability of the territorial planning finding, possibly, critical values of the parameters characterizing the territory itself. In Section 4 we use a nonlinear ordinary differential equation of Logistic-type with Harvesting ([10]), applied to a brownfield, ex-industrial area in the East side of Naples, subject to a Master Plan in the direction of improving the environment and we show the time-evolution trend of its ecological value and the existence of a critical value of a suitable environmental indicator linked to the geometrical setting of barriers to bioenergy fluxes.

2. Ecological Graph

In the modern discipline of Landscape Ecology, the landscape is defined as a heterogeneous land composed of interacting ecosystems that exchange energy and matter, and where natural and anthrop events coexist. In the present model, an environmental system is subdivided in a given number of different *landscape-units* separated from each other by natural or anthrop barriers. The bioenergy content of each unit may be represented by a circle (node) whose diameter is proportional to the magnitude itself. The barriers can have different degrees of permeability to bioenergy's flow ([3]). For example, an highway has almost zero value of permeability and determines, hence, a territorial fragmentation. We can represent the various levels of connection, among the units, by arcs whose width is proportional to the bioenergy flux shared among them. The collection of nodes and arcs is called Ecological Graph ([4]) of the environmental system. It can be drawn by means of a software GIS (Arch-View 3.x) by using the information, contained in suitable Shape Files furnished by local government, about land uses, presence and connection of road infrastructures (railroads, highways, government and provincial roads), system of water courses (natural and artificial) and administrative subdivision of the various urban territories within the studied landscape. We show, now, how to construct such a graph relative to National Park of Ci-

lento in Campania Region (Italy). Firstly we choose the subdivision in 18 units ([5]) as shown in **Figure 1**. Let us note that each landscape-unit, in turn, is composed of different *ecotopes*, *i.e.*, the smallest ecologically homogeneous distinct features in a landscape mapping, with a proper value of Biological Territorial Capacity, B_{TC}, that is the amount of energy (Mcal/m^2/year) that they need to dissipate in order to maintain their organizational level.

B_{TC} values can be computed on the basis of a standard classification ([3]), as reported in **Table 1**, once that it is known the kind of ecotopes.

We now define the bioenergy M_j of the landscape-unit j, j = 1, 2, ⋯ 18 as:

$$M_j = B_j\left(1 + k_j\right) \qquad (1)$$

where B_j is the average value of B_{TC} over all the ecotopes belonging to unit **j** and $k_j \in [0.1]$ is an environmental index computed as the average between three parameters

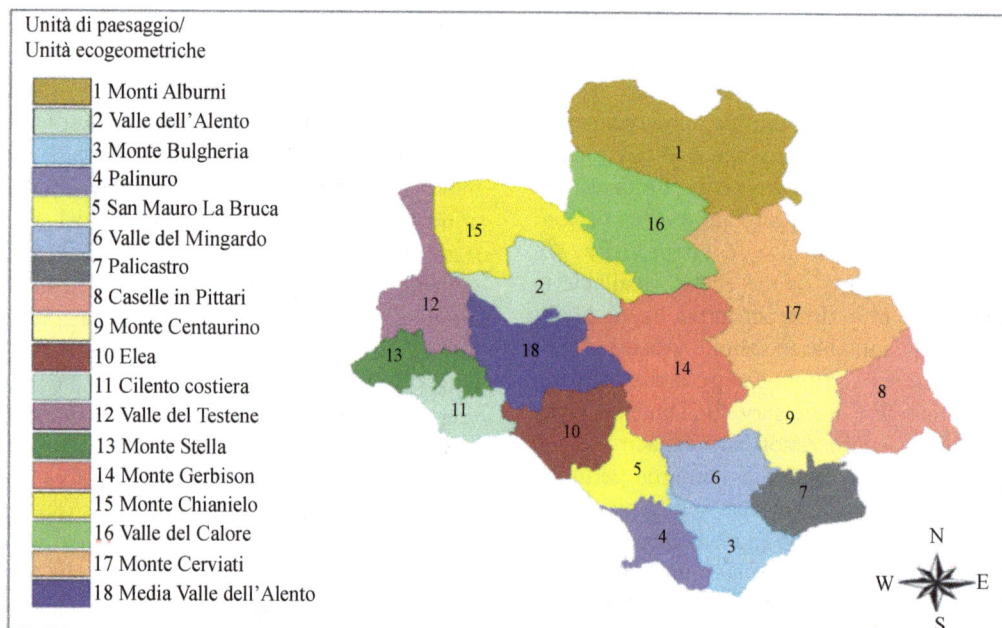

Figure 1. Landscape units for the national park of Cilento.

Table 1. Standard classification of B_{TC} values.

Class	Typology of ecotope	B_{TC} Mcal/m^2/year
A (Low)	Prevalence of systems that needs energy (industries, infrastructures, buildings, brownfields, rocky areas).	<0.5
B (Md-Low)	Prevalence of agricultural- technological systems or degraded ecotopes (sowed areas, shed built areas, inculted grassy areas, river corridords).	0.5 - 1.5
C (Medium)	Prevalence of agricultural seminatural systems (sowed areas, orchards, vineyards, hedges to medium resistance).	1.5 - 2.5
D (Md-High)	Prevalence of natural ecotopes (bushes area, pioneer vegetation, rows, poplar areas, reforestations areas, urban green).	2.5 - 3.5
E (High)	Prevalence of natural ecotopes that don't need a supply of energy (woods, mountains areas, damp zones).	3.5 - 5

k_{Fj}, k_{Pj}, k_{Dj}, each with values in [0, 1], as follows:

$$k_{Fj} = 1 - \frac{P_j^C}{P_j}, \quad k_{Pj} = \frac{\sum_{r=1}^{s} p_{rj} L_{rj}}{P_j} \text{ with } P_j = \sum_{r=1}^{s} L_{rj}$$

$$k_{Dj} = \frac{\sum_{i=1}^{5} \frac{n_i}{5} \log_{10} \frac{n_i}{5}}{\log_{10}(1/H)}$$

with P_j perimeter of unit j, P_j^C perimeter of a circle of area A_j, L_{rj} the perimeters of the s portions of P_j which have *permeability index* p_{rj}, r = 1, $\cdots$, s; n_k is the number of ecotopes of B_{TC} of class k among the whole number H of classes present in the unit j. Note that in our case the maximum number of classes is 5 as shown in **Table 1**.

The first index, k_{Fj}, is a parameter related to the shape of the patch borders, since their morphology influences strongly the energy exchanges between the patches themselves. Note that the most jagged is the hedge's shape the most favorable are the conditions for hiding and reproducing of wildlife.

The second one, k_{Pj}, again with the purpose of evaluating energy exchanges, takes into account the permeability of the barriers to energy flux, by following some standard classification ([3]) of values of the permeability parameter as reported in **Table 2**.

Finally, the third parameter K_{Dj} is related to biodiversity, determined by a Shannon entropy value, that takes into account the presence of different ecotopes inside each unit. High values of biodiversity contribute to more stable ecosystems.

Note that, being $\max\{k_j\} = 1$, we have that the maximum value of bioenergy, M_{max}, that a given landscape with *n units* can produce, will be:

$$M_{max} = 2B_{max} \text{ with } B_{max} = \max_{j=1,2,\cdots n}\{B_j\} \quad (2)$$

The last environmental indicator needed for evaluating the graph is the *bioenergy flux*, trough the borders of two consecutive units **i** and **j**, whose magnitude is proportional to the width of the link between the nodes of the

Table 2. Permeability of barriers.

Typology of barriers	p
Highways, principal net of communication	0.05
urban or secondary roads	0.4
artificial water net	0.4
railroad	0.5
White road	0.7
natural water net	0.85
principal river	1.0

units i and j:

$$F_{ij} = \frac{M_i + M_j}{2} \frac{L_{ij}}{P_i + P_j} p_{ij} \quad (3)$$

where all the quantities in (3) have already been defined.

For the construction of the ecological graph we have used the software ArcView 3.x of GIS (Geographical Information System).

The *shape files*, derived from the **regional land cover map** produced by Campania Region, furnish the information about the land uses, the presence and the connection of road infrastructures (railroads, highways, government and provincial roads), the system of water courses (natural and artificial) and the administrative subdivision of the various urban territories within the studied area.

In **Table 3** we have firstly reported the values found, by means of the spreadsheet application of Microsoft Excel, for the bioenergy M_j (normalized to 1), recalling that the diameters of the graph's nodes are proportional to these magnitudes.

Then, in **Table 4** we have reported the values of fluxes F_{ij} (see Equation (3)), normalized to one, between consecutive units, recalling that the width of graph's arcs are proportional to these values.

The final result of the construction of the Ecological Graph is shown in **Figure 2** that exhibits a picture of the actual state of ecological health of this territory.

From the graph of **Figure 2** we get, for example, the information that the energetic content of units 1, 14 and 17 (Monti Alburni, Monte Gerbison e Monte Cerviati) is high, hence they represent the territorial portions of greater ecological value and therefore deserve of more attention as they support the entire environmental system. Note from **Figure 3** that units 1 and 17 are Special Protection Areas (SPAs), while in unit 14 there is a Site of Community Importance (SCI). Even though, the flux between 1 and 17 is very weak due to the presence of a highway that is not permeable to energy flow. To strengthen this flow it might be possible to carry out structures that allow wildlife to cross above or below the roadway, as studied by the new discipline of Road Ecology. Moreover, the energy content of unit 9 (Monte Centaurino), even if it is characterized by SCI and SPA areas, is very low, as shown by the small size of its node, for the presence of agricultural areas used for annual crops associated with permanent, it would be appropriate to take action for environmental improvement works (hedges of natural vegetation) aimed to increase the level of biodiversity and thus the overall stability of the system. Unit 6 (Valle del Mingardo) is definitely a part of the territory on which it is advisable to aim for improve the system. It is in fact an area characterized by a low value of bioenergy but by a high number of links (5), charac-

Table 3. Bioenergy values of the 18 landscape-units.

Landscape-unit	area (mq)	B_j Average Value of B_{TC} over all the ecotopes in the unit j	k_j	Bioenergy $M_j = (1 + k_j)B_j$ normalized to 1
1_Monti Alburni	271795284.779	976415252.511	0.392	0.79
2_Valle dell'Alento	106677569.293	315876438.946	0.481	0.27
3_Monte Bulgheria	90479291.482	308154108.671	0.454	0.26
4_Palinuro	65705173.685	222999377.355	0.466	0.19
5_San Mauro La Bruca	73506910.281	207991143.863	0.536	0.18
6_Valle del Mingardo	88059522.990	256106446.029	0.429	0.21
7_Policastro	83583402.535	223884113.933	0.443	0.19
8_Caselle in Pittari	131787266.802	465534241.171	0.354	0.36
9_Monte Centaurino	117026977.046	315010972.459	0.482	0.27
10_Elea	104082537.911	305804409.005	0.526	0.27
11_Cilento Costiera	60436023.232	180283730.319	0.490	0.16
12_Valle del Testene	113333667.540	272387166.872	0.546	0.24
13_Monte Stella	69489471.921	214733859.953	0.503	0.19
14_Monte Gerbison	240080281.969	875649957.003	0.468	0.74
15_Monte Chianiello	173522652.513	469893200.610	0.513	0.41
16_Valle del Calore	183134923.836	522436272.450	0.512	0.46
17_Monte Cerviati	324094491.338	1223367667.853	0.412	1.00
18_Media Valle dell'Alento	142971752.086	430214999.459	0.531	0.38

Table 4. Bioenergy fluxes between consecutive landscape-units.

Consecutive units	L_{ij}	p_{ij}	$P_i + P_j$	Fluxes values $F_{ij} = \dfrac{M_i + M_j}{2} \dfrac{L_{ij}}{P_i + P_j} p_{ij}$	F_{ij} norm.
1 - 16	19092.000	0.200	157288.685	26088766.816	0.331
1 - 17	9513.000	0.200	182074.021	16128633.135	0.205
2 - 12	3914.000	0.200	130299.592	2670462.353	0.034
2 - 14	7330.000	0.200	141240.674	9100885.786	0.116
2 - 15	28216.000	0.200	139757.782	23793566.474	0.302
2 - 18	19773.000	0.200	129125.858	17247639.944	0.219
3 - 4	14800.000	0.200	111795.429	10256790.053	0.130
3 - 5	3000.000	0.200	107741.349	2136803.108	0.027
3 - 6	10500.000	0.200	101924.128	8384114.553	0.106
3 - 7	7970.000	0.200	106035.225	5794134.282	0.074
4 - 5	11187.000	0.200	102806.136	7033615.419	0.089
5 - 6	6039.000	0.200	92934.835	4454146.045	0.057
5 - 10	15200.000	0.600	104440.840	34324676.081	0.436
5 - 14	8163.000	0.200	131420.783	9971391.297	0.127
6 - 7	7550.000	0.200	91228.711	5701396.261	0.072
6 - 9	10078.000	0.200	103978.820	8072834.087	0.102
6 - 14	8246.000	0.200	125603.562	10844201.535	0.138
7 - 9	12600.000	0.600	108089.916	27624367.021	0.351
8 - 9	12624.000	0.200	118394.474	11701046.248	0.149
8 - 17	9600.000	0.200	153940.225	5943963.589	0.075
9 - 14	8525.000	0.200	142464.768	10488670.454	0.133
9 - 17	12684.000	0.200	156385.776	17798375.337	0.226
10 - 11	6127.000	0.200	105444.358	4272661.374	0.054
10 - 14	4173.000	0.200	318045.318	2299419.999	0.029
10 - 18	17636.000	0.300	124994.751	23814037.862	0.302
11 - 13	22500.000	0.200	109832.078	12114746.575	0.154
11 - 18	2741.000	0.200	580618.501	2919854.799	0.037
12 - 13	16089.000	0.350	130556.205	16043771.238	0.204
12 - 15	11245.000	0.350	151665.537	14687334.610	0.186
12 - 18	9200.000	0.600	141033.613	21131023.445	0.268
13 - 18	4729.000	0.200	129382.471	3586581.485	0.046
14 - 15	3647.000	0.200	162606.619	4478013.621	0.057
14 - 16	7747.000	0.600	153225.182	42264197.828	0.537
14 - 17	18611.000	0.500	178010.519	78760557.246	1.000
14 - 18	14830.000	0.200	151974.695	18973840.167	0.241
15 - 16	20334.000	0.200	151742.291	20108332.455	0.255
16 - 17	13905.000	0.600	167146.191	62825084.073	0.798

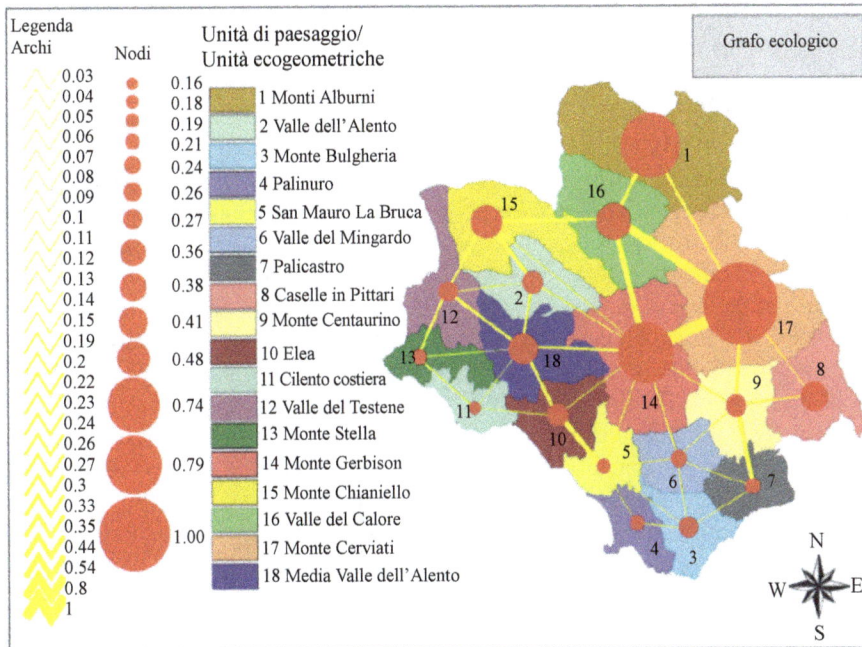

Figure 2. Ecological graph for the national park of Cilento.

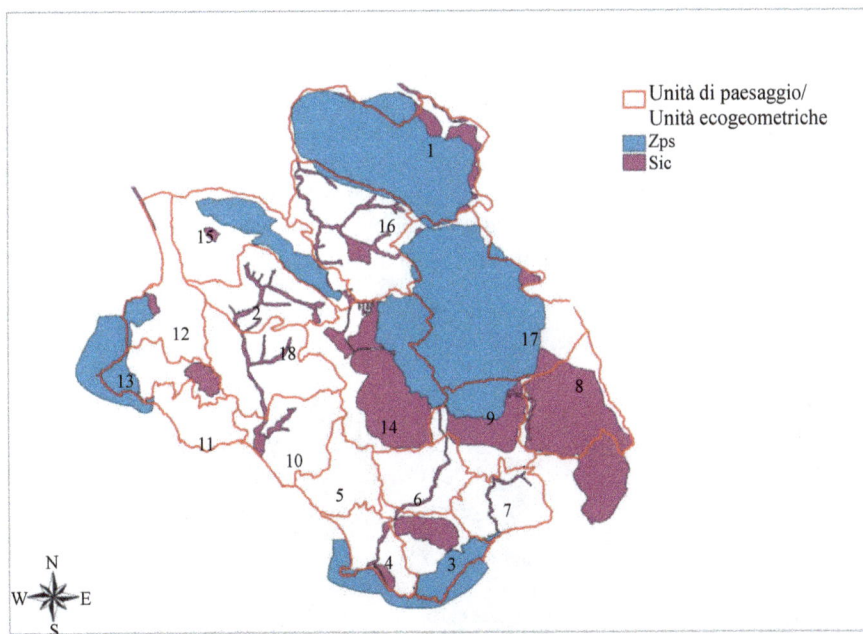

Figure 3. SCI (SIC) and SPAs (ZPS) areas in parco nationale del Cilento.

terized however by weak fluxes, even here, this weakness is linked to the widespread presence of agricultural areas on which to intervene in order to increase the biodiversity. We finally note that as decision-making tool we could match 2 graphs, one representing the actual situation, the second one representing the project solution in line with the intents of a Master Plan if any. The comparison among the graphs of the two sceneries allows, in decisional phase, to judge on the sustainability of the intervention on the area.

3. Ecological Network

As shown by the ecological graph in **Figure 2**, the studied area suffers of reduction and fragmentation of natural and semi-natural habitats as outcome of agricultural intensification, infrastructure networks and urbanization, even if it is a place of several SCI (Site of Community Importance) and SPAs (Special Protection Areas), as one can see from **Figure 3**. Recently, eco-regional planning is playing an increasingly important role on the acknowledgment that it is necessary to integrate, from both ecologically and socio-ecologically points of view, the protected areas in the landscape matrix of the entire territory. Hence the origin of *ecological networks*, characterized by their emphasis on biodiversity conservation at level of region.

The first step consists in creating a *resistance map*, that is, a map of resistance of the landscape matrix to the mobility of the selected species. In the language of ecological network the zones with the lowest value of resistance play the role of *core areas*. Then, the least-cost paths linking such areas, *i.e.*, the potential paths that minimize the cost of mobility, are computed ([6]). They represent the *corridors* of the ecological network that could be adopted as reference information in environmental evaluation of plans and projects in order to reduce as much as possible territorial fragmentation.

The design of corridors for integration in eco-regional planning demands to know the mobility requirements of certain target animal species with rather wide mobility ranges. In our Study Case we have chosen the Peregrine Falcon that lives on the cliffs of the coastal zone of Cilento and is renowned for its speed, reaching over 325 km/h during is hunting. The Peregrine Falcon requires open spaces in order to hunt, he often hunts over open water, marshes, valleys, fields, and tundra, searching for prey either from a high perch or from the air. While its diet consists almost exclusively of medium-sized birds, the Peregrine will occasionally hunt small mammals,

small reptiles, or even insects. Hence, we can say that its living space covers all over the Park.

In order to use the least-cost path algorithm, from GIS technology, we need to relate the specific ecological requirements of the chosen species to land uses of the Cilento Park, through a specific parameter called resistance that measures the degree of environmental opposetion to its spread and colonization.

The areas characterized by low value of resistance (resistance equal to zero to the most suitable area) are considered core areas, *i.e.*, natural areas with ecologically high value, for the potential ecological network specific to the species (they could be protected areas too). By using the GIS software used for the graph's construction we have built the map of resistance ([6]), as shown in **Figure 4**.

The resistance for the Peregrine Falcon runs from a zero value assigned to bare rocks, cliffs, ponds and meadows, to 60 value for woods, to 90 value for urban areas, rail and road networks. We then apply the PATHMATRIX tool ([7]), an implementation of the least-cost distance algorithm of the GIS software ArcVIEW 3.x, that is able to apply the *cost distance* algorithm in pair wise fashion among a set of sample locations. PATHMATRIX can also output the length of the least-cost path in geographical distance units. We recall that the least cost path minimizes the sum of resistances along the path (see **Figure 5**), hence, in our case, it corresponds to the corridor that minimizes the cost of mobility of the target species between the core areas, as shown in **Figure 6**.

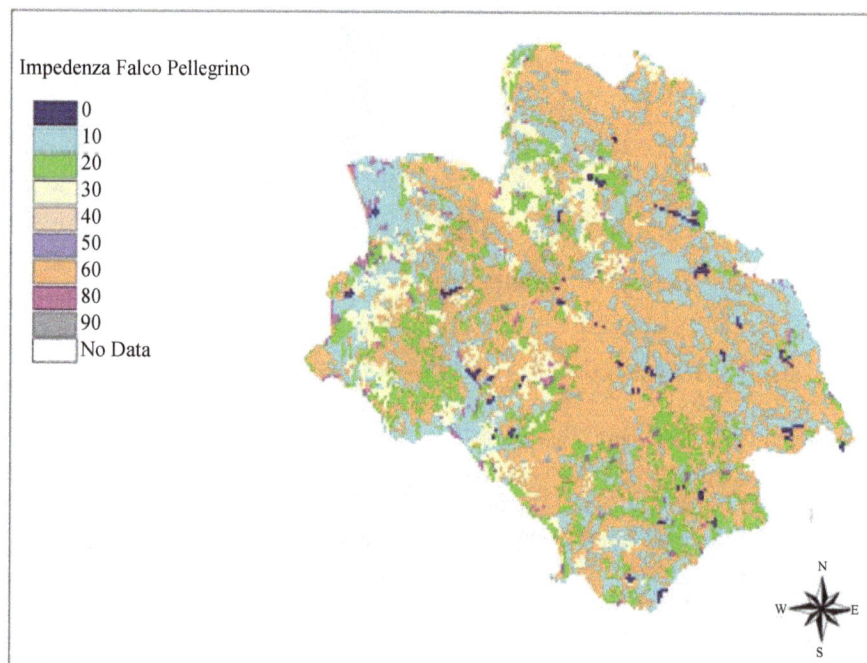

Figure 4. Resistance map for peregrine falcon in the national park of Cilento.

Figure 5. The least cost path minimizes the sum of resistances in going from A to B.

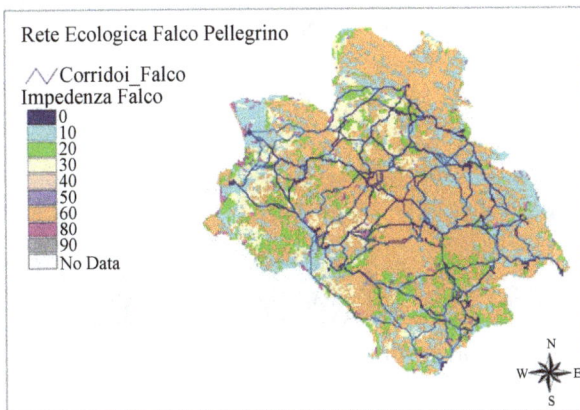

Figure 6. Potential ecological network for peregrine falcon.

The network suggests that along the corridors it would be better to avoid installing equipment in proximity of wetlands, places of wintering of waterfowl and frequented by several species of birds of prey. Also, not installing power lines or wind farms, sources of major impact in the fast flight of Peregrine Falcon. Moreover, rock climbing activity can have negative impact on species whose life is linked to cliffs (nesting, roosting food). Also agricultural activities that impact on the conservation of wildlife could pauperize Falcon's hunting.

These are some of the considerations coming from a rapid analysis of the ecological network that, hence, can be rightly considered as another decision-support tool in environmental planning and landscape management.

4. Logistic Equation with Harvesting

As said in the Introduction, the Ecological Graph furnishes the actual state of energy exchange in the territory,

hence, it would be interesting to investigate the time evolution of energy, starting from the actual settlement, in order to get information on possible future scenarios and check if the trend is toward a sustainable development of the territory or not. This can be made by studying the equilibrium solution of a suitable differential equation that models dynamics of the territory evolution under an ecological point of view. As said in the Introduction, the term "*bioenergy*" refers to the energy available in the environment, present in different forms like animals, seeds, plants, ruled by suitable metabolic processes, hence, its value must be limited by some carrying-capacity of the given environment. Then, we propose a simulation model based on a logistic-type differential equation ([8]) like that one approximating the evolution of population over time in presence of limited living resources of the environment. Moreover we add a harvesting term in order to simulate the growth of bioenergy over a landscape in spite of the obstacles coming from territory fragmentation.

Namely, if we denote by M(t) the average value of the bioenergy (see Formula (2)) over all the n Landscape Units constituting the entire system under study:

$$M(t) = \frac{1}{n}\sum_{j=1}^{n} M_j \, ,$$

the dynamical simulation model is given by the following nonlinear differential equation

$$M'(t) = cM(t)\left[1 - M(t)/M_{max}\right] - hS_o \qquad (4)$$

where M_{max}, the maximum value of bioenergy the given territory can produce, is given by formula (2) of Section 2 and the connectivity index, c is defined ([8]) by

$$c = \frac{1}{V}\sum_{s=1}^{V} \frac{F_s}{\max_s F_s}$$

with V the number of arcs present in the graph constructed for the given territory, F_s, s = 1, ···, V, are the values of energy fluxes given by Formula (3).

In Formula (4), the prime indicates the time derivative, t the time variable, the *harvesting term*, $-hS_o$, is given by the product of h, the ratio between the sum of the impermeable barrier lengths and the total external perimeter of the territory, and S_o, the ratio between the sum of the territory surfaces with low values of B_{TC} and the total surface of the system.

Let us note that, from its definition, the connectivity c represents the territorial ability to spread the bioenergy, furnishing, hence, a measure of territorial fragmentation (the flux F_S through an impermeable barrier is equal to zero). It plays the role of the constant growth rate as in population dynamics.

By using the normalized bioenergy $M(t) = M(t)/M_{max}$,

the time evolution equation for M becomes:

$$M'(t) = cM(t)[1 - M(t)] - hS_o \qquad (5)$$

In the application of this model to a Brownfield at the East zone of Naples (**Figure 7**), subject to a Master Plan in the direction of improving the environment, S_o will represent the percentage of edified areas, while h will be the ratio between the sum of the perimeters of edified areas and the total perimeter of the area ([9]).

With a similar procedure as before we can construct the ecological graph for this area. We shall not give the details but, instead, we shall provide the main and significant results coming from the mathematical approach based on the study of the equilibrium solutions of Equation (5).

We outline that the mathematical model basic assumption of Equation (5) is that the time evolution of bioenergy will consist of the balance between two quantities with opposite signs. The first one, positive, describes the bioenergy growth by following a logistic law, driven by the connectivity parameter c; the second one, negative, $-hS_o$, the *harvesting term*, opposes to bioenergy growth due to the presence of barriers related to the edified areas that hamper the flux of energy.

Note that, once subdivided the territory in units as in Section 2, from the relative ecological graphs, as those in **Figures 8** and **9**, we see that in the project plan the diameters of the nodes, together with the width and the number of the arcs, are increased in line with the planned environmental improvements.

Figure 7. Brownfield-east zone of Naples.

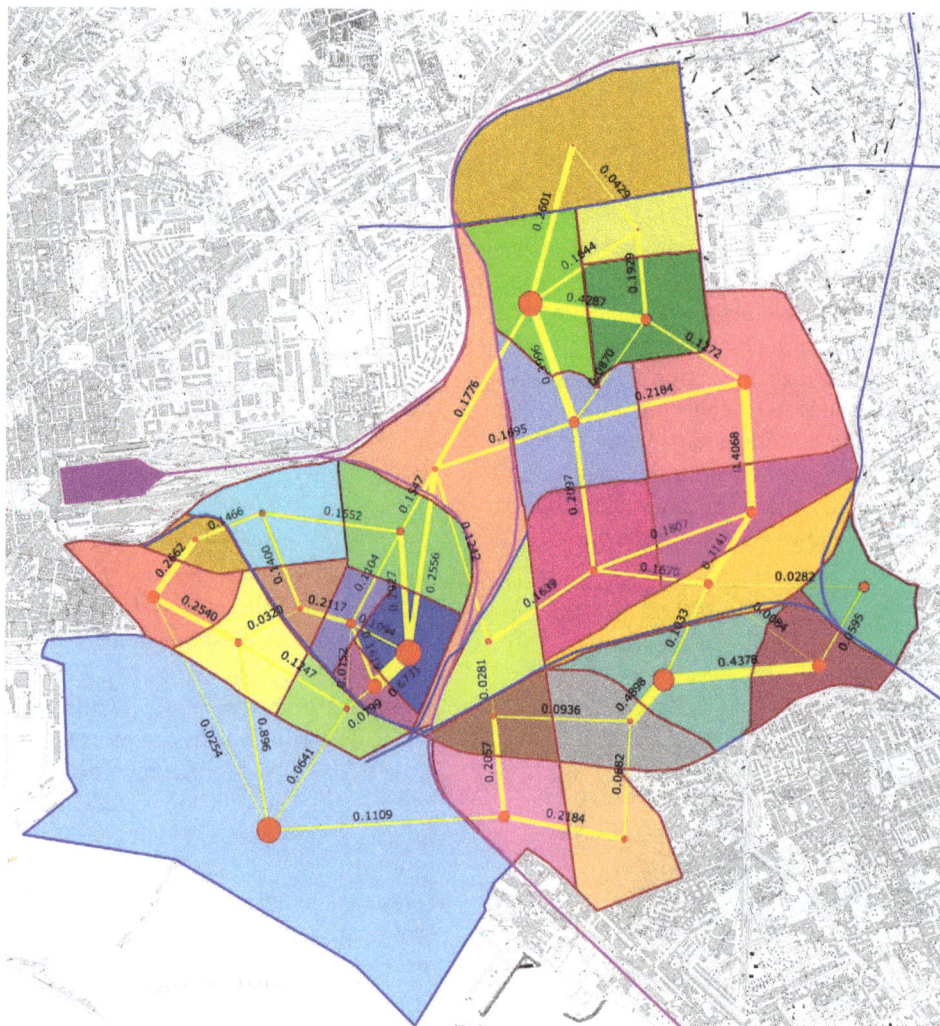

Figure 8. Ecological graph of brownfield actual state.

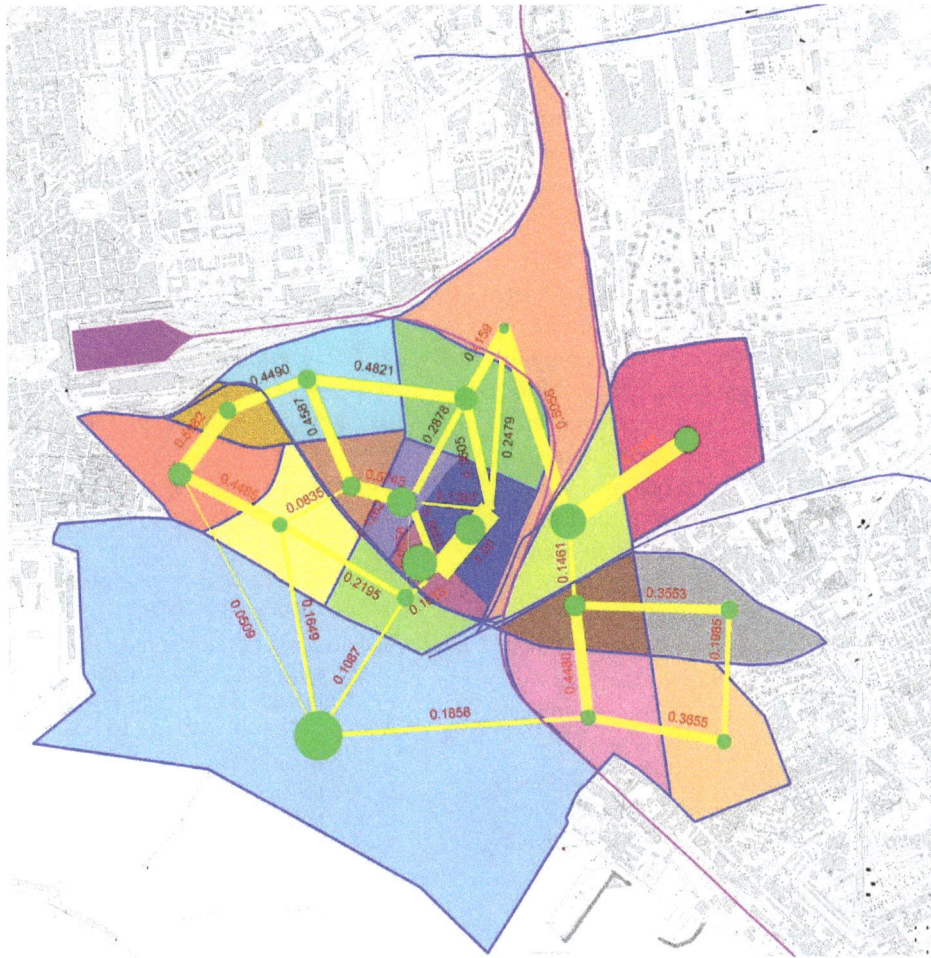

Figure 9. Master plan project ecological graph.

This happens because between the two situations there will be different typologies of intended use of the ground and barriers that will make changes in the analysis and in the calculation of the environmental indices (the Master Plan foresees in fact, among other, the presence of a Urban Park). As a consequence the initial value of M, $M(0)$, as well as the values of c, h and S_o will change.

If we now turn our attention to the graphs of the solutions ([10]) to the differential Equation (5) of **Figure 10**, we can see three different possible future sceneries: the first one shows that, starting from the actual state of the area, *i.e.* with initial value of bioenergy $M(0) = 0.048$, c $= 0.58$, $h = 0.28$, $S_o = 0.79$, there is a quick trend to the environmental collapse corresponding to $M = 0$; in the second one, starting from the project plan, $M(0) = 0.086$, c $= 0.65$, $h = 0.069$, $S_o = 0.66$, the value of the bioenergy grows visibly, due to the environmental improvements given by the interventions in line with the Master Plan and tends to a stable good value; the third one furnishes a critical value of the parameter, $h = 0.08$, at which the system tends to collapse even if it starts from the project value of bioenergy, showing the important role played by

the geometrical configuration of impermeable barriers of the buildings.

AREA IN THE ACTUAL STATE	PROJECT	Criticality
$M(0) = 0.048$	$M(0) = 0.086$	$M(0) = 0.086$
$c = 0.58$	$c = 0.65$	$c = 0.65$
$S_o = 0.79$	$S_o = 0.66$	$S_o = 0.66$
$h = 0.28$	$h = 0.069$	$h = 0.08$
$t = 1$ year	$t = 50$ years	$t = 10$ years

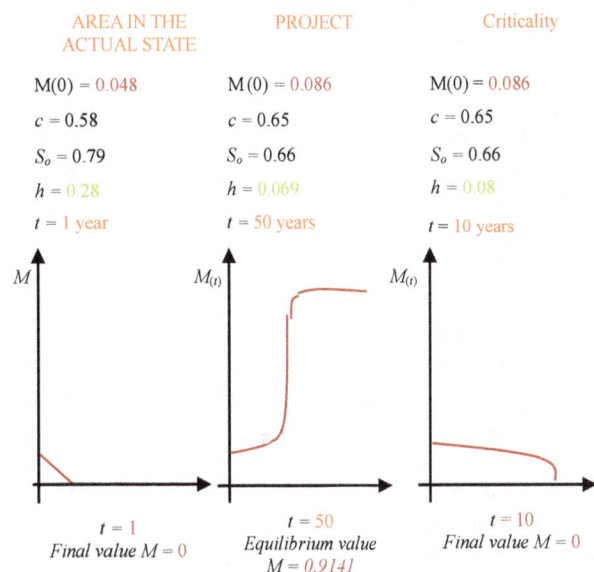

t = 1
Final value M = 0

t = 50
Equilibrium value
M = 0.9141

t = 10
Final value M = 0

Figure 10. Solution graphs.

We would like to stress that the simple simulation models here presented do not pretend to perform quantitative predictions, but to estimate the goodness of a territorial plan, getting an insight to possible criticalities and hence serving as a decision support in sustainable environmental planning and landscape management.

5. Conclusions

Strategic Environmental Assessment, nowadays, has been adopted in Europe in landscape planning, whose task is to verify the compatibility of territory transformations with respect to their levels of criticality and vulnerability, to evaluate possible future scenarios as consequence of interventions by checking if they are in line with presservation and valorization of environmental quality.

This evaluation must be based, hence, on the knowledge of the mechanisms that rule a territorial transformation. In order to assess the ecological functioning of an environmental system is necessary to single out the energetic contents of the units composing the territory, their connections carrying energy and material fluxes, as well as their breaking points, due to the presence of impermeable barriers, that produce territory fragmentation.

We have presented three different simulation models in the framework of Landscape Ecology, all of them based on GIS technology, which can be used as decision support in environmental planning, such as:

1) **Planar graph**, the so called *ecological graph*, whose construction needs the computation of suitable indices of environmental control, proper of Landscape Ecology, such as *biodiversity, Biological Territorial Capacity, connectivity*. The planar graph for the considered environmental system returns a picture of its actual ecological health conditions and provides very detailed indications and operational assistance to guide toward ecological sustainable interventions.

2) **Ecological network**, based on the *resistance map* of target animal species, that gives useful information to prevent and to reduce territorial *fragmentation* caused by intense processes of urbanisation and industrialisation

3) A mathematical model based on nonlinear differential equation of **logistic-type with harvesting** used to investigate the time evolution of the ecological value of a given territory, by starting from a given settlement and looking for the trend to consequent future scenarios, hence, furnishing qualitative predictions on the sustainability of a given territorial plan; a recent implementation of this model can be found in [11,12].

These mathematical and GIS interfaced models can help in understanding environment response and dynamic change in time to correctly manage and preserve natural resources; they can represent a powerful decision support to compare effects and impacts of possible alternative future scenarios.

6. Acknowledgements

The author wishes to acknowledge the support given by the Department of Industrial and Information Engineering of the Second University of Naples.

REFERENCES

[1]   R. T. T. Forman, "Land Mosaics. The Ecology of Landscape and Regions," Cambridge Press, Cambridge, 1995.

[2]   A. Farina A., "Ecologia del Paesaggio," UTET Libreria, Torino, 2001.

[3]   V. Ingegnoli, "Landscape Ecology: A Widening Foundation," Springer-Verlag, New York-Berlin, 2002.

[4]   P. Fabbri, "Paesaggio, Pianificazione, Sostenibilità," Alinea Editrice, Firenze 2003.

[5]   G. Lauro and R. De Martino, "Environmental assessment of the meta-ecosystem Cilento with the tools of Landscape Ecology," In: C. Gambardella, Ed., *Atlante del Cilento*, ESI, Napoli, 2009, pp. 533-538.

[6]   G. Lauro and R. De Martino, "The Ecological Network as a Tool for the Protection of Coastal Ecosystems. Workshop: The Mediterranean Coastal Monitoring: Issues and Measure Techniques," Livorno, 15-16-17 Giugno 2010, Firenze: CNR-IBIMET, pp. 163-170.

[7]   N. Ray, "PATHMATRIX: A Geographical Information System Tool to Compute Effective Distances among Samples," Molecolar Ecology Notes, Wiley Online Library, 2005.

[8]   G. Lauro, R. Monaco and G. Servente, "A Model for the Evolution of Bioenergy in an Environmental System," In: T. Ruggeri and M. Sammartino, Eds., *Asymptotic Methods in Nonlinear Wave Phenomena*, World Scientific, 2007, pp. 96-106.

[9]   G. Lauro and M. Musto, "Simulation Models for Environmental Control," *Proceedings of Sixth EAAE-ENSHA Construction Teachers' Network Workshop*, Mons, 22-24 November 2007, pp. 178-185.

[10] A. D. Bazykin, "Nonlinear Dynamics of Interacting Populations," World Scientific, River Edge, 1998.

[11] F. Gobattoni, R. Pelorosso, G. Lauro, A. Leone and R. Monaco, "A Procedure for Mathematical Analysis of Landscape Evolution and Equilibrium Scenarios Assessment," *Landscape and Urban Planning*, Vol. 103, No. 3, 2011, pp. 289-302.

[12] F. Gobattoni, G. Lauro, R. Monaco and R. Pelorosso, "Mathematical Models in Landscape Ecology: Stability Analysis and Numerical Tests," *Acta Applicandae Mathematicae*, Vol. 125, No. 1, 2013, pp. 173-192.

8

A Web-Based Cancer Atlas of Saudi Arabia

Khalid Al-Ahmadi[1*], Ali Al-Zahrani[2], Atiq Al-Dossari[1]
[1]Space Research Institute, King Abdulaziz City for Science and Technology, Riyadh, Saudi Arabia
[2]King Faisal Specialist Hospital and Research Centre, Riyadh, Saudi Arabia

ABSTRACT

There is a distinct lack of online atlases to visualize and explore cancer incidence in Middle Eastern countries despite the clear benefit that such tools can deliver. This paper describes the development and implementation of a cancer Atlas of Saudi Arabia, which is a web-based client-server application with built-in analysis functions for analyzing patterns of cancer incidence. Built using ESRI's ArcGIS Server API and ASP.NET, the atlas contains 45,532 incidences of cancer for the period from 1998 to 2004, which were provided by the Saudi Arabian National Cancer Registry. This tool is aimed at health care practitioners and researchers, who can use this tool for exploring cancer distribution and investigating trends, and as a decision support tool for service allocation. The tool allows users to map cancer incidence and undertake analyses at four spatial scales from city to national level.

Keywords: Cancer Incidence; Web-Based Mapping; GIS; Colorectal Cancer; Saudi Arabia

1. Introduction

Cancer is a leading cause of death in the world, affecting populations in countries of varied levels of industrialization and wealth. Although 70% of cancer deaths are in low and middle income countries, the incidence in all countries, including high income countries, is expected to increase over the next few decades. In 2008, there were 7.8 million deaths attributable to cancer, and this number is expected to increase to more than 11 million in 2030 [1]. With such a global presence in public health, research is ongoing to explore not only the aetiology of the disease but also the spatial variation in the incidence of cancer. A greater understanding of the spatial distribution of different types of cancer can be used to target early screening efforts and treatments in those areas where the need is greatest.

The use of Geographic Information Systems (GIS) and spatial analysis techniques is a relatively new area of investigation in cancer research. Elevated rates of breast cancer in Cape Cod between 1982 and 1990, when compared to the state of Massachusetts as a whole, prompted Brody et al. [2] to consider the challenges of using GIS to investigate the potential environmental influences of breast cancer from hormone disrupters in drinking water or from exposure to pesticides. Subsequent analyses were

reported in Brody et al. [3,4]; no evidence was found for a relationship between breast cancer and drinking water contaminated by wastewater but a small increased risk was found for some pesticides. Around the same time, Biggeri et al. [5] investigated the effect of four sources of air pollution on lung cancer, including the effect of distance. They found strong relationships between lung cancer and proximity to the city centre and an incinerator, with decreasing risk as distance from these places increased. Since then, a number of cancer studies have appeared in the literature and utilized spatial analysis and GIS for disease mapping, exposure and risk assessment [e.g. 6-10]. Beyond the methods used by these authors to map and visually assess patterns in cancer incidence/ mortality, advanced spatial analysis methods have been applied to identify spatial clusters of different cancers. For example, Wang [11] applied the R statistic of Rogerson [12] to look for spatially significant clusters of breast, lung, colorectal and prostate cancers in Illinois for the period from 1986 to 2001. Vieira et al. [13] used a similar approach to assess potential clusters in the incidence of breast, lung and colorectal cancers in Massachusetts. In this study, the authors considered length of residence (up to forty years prior to diagnosis) to adjust for population movement. Using generalized additive models, they found that after adjusting for confounders, breast cancer hotspots increased and were statistically signifi-

---

[*]Corresponding author.

cant near a military base and groundwater plumes.

Interest in GIS as an analytic tool for cancer incidence analysis remains strong, as evidenced by recent publications (see, for example, [14,15]). However, the focus to date has been largely on breast and lung cancer, perhaps due to larger numbers of diagnosed cases and the interest in potential environmental determinants of these cancer types.

All of the above examples have been concerned with the investigation of case studies based on a small number of cancers or they have a limited spatial focus. To provide a more comprehensive overview of cancer incidence or mortality, a number of different cancer atlases have been developed (see [16] for an overview of selected atlases). The Atlas of Cancer Mortality of the United States for the period from 1950-1994 was originally published in hard copy, but with rapid changes in IT and the internet, web-based applications have become a more common way to rapidly disseminate this type of information to a wider audience. Initially a method of static map dissemination, recent advances in Web 2.0 technology [17] mean that advanced spatial analysis functionality can now be embedded into a web-based GIS as a lightweight browser application. The Atlas of Cancer Mortality is now available as an online atlas with both static and dynamic mapping capabilities [18]. Other online atlases include the Pennsylvania Center Atlas (PACA) [19,20], the Interactive Cancer Atlas (InCA) for the US, [21], the Cancer e-Atlas, covering Yorkshire and Humberside in the UK [22], and the Atlas of Cancer in India [23]. Some of the online versions allow for the dynamic production of maps that can illustrate the type of cancer, age and gender of patients by area as well as basic statistics on cancer incidence. Online tools with a broader health focus are also available, for example to assess local health outcomes in the city of London, such as the London Healthcare Benchmarking Tool [24] and the Health Needs Assessment Toolkit [25]. These sites allow healthcare professionals, decision makers and the public to create maps or charts for downloading and to show how the various health outcomes in their local area compared to other areas in London. The geographic focus of existing online cancer atlases and broader health online applications has remained on more developed countries, with other highly populated regions in the world underrepresented.

Reflecting this unintentional "global north" perspective, there is a distinct lack of online atlases to visualize and analyze cancer incidence in Middle Eastern countries despite the clear benefit that such tools can deliver. This paper describes the development and implementation of the Cancer Atlas of Saudi Arabia (CASA), which is a web-based client-server application with built-in analysis functions for analyzing patterns of cancer incidence. Built using ESRI's ArcGIS Server API and ASP.NET, the atlas

contains 45,532 incidences of cancer for the period from 1998 to 2004, which were provided by the Saudi Arabian National Cancer Registry (SCR). This tool is aimed at health care practitioners and researchers, who can use it to identify cancer "hotspots" and investigate trends, and also as a decision support tool for service allocation. The tool allows users to map cancer incidence and undertake analyses at four spatial scales from city to national level. The CASA extends the functionality of previous online cancer atlases through the provision of additional mapping types, animation, the ability to export the maps and graphs as high quality graphics, and the calculation of cancer rates that adjust for population and age. The technology used to build the CASA is also sufficiently generic that it could be easily transferred to other areas and health outcomes beyond the one presented here. The functionality of the atlas is illustrated using colorectal cancer in Saudi Arabia as an example.

2. Development of the Atlas

The functionality of the Cancer Atlas of Saudi Arabia (CASA) was driven by the requirements of the end-user, *i.e.* researchers and health practitioners. The literature on existing cancer atlases and cancer analysis was reviewed and the end-users were interviewed to undertake a needs assessment. It was clear from both these exercises that the atlas must be capable of both statistical and spatial analysis. As well as allowing the user to create and export maps, figures and tables showing counts of people diagnosed with cancer by cancer site, gender, age groups, stage distribution, morphological distribution, region and geography, users wanted to visualize the data at different spatial scales and over time. To analyze spatial and temporal changes simultaneously, animation capabilities were added. **Table 1** lists the specifications of the CASA in comparison with those of the main online cancer atlases. Although this list is not meant to be fully comprehensive, we did not find comparable atlases for other Middle Eastern countries (in English or Arabic). From **Table 1**, it can be seen that the CASA was developed with a greater level of functionality and flexibility. The architecture of the system is outlined in the next section followed by a description of the data, how cancer incidence is reported and an overview of the statistical and spatial interface of the atlas.

2.1. Architecture of the System

The atlas was developed using ESRI's ArcGIS Server API software for Flex [26]. The ESRI API provides a comprehensive set of functions for designing and creating professional cartographic products from a client computer or mobile device using a browser. The CASA is comprised of four main components, as shown in **Figure 1**:

Table 1. Comparison of the features of different cancer atlases with the CASA.

Feature	Pennsylvania Cancer Atlas	Cancer e-Atlas	Interactive Cancer Atlas (CDC)	Atlas of Cancer Mortality	Cancer Atlas of Saudi Arabia (CASA)
Country applied to	USA	UK	USA	USA	Saudi Arabia
Interactive	Yes	Yes	Yes	Yes	Yes
Dynamic Composite Analysis	Yes	No	Yes	No	Yes
Cancer Site	Yes	Yes	Yes	Yes	Yes
Gender	Yes	Yes	Yes	Yes	Yes
Period of cancer	Yes	No	Yes	Yes	Yes
Age at diagnosis	Yes	No	Yes	Yes	Yes
Stage of cancer	Yes	No	Yes	No	Yes
Morphology	No	No	No	No	Yes
Cancer Incidence Rate	No	No	No	No	Yes
AIR	No	No	No	No	Yes
ASR	Yes	Yes	Yes	Yes	Yes
95% Confidence Interval	Yes	Yes	Yes	Yes	Yes
Animation Player	No	Yes	Yes	No	Yes
Map Types	Choropleth Map	Choropleth Map	Choropleth Map	Choropleth Map	Choropleth, Graduated Symbol Map, Pie Chart Map, Density Map
Map Classification Scheme	Equal Interval, Quantile	Equal Interval, Quantile, Natural Break, Standard Deviation, Continuous	Equal Interval, Quantile	No	Equal Interval, Quantile
Change Color of Map	No	Yes	No	No	Yes
Export Data (format)	Excel	Excel	Excel	Excel	Excel
Export Figures and Maps	No	JPEG PNG	No	No	JPEG PNG PDF
Figure Types	Line, Column				Line, Column, Pie
2D-3D visualisation	2D	2D	2D	2D	2D-3D

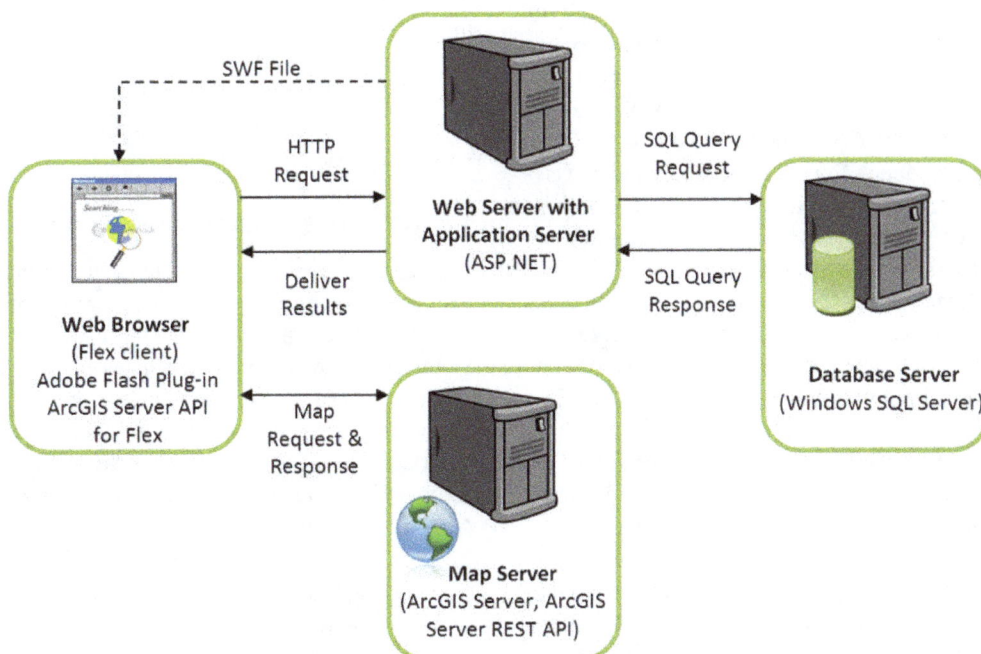

Figure 1. Architecture of the web-based cancer atlas.

- *The Client*: Flex was used to build the web client for the atlas, where MXML and Action Script define the layout, appearance and behavior of the application. These were compiled into a single SWF file that makes up the Flex client CASA application. Maps and analyses are displayed in the client using the functionality from the ArcGIS Server API for Flex.
- *The Web and Application Server*: The web server responds to client requests from the databases or the map server. ASP.NET was used to build the web application server, since it is a powerful tool for creating dynamic and interactive web applications. The cancer database is maintained in a Microsoft SQL Server and the operating system is Windows Server, so the application server interfaces with these other components seamlessly.
- *The Map Server*: The map server fulfills spatial queries, conducts spatial analysis, and generates and delivers maps to the client based upon requests by the users. The output from the map server can be a simple map image in a graphic format or map elements served by ArcGIS Server.
- *The Database Server*: The database server houses the cancer data in a relational database structure stored in Windows SQL Server 2008 (Enterprise Edition).

The design of the client interface was the single most important consideration, given that the target end-users were unlikely to have much technical knowledge and skills in computing or the principles of GIS. The design was partly driven by a review of existing atlases (**Table 1**), which were critiqued from a design as well as a functionality perspective. Users were also involved in the development process in order to offer the right level of GIS functionality while still providing the opportunity to create high-quality maps, charts, tables and commentaries on cancer incidence. The system is also flexible enough to allow for additional functionality to be added in the future based on feedback from the users. Choice of variables and options was designed to be as intuitive as possible. The design was implemented and tested with a set of potential users. The system provided considerable flexibility in terms of satisfying the needs of the users and high levels of responsiveness and performance during tests in an office where client computers were connected over a local network to a server in the office. The result was a web-based client-server system which is relatively straight-forward to use, offers the option of an Arabic or an English interface and worked well during feedback sessions with staff who work in the Saudi Cancer Registry. Only the English interface is shown in this paper. The atlas is currently only available on an internal network at the Saudi Cancer Registry while testing continues for the next year and feedback continues to be collected. The plan is to then open up the atlas to relevant health agencies and organizations working at the regional level in Saudi Arabia.

2.2. Data

The data were supplied by the Saudi Cancer Registry (SCR), a government agency which collects, collates and evaluates statistics on cancer incidence, survival and mortality in Saudi Arabia. The statistics on cancer incidence included the most common types of cancers which had been reported and recorded in Saudi Arabia from January 1998 to December 2004. The total number of patients diagnosed with cancer in this period was 45,532. The SCR has detailed medical information about each individual cancer case, but due to confidentiality, this data has not been made available for the atlas.

After georeferencing and cleaning the original cancer database, there were fourteen variables for each individual cancer case, including: gender, age, birth date, marital status, region, city, diagnosis date, site, topography, morphology, behavior and stage of diagnosis. The data were then aggregated for use at four spatial levels: national, regional, governorate and cities. At present there are no plans to analyze the data at a finer scale.

To estimate the annual population between 1998 and 2004, the 2004 Saudi census was used, along with annual population growth estimates [27]. Estimates were compiled for the four levels of geographical resolution. Growth rates were obtained from the Central Department of Statistics and Information (CDSI) for the whole of Saudi Arabia and the thirteen regions; as gender is not distinguishable in this data, estimates were used. Growth rates were not available at the governorate or city region so they were assigned the growth rates of the regions in which they are contained. This could be subject to the ecological fallacy [28], as areas within a region are unlikely to all be the same. However, given the data available, this was the best possible solution.

2.3. Reporting of Cancer Incidence

The rates of cancer incidence in each administrative zone over a geographical region provide an important measure of the relative risks of individual cancers in each zone. Such rates provide direct estimates of the probability or risk of cancer or other illnesses and are particularly important for epidemiological studies. The absolute or relative rates of cancer incidence are especially important for comparative studies, e.g. comparing rates of cancer incidence at two or more different times, for measuring change over time or for comparing rates in different zones. Three of the most common indices for cancer research are the crude incidence rate (CIR), the age-specific incidence rate (AIR) and the age standardized incidence rate (ASR) [29-31], which are calculated in the CASA. The CIR is expressed as the total number of incidences of cancer for each 100,000 people in the population. However, rates of cancer incidence vary greatly with age, and the crude rate is strongly influenced by the demographic structure of a population. Hence, if the population structure changes over a period, the crude rate over that period may be artificially altered. For similar reasons, one cannot compare crude rates across geographical areas with different population age structures. Therefore, in order to assess trends in the incidence of a particular type of cancer or compare the incidence over geographical areas or between different cancer registries, it is necessary to standardize the rates with respect to age through computing the AIR. The AIR is the number of particular types of cancer incidences occurring during a specific period in a population of a specific age and gender, divided by the number of the mid-year population of that age and gender. The AIR for age class i is calculated as:

$$AIR_i = \frac{r_i}{n_i} * 100,000 \qquad (1)$$

where r_i is the number of incidences in the age class i and n_i is the corresponding person-years of observation. The Age-Standardized Incidence Rate (ASR) is a summary measure of the rate of cancer incidence which a population would have if it had a standard age structure. Standardization is necessary when comparing several populations which differ with respect to age structure; in effect, this allows the researcher to keep age constant to see how a rate varies across areas. The most frequently used standard population is the World Standard Population. The calculated incidence of a cancer is known as the World Standardized Incidence Rate. The populations in each age-class of the Standard Population are used as weights in the standardization process. Expressed per 100,000 people, the ASR can be computed as [29]:

$$ASR = \frac{\sum_{i=1}^{A} AIR_i w_i}{\sum_{i=1}^{A} w} \qquad (2)$$

where w_i is the Standard Population of the age class i where $i = 1$, n age classes. In a statistical sense, the ASR represents an estimate of some true parameter value (which could only be known if the units of observation were infinitely large). Therefore, it is usual to give some measure of uncertainty of the estimated rate, such as the standard error (SE) of the rate. The standard error can also be used to calculate confidence intervals (CI) for the rate, which are intuitively rather easier to interpret. Both of these measures are computed in the atlas.

2.4. Statistical and Spatial Analyses in the CASA

One of the most useful aspects of the CASA is the range of functionality beyond simple mapping and statistical

analysis available in similar atlases. With the CASA, users are able to collate data over time, choose from a range of available map styles (e.g. choropleth, graduated symbol, pie charts and density maps) and calculate a range of cancer incidence rates to create exportable figures and maps. This section provides an overview of the functionality contained in the two main interfaces in the atlas, i.e. the statistical and spatial analysis windows. A detailed help file and manual are available for users of the atlas.

2.4.1. The Statistical Analysis Interface

The CASA's statistical interface is divided into the following five sections, as shown in **Figure 2**: 1) analysis panel; 2) cancer site and time period panel; 3) figure panel; 4) table panel; and 5) legend panel. These five panels are linked to each other and work in a dynamic and interactive way, i.e. any action taking place in any one panel affects the results in the other panels.

The analysis panel is used to select the type of cancer analysis, which is then displayed in the figure and table panels on the right hand side. The user first selects the spatial level for analysis from a drop-down list where the options are National, Regional, Governorate or Cities. At the national level, a user can analyze and explore cancer distributions over the whole of Saudi Arabia, while at the regional level, the exploration and analysis of cancer distributions, variations, trends and patterns are conducted over the thirteen administrative regions. As for the governorate level, a user can view and analyze cancer distributions over the 118 governorates. At the city spatial level, the analysis is undertaken on cancer data over the 240 cities and towns.

Radio buttons then allow the user to select all cancers for analysis or just the most common cancers, followed by an option to choose standard or advanced analyses. This is followed by a list of types of different analyses that the user can select. When a user selects all cancers, the results will be displayed for approximately sixty types of cancer according to the International Classification of Diseases [32]. A query of the most common cancers will provide results for the top ten cancers, updated dynamically upon selection of the gender and period. Selection of the standard analysis will provide the user with thirty-two pre-defined types of analysis, which will appear as a list in the analysis panel (**Supplementary Table 1**). **Figure 3** illustrates the results from four different analysis types.

The advanced analysis option allows users to create customized queries by adjusting different parameters, as listed in **Supplementary Table 2**. The statistics are then computed based on this user-defined query, which provides much more flexibility in undertaking specific analyses. An example of a user-defined query might be to display the incidence of liver cancer in males aged between

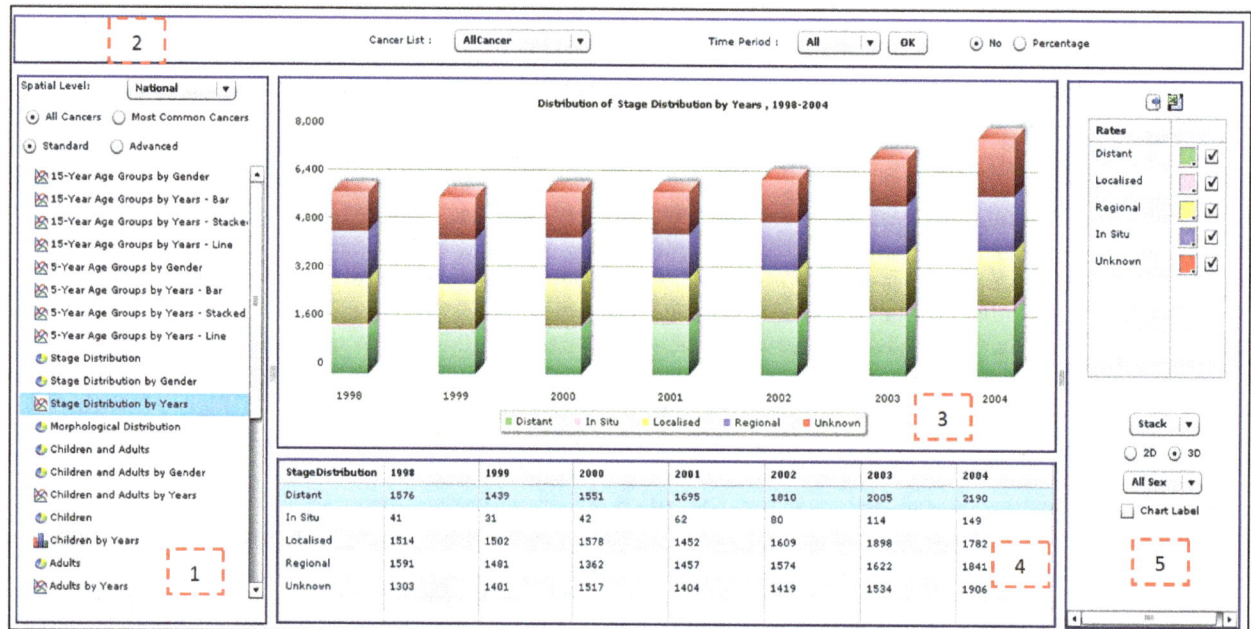

Figure 2. The statistical analysis interface of the CASA.

fifteen and twenty-nine between 1999 and 2003 with distant and regional stage distribution for all morphological distributions. By default, the CASA chooses all cancer types for analysis. The cancer site and period panel (labeled as item 2 in **Figure 2**) allows the user to select a specific cancer type. The user can also select all the years for analysis (1998-2004), an individual year or a subset of years. The absolute number of percentage option specifies how the data are displayed in tables and figures but does not apply to rates such as CIR, AIR and ASR. Finally, the export functions enable users to export a table as text or in Microsoft Excel format, and figures in image format such as jpeg, png or pdf.

The figure panel (item 3 in **Figure 2**) presents the result of an analysis in the form of a figure. This includes the figure itself, a dynamic figure title and a dynamic figure legend. The table panel (item 4 in **Figure 2**), which appears below the figure panel, presents the results of an analysis in tabular format. The legend panel (item 5 in **Figure 2**) provides the user with some interactive functions for controlling colors, figure types (bar chart, pie chart, stacked bar chart or lines), 2D or 3D and labels. A user can select the type of figure to be used to present the data.

2.4.2. The Spatial Analysis Interface
The spatial analysis tools are available in a similar interface to that of the statistical analysis (**Figure 4**). The main difference is that the results are displayed as maps, although the user can display figures and tables through the legend panel. The statistical analysis interface is divided into four sections: 1) analysis panel; 2) cancer site

and time period panel; 3) map panel; and 4) legend panel. The function of the analysis panel is similar to the one found in the statistical analysis interface, *i.e.* to allow users to select across the spatial level of analysis, cancer types, standard or advanced analysis and a list of predefined analyses. The results of the analysis are displayed as a map.

The cancer site and period panel (item 2 in **Figure 4**) is the same in both the statistical and spatial analysis windows. The map panel (item 3 in **Figure 4**) presents the result of the analysis selected by the user. Different maps can be generated, such as choropleth maps, density maps, graduated symbol maps, pie chart maps and bar chart maps. Examples of different map types are shown in **Figure 5**. The function of the legend panel is to provide the user with some interactive capabilities to adjust map parameters such as color scheme, classification method (equal interval, quantile, natural break or standard deviation), number of classes, size of symbols and transparency level. To help researchers become familiar with these different mapping types, a manual was written which describes them in more detail. Moreover, training sessions have been held with staff to explain the different mapping types available. These sessions have also allowed us to test the atlas and gather feedback.

3. Analyzing Colorectal Cancer (CRC) Using the Cancer Atlas

It has been reported that presently very few reports deliver a descriptive epidemiology of colorectal cancer (CRC) in Saudi Arabia [33-35]. Hence, this section provides the results of an analysis of CRC in Saudi Arabia, to illus-

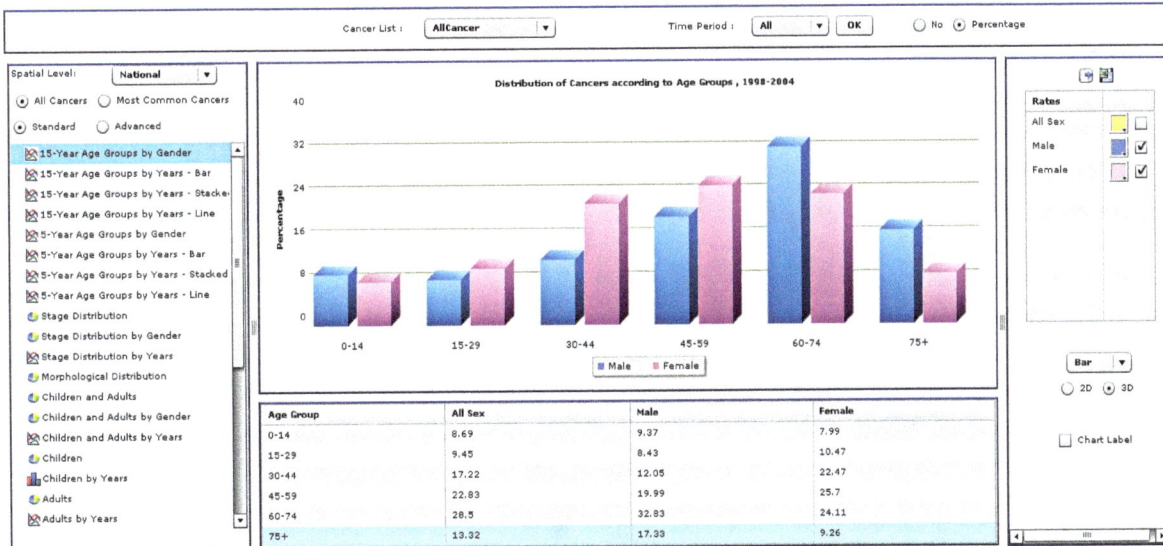

Distribution of Cancers according to Age Groups, 1998-2004

Age Group	All Sex	Male	Female
0-14	8.69	9.37	7.99
15-29	9.45	8.43	10.47
30-44	17.22	12.05	22.47
45-59	22.83	19.99	25.7
60-74	28.5	32.83	24.11
75+	13.32	17.33	9.26

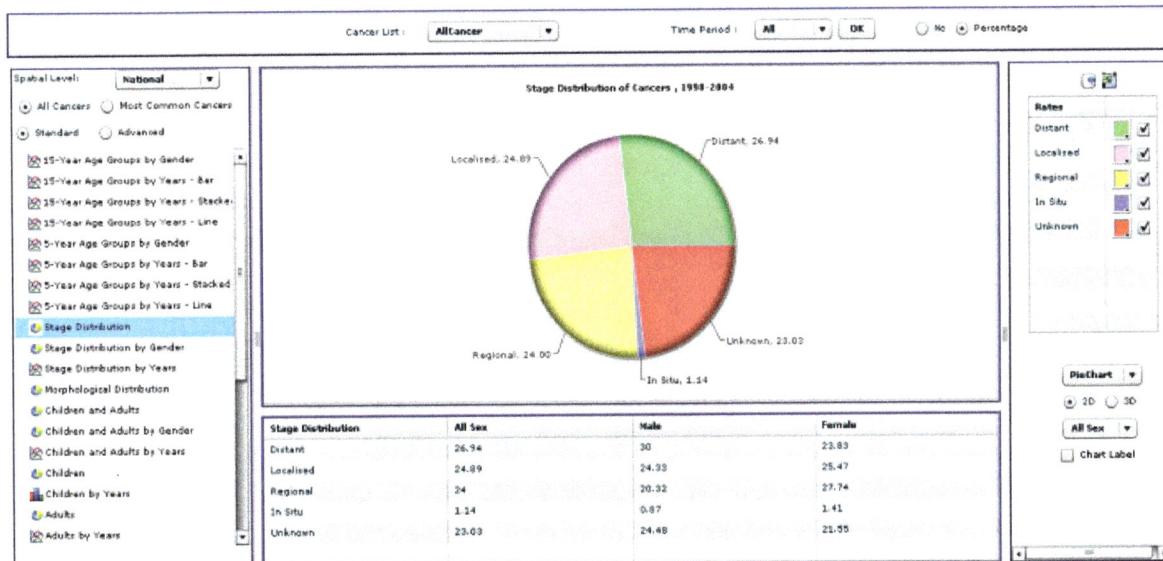

Stage Distribution of Cancers, 1998-2004

Stage Distribution	All Sex	Male	Female
Distant	26.94	30	23.83
Localised	24.89	24.33	25.47
Regional	24	20.32	27.74
In Situ	1.14	0.87	1.41
Unknown	23.03	24.48	21.55

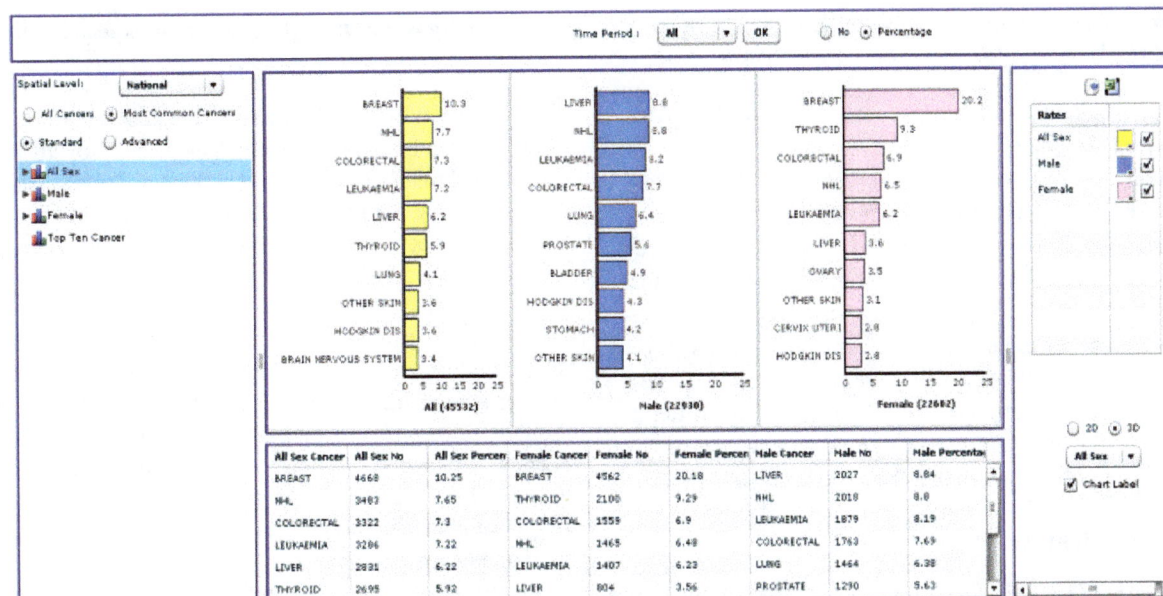

All Sex Cancer	All Sex No	All Sex Percen	Female Cancer	Female No	Female Percen	Male Cancer	Male No	Male Percenta
BREAST	4668	10.25	BREAST	4562	20.18	LIVER	2027	8.84
NHL	3483	7.65	THYROID	2100	9.29	NHL	2018	8.8
COLORECTAL	3322	7.3	COLORECTAL	1559	6.9	LEUKAEMIA	1879	8.19
LEUKAEMIA	3286	7.22	NHL	1465	6.48	COLORECTAL	1763	7.69
LIVER	2831	6.22	LEUKAEMIA	1407	6.23	LUNG	1464	6.38
THYROID	2695	5.92	LIVER	804	3.56	PROSTATE	1290	5.63

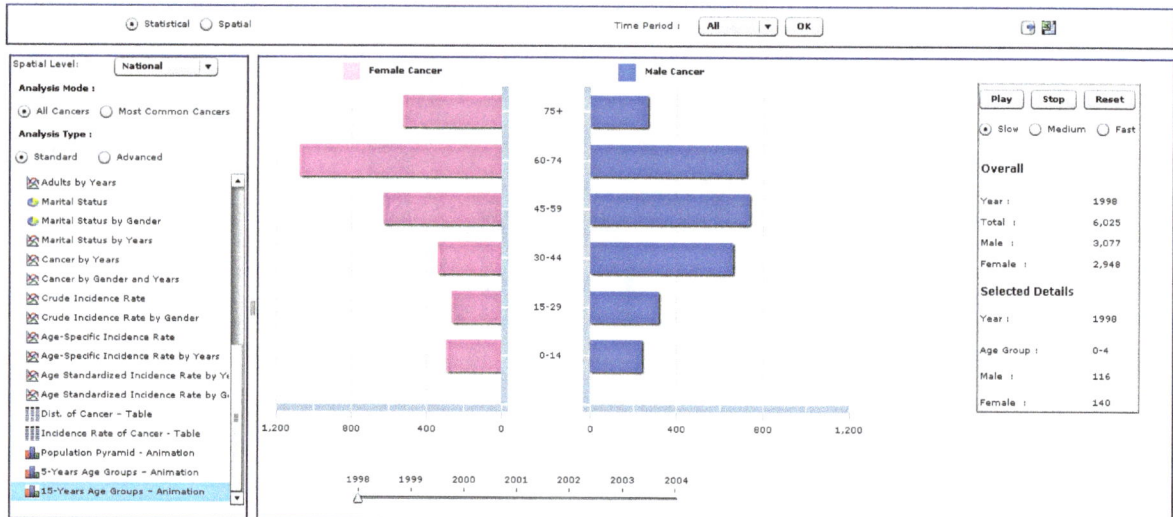

Figure 3. Four different analysis types from the standard analysis list.

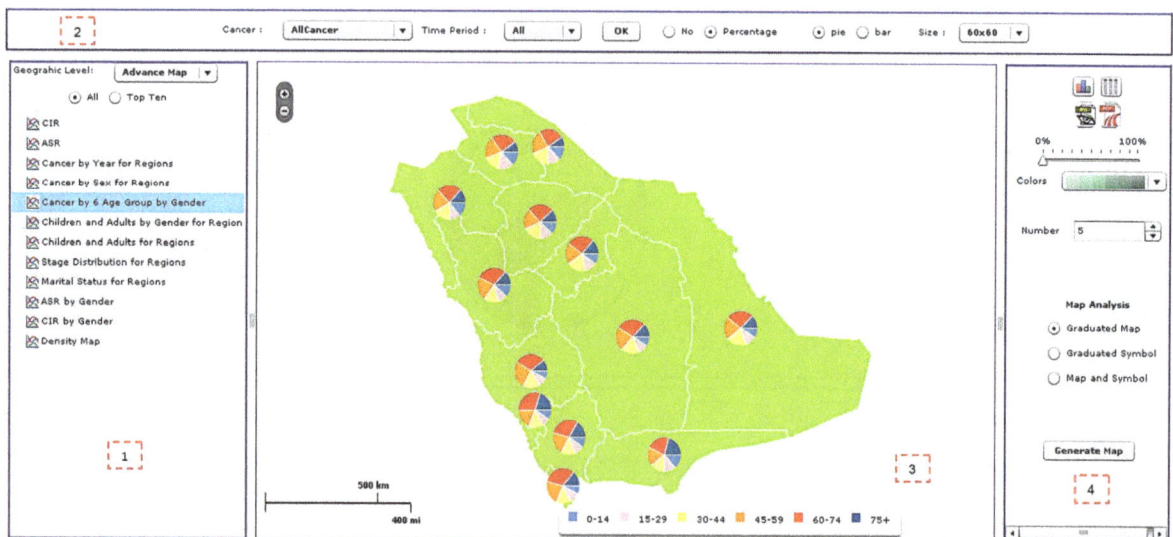

Figure 4. The spatial analysis interface of the CASA, including (1) the analysis panel; (2) the cancer site and time period panel; (3) the map panel; and (4) the legend panel.

trate the statistical and spatial functionality of the CASA at varying geographical scales.

Colon cancer is cancer of the large intestine (colon), the lower part of the digestive system. Rectal cancer is cancer of the last six inches of the colon. Together, they are often referred to as colorectal cancers (CRC). Most cases of colon cancer begin as small, noncancerous (benign) clumps of cells called adenomatous polyps. Over time, some of these polyps become colon cancers. Polyps may be small and produce few, if any, symptoms. Regular screening tests can help to prevent colon cancer by identifying polyps before they become cancerous [36].

3.1. Statistical Analysis

Between 1998 and 2004, a total of 3322 cases of CRC

were diagnosed in Saudi Arabia. This accounts for 7.3% of the total number of cancer cases, 7.69% of all male cancers and 6.90% of all female cancers. CRC was ranked as the fourth and third most common cancer in males and females respectively. The overall ASR was 4.28 per 100,000 people between 1998 and 2004, with a higher ASR for males of 4.44 compared to 4.03 for females. Higher ASR in males is consistent with the majority of other countries. The reasons why colorectal cancer is greater among males than females are not clear [35], although factors such as diet, body size, physical activity, hormones and family history of CRC could be accountable for the higher incidence amongst males than females [37]. In Saudi Arabia, progressively increasing exposure to risk factors and the lack of a nationwide screening program, along with an aging and growing population,

(a)

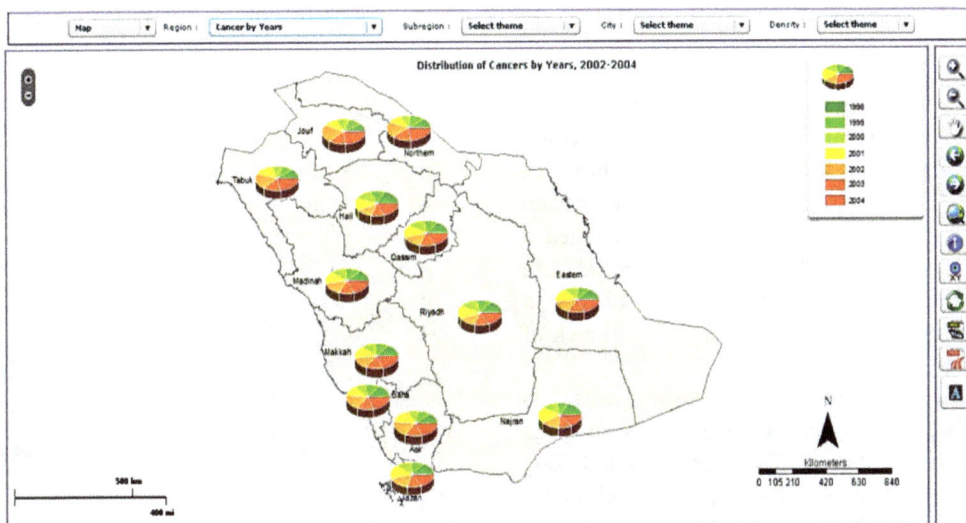

(b)

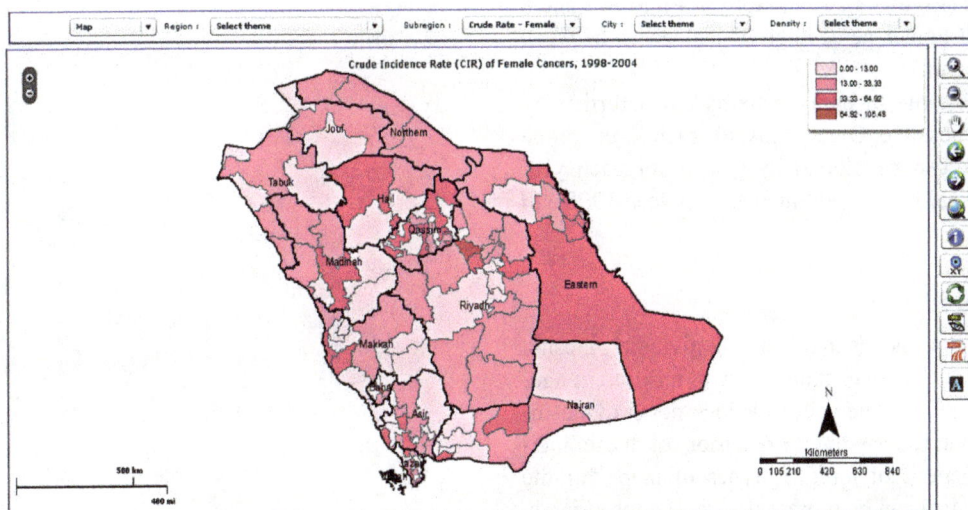

(c)

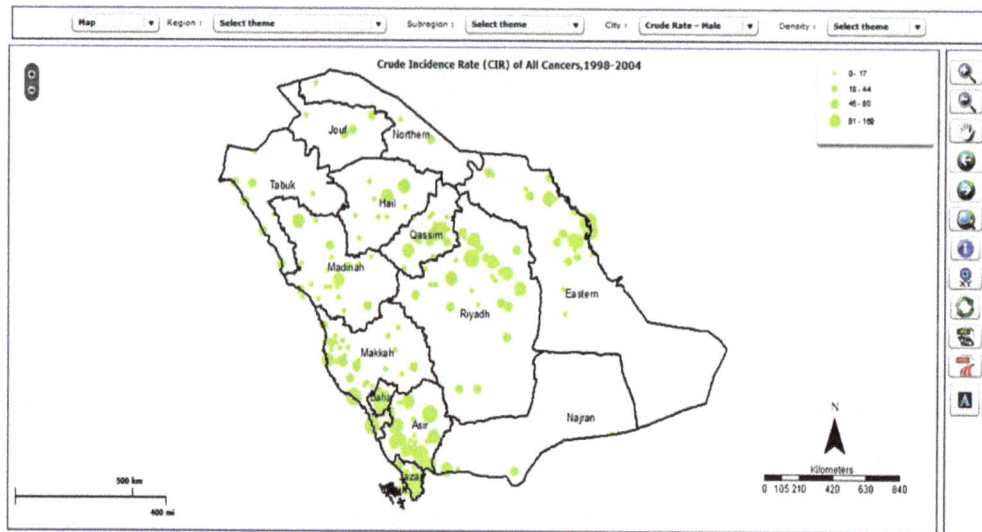

(d)

Figure 5. Examples of mapping types in the CASA: (a) choropleth map; (b) map with pie charts; (c) density map; and (d) graduated symbol map.

probably explain the rising CRC rates. These findings, in addition to the possible hidden familial risk for colon cancer, highlight the need for a mass screening program for CRC in Saudi Arabia, preferably for individuals aged forty years and above. Ibrahim *et al.* [34] predicted a significant increase of colorectal cancer incidence in both sexes by almost four-fold by 2030 in Saudi Arabia due to possible westernization of our dietary habits and lack of proper screening.

A number of useful graphs can be generated very easily from the atlas. For example, **Figure 6** shows that the number of new cases diagnosed between 1998 and 2004 by gender, which shows a steady increase in both males and females over this time period. CRC can also be viewed by fifteen-year age groups for all years (**Figure 7**) or single years (**Figure 8**). The graphs in **Figures 7** and **8** clearly show a higher percentage of incidences between the ages of 60 and 74.

It is also possible to break patterns down further by gender is also possible. Other statistics, such as the crude incidence rate, can be plotted by gender for each year (**Figure 9**), revealing a steady increase between 1998 and 2004.

3.2. Spatial Analysis

The statistical interface is invaluable at giving a snapshot of the rates of cancer over time and within different age-gender classifications; however, it does not provide information about the spatial distribution of these incidences. The benefit of multiple types of maps for the outputs is that data can be represented most appropriately for each situation. For instance, a choropleth map may be

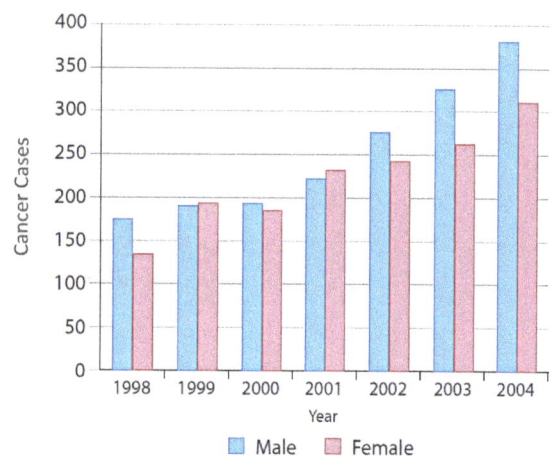

Figure 6. Distribution of CRC by gender and years.

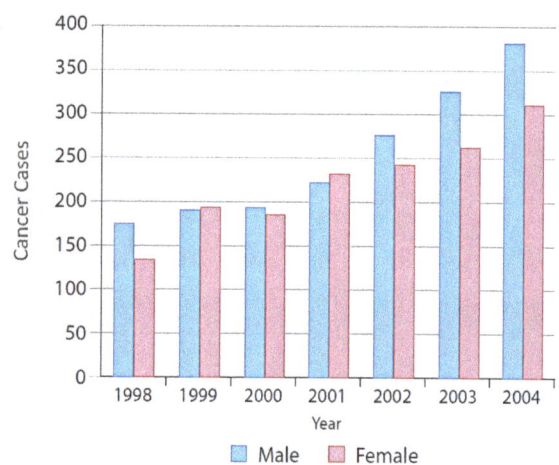

Figure 7. Overall distribution of CRC by gender and 15-year age groups for 1998.

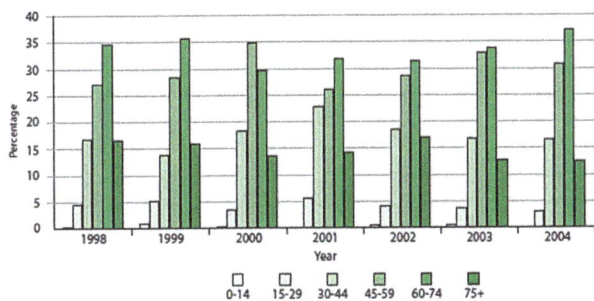

Figure 8. Distribution of CRC by 15-year age groups across all years.

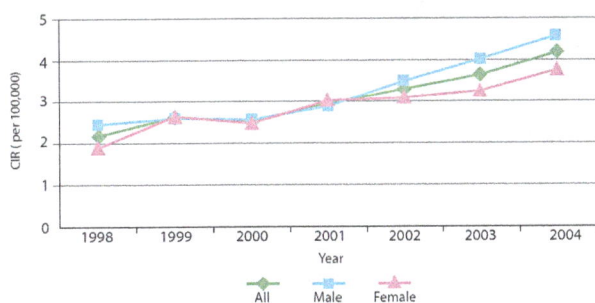

Figure 9. Crude Incidence Rate of CRC over time.

created and downloaded for inclusion in a report for local city officials who need to know general statistics about cancer incidence by local area. A more detailed output, such as a pie chart illustrating the distribution of cancer cases by age in each area, may be suitable for the health practitioners in a region so that they may better target the local population for screening. The graduated symbols are useful for finer spatial scales and provide a valuable alternative where color reproduction of the maps may not be feasible so choropleth maps could hinder clear interpretation of the outputs.

Starting at the regional level, the CASA allows one to examine spatial patterns across the thirteen regions of Saudi Arabia using a combination of tables and maps. For 1998 to 2004, the Riyadh region reported the highest number of colorectal cases (990 cases; 29.9%), followed by the Makkah region (822 cases; 24.8%) and the Eastern region (558 cases 16.9%). These three regions showed significant increased trend in the number of colorectal cases diagnosed between 1998 and 2004. The prominence of these regions was also found through mapping the CIR, as shown in **Figure 10**. Regions with the highest CIRs were Riyadh with 4.16, Makkah with 3.49 and Eastern Regions with 3.37 per 100,000 population for both genders. Qassim, Madinah, Hail, Baha and Asir reported medium CIR that ranged between 2.26 and 2.97 per 100,000 population. In contrast, the lowest CIRs were reported from Jouf, Jazan, Najran, Northern province and Tabuk, ranging between 1.51 and 1.73 per 100,000 population.

Overall, colorectal cancer is mainly a disease of the middle aged and elderly: 78% of cases (80.1% males and 75.6% females) occurred in people aged above forty-five years and only 4.4% occurred in people less than thirty years old (3.7% males and 5.1% females). However, distribution of colorectal cancers by fifteen-year age groups showed high percentages of colorectal cases reported among younger males in the Jouf, Jazan, Tabuk and Northern regions and younger females in the Qassim, Najran, and Hail regions (**Figure 11**).

This section has illustrated the various options for spatial display of data at a range of scales and levels of data specificity. This functionality is superior to many atlases which are dominated by choropleth maps, which may not be the best option for data display or for exploring data at the city level. Pie charts provide an option to include age and gender-specific data for larger geographic regions, as shown in **Figure 11**. Although these data may be exported to a table, the spatial display on a map is easier to interpret and enables the user to identify areas where more CRC screening services may need to be offered. To find specific data for any area (rather than ranges or relative densities), users can refer back to the tables. Our illustrations have used CIR rather than age-adjusted rates, but all maps may be created with AIR or ASR data.

4. Conclusions

This paper has described the development and implementation of an interactive web-based Statistical Spatial Cancer Atlas for Saudi Arabia. Users are able to select sets of cancer statistics and then select criteria such as type of cancer, stage, site, gender, age, ethnicity, etc., in order to create custom maps, charts, corresponding tables of statistics and commentaries of cancer incidence at regional, governorate or city level in Saudi Arabia. The functionality, ease of use and high quality of the output maps, charts, tables and commentaries were designed to meet the requirements of a target group of users working in cancer medicine, cancer care, public health, hospital management, health economics, cancer screening, prevention and awareness and training.

The design of the atlas was based on the use of ESRI ArcGIS Server API software for Flex running on a server in order to access functions for mapping, chart creation, and the generation of tables and commentaries on client computers, which access the server over the Internet using a web browser. The design was implemented and tested using standard sets of statistical and vector map data. The final implementation gave considerable flexibility and high levels of responsiveness and performance during tests where client computers were connected over a local network to a server in the office.

In the future, the atlas will be used to determine whether observed geographic variation in the cancer incidence rates for the most common cancers, such as breast, liver,

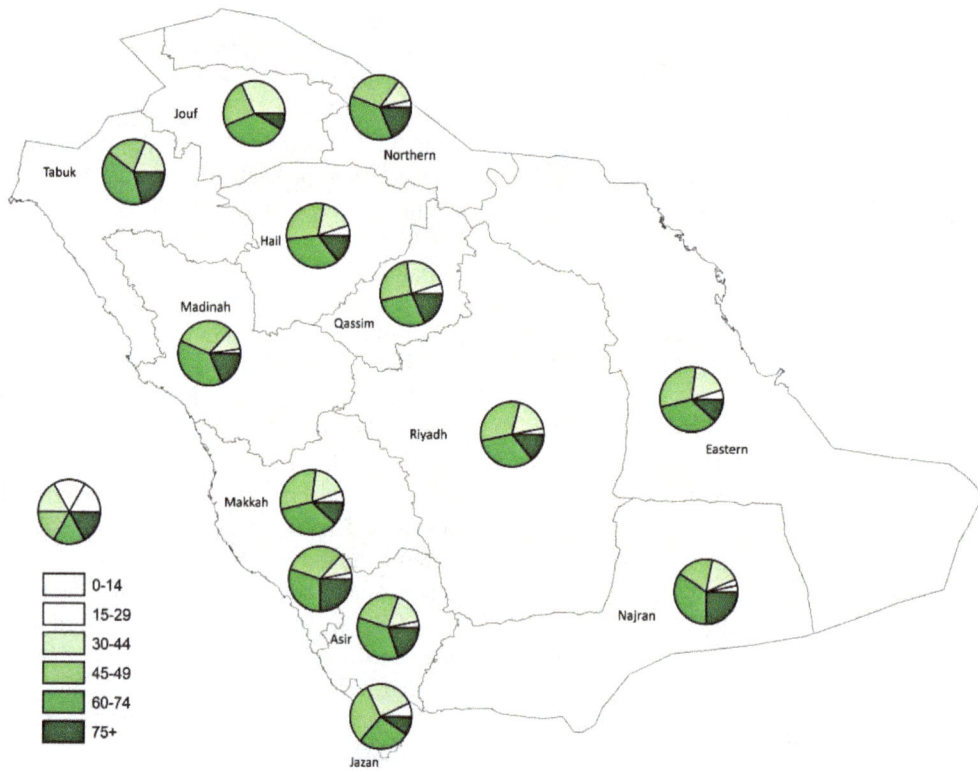

Figure 10. Crude Incidence Rate of Colorectal Cancer, 1998-2004.

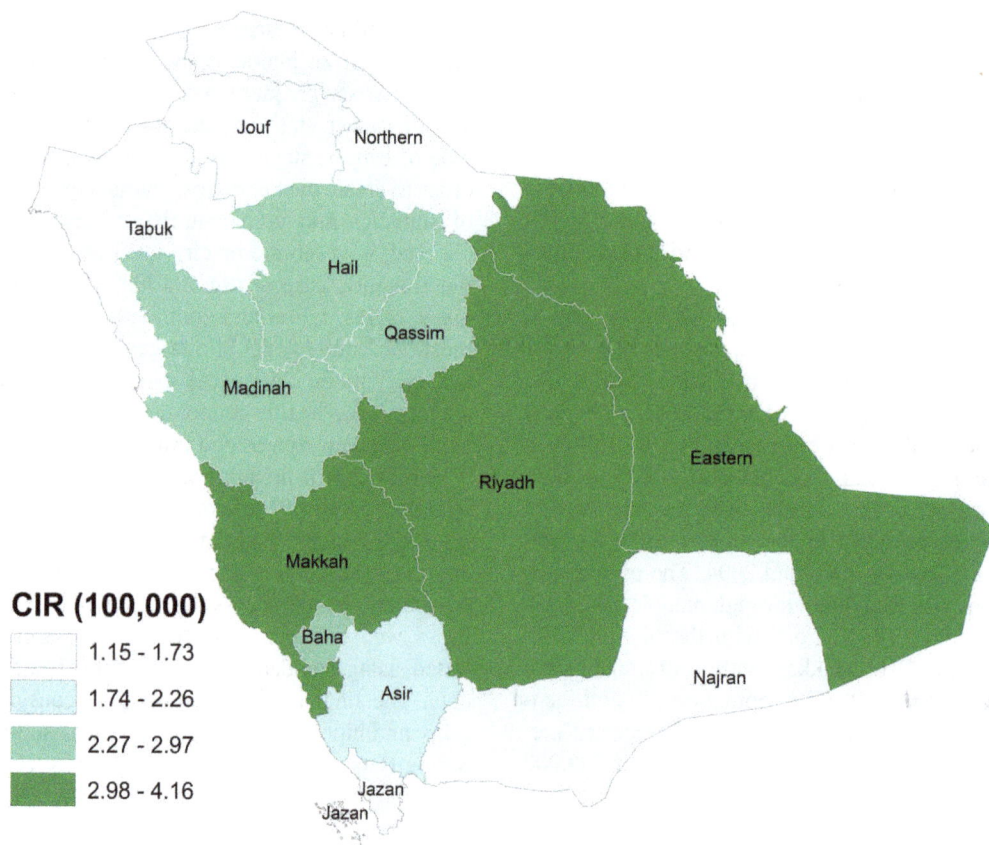

Figure 11. Distribution of CRC by 15-year age groups for both genders.

thyroid, and colorectal cancers, are random or statistically significant. If there are areas of excess, then research questions can be asked regarding whether that excess is stable or temporary over the seven-year study period (1998-2004), whether excesses are consistent across all diagnostic stages, or whether they might be due to excesses in early or late stage diagnoses, and whether they can be attributed to covariates such as age, sex, and urban/rural status. Moreover, there are plans to add other datasets (e.g. socio-economic data, exposure to solar radiation, nitrogen dioxide, fine particle air pollution, etc.) as well as more advanced functionality in the form of various spatial statistical methods, e.g. to help in finding spatial correlations and relationships between cancer and different risk factors using techniques such as geographically weighted regression. Spatial aggregation errors will then need to be carefully considered, as outlined clearly by Luo et al. [38], particularly when using large zones like regions or where health care services are sparsely located or spatially clustered if access to services is considered in the future.

5. Acknowledgements

This research was funded by King Abdulaziz City for Science and Technology (KACST), Saudi Arabia. The authors acknowledge the Saudi Cancer Registry and its board members.

REFERENCES

[1] World Health Organization (WHO), Cancer Fact Sheet (N 297), updated Feb 2011. 2011. http://www.who.int/mediacentre/factsheets/fs297/en/

[2] J. G. Brody, R. Rudel, N. I. Maxwell and S. R. Swedis, "Mapping out a Search for Environmental Causes of Breast Cancer," *Public Health Reports*, Vol. 111, No. 6, 1996, pp. 494-507. http://www.ncbi.nlm.nih.gov/pmc/articles/PMC1381895/

[3] J. G. Brody, A. Aschengrau, W. McKelvey, R. A. Rudel, C. H. Swartz and T. Kennedy, "Breast Cancer Risk and Historical Exposure to Pesticides from Wide-Area Applications Assessed with GIS," *Environmental Health Perspectives*, Vol. 112, No. 8, 2004, pp. 889-897.

[4] J. G. Brody, A. Aschengrau, W. McKelvey, C. H. Swartz, T. Kennedy and R. A. Rudel, "Breast Cancer Risk and Drinking Water Contaminated by Wastewater: A Case Control Study," *Environmental Health*: *A Global Access Science Source* 2006, Vol. 5, No. 28, 2006, p. 28.

[5] A. Biggeri, F. Barbone, C. Lagazio, M. Bovenzi and G. Stanta, "Air Pollution and Lung Cancer in Trieste, Italy: Spatial Analysis of Risk as a Function of Distance from Sources," *Environmental Health Perspectives*, Vol. 104, No. 7, 1996, pp. 750-751.

[6] M. B. Toledano, L. Jarup, N. Best, J. Wakefield and P.

Elliott, "Spatial Variation and Temporal Trends of Testicular Cancer in Great Britain," *British Journal of Cancer*, Vol. 84, No. 11, 2001, pp. 1482-1487.

[7] G. Rushton, I. Peleg, A. Banerjee, G. Smith and M. West, "Analyzing Geographic Patterns of Disease Incidence: Rates of Late-Stage Colorectal Cancer in Iowa," *Journal of Medical Systems*, Vol. 28, No. 3, 2004, pp. 223-236.

[8] J. R. Meliker, M. J. Slotnick, G. A. AvRuskin, A. Kaufmann, G. M. Jacquez and J. O. Nriagu, "Improving Exposure Assessment in Environmental Epidemiology: Application of Spatio-Temporal Visualization Tools," *Journal of Geographical Systems*, Vol. 7, No. 1, 2005, pp. 49-66.

[9] L. M. DeChello and T. J. Sheehan, "Spatial Analysis of Colorectal Cancer Incidence and Proportion of Late-Stage in Massachusetts Residents: 1995-1998," *International Journal of Health Geographics*, Vol. 6, 2007, p. 20.

[10] P. Goovaerts, "Combining Area-Based and Individual-Level Data in the Geostatistical Mapping of Late-Stage Cancer Incidence," *Spatial and Spatio-temporal Epidemiology*, Vol. 1, No. 1, 2009, pp. 61-71.

[11] F. Wang, "Spatial Clusters of Cancers in Illinois 1986-2000," *Journal of Medical Systems*, Vol. 28, No. 3, 2004, pp. 237-256.

[12] P. A. Rogerson, "The Detection of Clusters Using a Spatial Version of the Chi-Square Goodness-of-Fit Statistic," *Geographical Analysis*, Vol. 31, No. 2, 1999, pp. 130-147.

[13] V. Vieira, T. Webster, J. Weinberg, A. Aschengrau and D. Ozonoff, "Spatial Analysis of Lung, Colorectal, and Breast Cancer on Cape Cod: An Application of Generalized Additive Models to Case-Control Data," *Environmental Health*: *A Global Access Science Source*, Vol. 4, 2005, p. 11.

[14] J. C. McEntee and Y. Ogneva-Himmelberger, "Diesel Particulate Matter, Lung Cancer, and Asthma Incidences along Major Traffic Corridors in MA, USA: A GIS Analysis," *Health & Place*, Vol. 14, No. 4, 2008, pp. 815-826.

[15] N. Tian, J. Gaines Wilson and F. Benjamin Zhan, "Female Breast Cancer Mortality Clusters within Racial Groups in the United States," *Health Place*, Vol. 16, No. 2, 2010, pp. 209-218.

[16] S. M. Cramb, K. L. Mengersen and P. D. Baade, "Developing the Atlas of Cancer in Queensland: Methodological Issues," *International Journal of Health Geographics*, Vol. 10, 2011, p. 9.

[17] M. Haklay, A. Singleton and C. Parker, "Web Mapping 2.0: The Neogeography of the GeoWeb," *Geography Compass*, Vol. 2, No. 6, 2008, pp. 2011-2039.

[18] National Cancer Institute, Atlas of Cancer Mortality in the US, 1950-1994. http://ratecalc.cancer.gov/

[19] M. MacEachren, S. Crawford, M., Akella and G. Lengerich, "Design and Implementation of a Model, Web-Based, GIS-Enabled Cancer Atlas," *The Cartographic Journal*, Vol. 45, No. 4, 2008, pp. 246-260.

[20] E. Lengerich, A. MacEachren, R. Parrott, G. Chase, K. Brenda and S. Crawford, Pennsylvania Cancer Atlas, 2010.
http://www.geovista.psu.edu/resources/flyers/GV_PA_At las.pdf

[21] CDC, Interactive Cancer Atlas (InCA), 2010.
http://www.cdc.gov/Features/CancerAtlas/

[22] NCIN, Cancer e-Atlas, 2010.
http://www.ncin.org.uk/cancer_information_tools/eatlas.a spx

[23] A. Nandakumar, P. C. Gupta, P. Gangadharan, R. N. Visweswara and D. M. Parkin, "Geographic Pathology Revisited: Development of an Atlas of Cancer in India," *International Journal of Cancer*, Vol. 116, No. 5, 2005, pp. 740-754.

[24] NHS, "London Healthcare Benchmarking Tool," 2010.
http://lhbt.london.nhs.uk/

[25] NHS, "London Health Programmes: Health Needs Assessment Toolkit," 2011. http://hna.londonhp.nhs.uk/

[26] ESRI, "ArcGIS API for Flex," 2010.
http://help.arcgis.com/en/webapi/flex/index.html

[27] SCR, "Cancer Incidence Report Saudi Arabia 2004," Saudi Cancer Registry, Saudi Arabia, 2004.

[28] S. Openshaw, "Ecological Fallacies and the Analysis of areal Census Data," *Environment and Planning A*, Vol. 16, No. 1, 1984, pp. 17-31.

[29] P. Boyle and D. M. Parkin, "Statistical Methods for Registries," In: O. M. Jensen, D. M. Parkin, R. MacLennan, C. S. Muir and R. G. Skeet, Eds., *Cancer Registration: Principles and Methods*, IARC Scientific Publication No. 95, International Agency for Research on Cancer, Lyon, 1991, pp. 126-158.

[30] A. Nandakumar, P. C. Gupta, P. Gangadharan, R. N. Visweswara and D. M. Parkin, "Geographic Pathology Revisited: Development of an Atlas of Cancer in India," *International Journal of Cancer*, Vol. 116, No. 5, 2005, pp. 740-754.

[31] M. Quinn, H. Wood, N. Cooper and S. Rowan, "Cancer atlas of the United Kingdom and Ireland 1991-2000," National Statistics Publication, Ashford Colour Press Ltd. Gosport, Great Britain, 2005, pp. 211-217.

[32] WHO, World Health Organization, "International Statistical Classification of Diseases and Related Health Problems," 10th Revision, 2007.

[33] I. Mansoor, I. H. Zahrani, S. Abdul Aziz, "Colorectal cancers in Saudi Arabia," *Saudi Medical Journal*, Vol. 23, No. 3, 2002, pp. 322-327.

[34] E. M. Ibrahim, A. A. Zeeneldin, T. R. El-Khodary, A. M. Al-Gahmi and B. M. Bin Sadiq, "Past, Present and Future of Colorectal Cancer in the Kingdom of Saudi Arabia," *Saudi Journal of Gastroenterology*, Vol. 14, No. 4, 2008, pp. 178-182.

[35] M. H. Mosli, M. S. Al-Ahwal, "Colorectal Cancer in the Kingdom of Saudi Arabia: Need for Screening," *Asian Pacific Journal of Cancer Prevention*, Vol. 13, No. 8, 2012, pp. 3809-3813.

[36] ACS, American Cancer Society. http://www.cancer.org

[37] T. T. Fancher, J. A. Palesty, L. Rashidi and S. J. Dudrick, "Is Gender Related to the Stage of Colorectal Cancer at Initial Presentation in Young Patients?" *Journal of Surgical Research*, Vol. 165, No. 1, 2011, pp. 15-18.

[38] L. Luo, S. McLafferty and F. H. Wang, "Analyzing Spatial Aggregation error in Statistical Models of Late-Stage Cancer Risk: A Monte Carlo Simulation Approach," *International Journal of Health Geographics*, Vol. 9, 2010, p. 51.

Supplementary Material

Supplementary Table 1: Name of the standard analysis type in the list and the description.

No.	Name in the Analysis List	Description
1	Cancer by Years	Distribution of cancer cases by years
2	Cancer by Gender and Years	Distribution of cancer cases by years and gender
3	15-Year Age-Groups by Gender	Distribution of cancer cases according to fifteen-year age-groups by gender
4	15-Year Age-Groups by Years—Column	Distribution of cancer cases according to fifteen-year age-groups by years
5	15-Year Age-Groups by Years—Stack	Distribution of cancer cases according to fifteen-year age-groups by years
6	15-Year Age-Groups by Years—Line	Distribution of cancer cases according to fifteen-year age-groups by years
7	5-Year Age-Groups by Gender	Distribution of cancer cases according to five-year age-groups by gender
8	5-Year Age-Groups by Years—Column	Distribution of cancer cases according to five-year age-groups by years
9	5-Year Age-groups by Years—Stack	Distribution of cancer cases according to five-year age-groups by years
10	5-Year Age-groups by Years—Line	Distribution of cancer cases according to five-year age-groups by years
11	Stage Distribution	Stage distribution of cancer cases
12	Stage Distribution by Gender	Stage distribution of cancer cases by gender
13	Stage Distribution by Years	Stage distribution of cancer cases by years
14	Morphological Distribution	Morphological distribution of cancer cases
15	Children and Adults	Distribution of cancer cases among children and adults
16	Children and Adults by Gender	Distribution of cancer cases among children and adults by gender
17	Children and Adults by Years	Distribution of cancer cases among children and adults by years
18	Children by Gender	Distribution of cancer cases among children by gender
19	Children by Years	Distribution of cancer cases among children by years
20	Adults by Gender	Distribution of cancer cases among adults by gender
21	Adults by Years	Distribution of cancer cases among adults by years
22	Crude Incidence Rate by Years	CIR per 100,000 by years
23	Crude Incidence Rate by Gender	CIR per 100,000 by gender
24	Age-Specific Incidence Rate	AIR per 100,000
25	Age-Specific Incidence Rate by Years	AIR per 100,000 by years
26	Age-Standardized Incidence Rate by Years	ASR per 100,000 by years
27	Age-Standardized Incidence Rate by Gender	ASR per 100,000 by gender
28	Distribution of Cancer Cases (Table)	Distribution of cancer cases according to five-year age-groups and ICD-10 Sites
29	Incidence Rate of Cancer Cases (Table)	Incidence rate (per 100,000) of cancer cases according to 5-year age-groups and ICD-10 sites
30	Population Pyramid—Time Animation Player	Time animation pyramid figure depicts population distribution of five-year age-groups by gender over a seven-year period. The animated figure shows how the population changed over the years from 1998 to 2004
31	5-year Age-groups—Time Animation Player	Animation of cancer cases distribution of five-year age-groups by gender over a seven-year period. The animated figure shows how the distribution of cancer cases changed over the years from 1998 to 2004
32	15-year Age-groups—Time Animation Player	Animation of cancer cases distribution of fifteen-year age-groups by gender over a seven-year period. The animated figure shows how the distribution of cancer cases changed over the years from 1998 to 2004

Supplementary Table 2: Parameters that can be adjusted using the advanced analysis option in the CASA. Note that since the atlas is geared to researchers in the public health sector, many are familiar with the terminology used here. However, a more detailed manual, which is available for the atlas, explains this terminology in more detail.

Parameter Name	Description
Cancer Site	All cancer sites coded according to ICD-10 (WHO, 2007)
Period	Year of cancer diagnosis. Options includes single year basis, multiple years and entire period. Period list: All (1998-2004), 2004, 2003, 2002, 2001, 2000, 1999 and 1998
Gender	Includes three options: all sex, male and female.
Age	Age at diagnosis. Includes five options: all ages, fifteen-year age-groups, five-year age-groups, user-defined age-group and user-defined age
Stage	Stage distribution of cancer. Options include: localized, regional, distant, in-situ and unknown.
Morphology	Morphological distribution. This option will only work if one cancer site is selected.

9

Accuracy of Stream Habitat Interpolations Across Spatial Scales

Kenneth R. Sheehan[1], Stuart A. Welsh[2]

[1]Water Systems Analysis Group, University of New Hampshire, Durham, USA
[2]US Geological Survey, West Virginia Cooperative Fish and Wildlife Research Unit, Morgantown, USA

ABSTRACT

Stream habitat data are often collected across spatial scales because relationships among habitat, species occurrence, and management plans are linked at multiple spatial scales. Unfortunately, scale is often a factor limiting insight gained from spatial analysis of stream habitat data. Considerable cost is often expended to collect data at several spatial scales to provide accurate evaluation of spatial relationships in streams. To address utility of single scale set of stream habitat data used at varying scales, we examined the influence that data scaling had on accuracy of natural neighbor predictions of depth, flow, and benthic substrate. To achieve this goal, we measured two streams at gridded resolution of 0.33×0.33 meter cell size over a combined area of 934 m^2 to create a baseline for natural neighbor interpolated maps at 12 incremental scales ranging from a raster cell size of 0.11 m^2 to 16 m^2. Analysis of predictive maps showed a logarithmic linear decay pattern in RMSE values in interpolation accuracy for variables as resolution of data used to interpolate study areas became coarser. Proportional accuracy of interpolated models (r^2) decreased, but it was maintained up to 78% as interpolation scale moved from 0.11 m^2 to 16 m^2. Results indicated that accuracy retention was suitable for assessment and management purposes at various scales different from the data collection scale. Our study is relevant to spatial modeling, fish habitat assessment, and stream habitat management because it highlights the potential of using a single dataset to fulfill analysis needs rather than investing considerable cost to develop several scaled datasets.

Keywords: Natural Neighbor Interpolation; Residuals; Ordinary Least Squares; Stream Modeling; Habitat; Benthic Substrate

1. Introduction

Stream habitat data at varying spatial scales provide integral information for lotic management and broad ecologic study. Typically, stream data are collected at multiple spatial scales to provide more complete representation of habitat and allow additional analysis power and ecologic insight [1-3]. The spatial scale at which stream habitat data are collected is important due to connectivity among habitat patches, species occurrence, and life history [3-8]. Because of ecological links between scales, spatial analysis in varying forms has become a staple tool for examining multi-scale stream habitat data [4,9,10]. Collection of stream variables at multiple scales is also necessary for complex analysis of macroinvertebrates, fish habitat relationships, ecological processes, and stream habitat [3,4,7,10,11].

Inability to make inferences at scales other than those collected is linked directly to the unknown amount of accuracy lost when scaling between fine and coarse stream habitat scales [12,13]. Due to the inability of data sets to be scaled for comparative purposes, several data sets are often required for stream habitat spatial analysis at great expense [14]. Data analysis may only be as accurate as the finest scale of data collected [12,15], leading scientists to collect data at the finest scale possible for each study. Unfortunately, an inverse relationship exists between the spatial scale of data and cost to acquire it; the finer the data scale is required, the smaller the area is able to be examined for a given amount of funding.

Utilizing data at various scales has long presented further problems such as pattern analysis and combination of data at varying scales [12,16,17]. Interpolation methods represent a family of spatial statistics able to create map products at multiple scales to aid in ecological pattern analysis and presentation of spatial data [9,18,19]. There are various interpolative methods including inverse distance weighted, several forms of kriging, natural neighbor, point interpolation, trend, and spline to create predictions of stream habitat data [21,22]. Many of these methods have been directly compared on environmental data [23-25]. Comparisons have shown that each method of interpolation has its own strengths in dealing with data of different types and number [19,22,26-30]. Natural neighbor interpolation has shown promise in producing practical maps of streams from small amounts of spatial data [19]. Demonstration of the ability of natural neighbor interpolation to accurately model various scales from a single stream habitat dataset may provide avenues to make multiple scale data collection redundant and opportunity for substantial cost and time savings.

Interpolation creates continuous surfaces from spatial data [20], offering opportunity to alleviate the problem of data gaps in spatial environmental data (e.g. trying to use data across scales). Interpolation provides predictive values of variables in regions which have no data by using information from adjacent regions. This ability provides better potential to use datasets at different scales than those they were initially collected.

Specifically, when selecting stream habitat data variables of depth, flow velocity, and benthic substrate at known locations, natural neighbor interpolation has shown to be accurate [19]. Natural neighbor works well with large datasets and has a nearly identical algorithm as inverse distance weighted interpolation. Natural neighbor interpolation is based on Theissen polygon networks, and weights adjacent data within a specified search radius. It takes a set of spatially located points and creates a grid (raster map) of the area based on the input points at the centroid of each cell. Natural neighbor interpolation works well with stream habitat data such as substrate because depositional patterns in rivers are typically well ordered, and not random [31-33]. Depth, flow velocity, and substrate have a high degree of spatial auto-correlation which further helps prediction accuracy [22,34].

While stream habitat variables of depth, flow velocity, and substrate have been recreated accurately by using natural neighbor interpolation when applied to small amounts of data [19]. There has been no evaluation of the role of spatial scale on stream habitat model accuracy with this type of interpolation. Scaling of stream habitat data typically involves using data across scales in an attempt to understand links between habitat patches, species occurrence, or other environmental variable [11,35-38]. Such studies often highlight problems caused by aggregating data across ecosystem scales [11,13,16,35,36, 39,40].

This study evaluates accuracy loss of predictive stream models when moving to coarser scales using natural neighbor interpolation. We hypothesize that accuracy retention will be high enough to create practical maps for analysis purposes at scales well removed from the initially collected data scale. A further objective of the study is to examine accuracy of natural neighbor interpolation predictive models at stream sites using data on water depth, water velocity, and benthic substrate at multiple spatial scales. This study will help establish potential for the use of a single dataset across scales in stream habitat modeling. To our knowledge, there has been no such study on scalability of stream habitat data when using natural neighbor interpolation. Our study is relevant to spatial modeling, fish habitat assessment, and stream habitat management because it examines the potential of a single dataset to fulfill analysis needs which would otherwise require multiple datasets at varying spatial scales at an increased cost of time and money. Further, this study emphasizes the rate of accuracy loss between data scales while creating visual maps of stream habitat, which could potentially aid and streamline both data and stream habitat management.

2. Methods

Study Sites and Data Collection

Benthic substrate data were collected from two wadeable streams which were located in the Greater Yellowstone Ecosystem, Gallatin National Forest, Montana, USA. The first of our two sites was located on Little Wapiti creek (111°16'53.546"W, 45°2'20.639"N). The Little Wapiti creek site measured nearly 34 meters long by 12 meters wide. The second study site was located on Grayling creek (111°6'16.407"W, 44°48'16.878"N). The Grayling creek site measured nearly 29 meters long by 19 meters wide.

Study sites were delineated by grid cells which measured 0.33 by 0.33 meters, or an area of 0.11 m^2 resolution per cell, using a fifty meter tape measure, laser rangefinder, and flagging (later removed). For purposes of this study, 0.33 × 0.33 meter grid cells will be referred to by its area, 0.11 m^2, or one third (1/3) of a meter squared. This is also referred to as base scale, the finest scale in the study. One third of a meter squared cells were chosen as the base resolution because stream habitat patches could be adequately captured at this scale on a wide variety of stream sizes, including those found in this study. A single piece of rebar was inserted into the bank material on each stream bank and high tensile line was secured to the rebar to guide the tape measure. As each row of data collection was finished the rebar was repositioned upstream to provide support for the next. Starting at the downstream left of each site, values for benthic substrate size, depth, and flow velocity were recorded for each x,y coordinate. Substrate was recorded along a continuous scale in millimeters from 0.05 to >300 mm based on the intermediate axis diameter [41]. Thus, actual values of substrate size were recorded for each 0.11 m^2 cell for each study site. Stream depth (cm, top-setting wading rod) and mean water velocity (m/s at 60% depth, Marsh-McBirney Flowmate 2000) were measured at the center of each cell. This was repeated until the site was captured in a complete grid of x,y coordinate points (**Figure 1**, Grayling creek example, upper left inset). All values were recorded in Microsoft Excel. Corner points for each study site were recorded and exported to ArcMap 10. In ArcMap 10, corner points for each study site were geo-referenced and exported to Microsoft Excel. Next, x,y coordinates were calculated for the remainder of cells in the site grid and appended to the initial Excel dataset of water depth, flow, and benthic substrate size. Little Wapiti creek had 3630 x,y coordinate points at the base scale,

and Grayling creek had 4950 x,y coordinate points at the base scale. The final base scale datasets were imported back to ArcMap 10.

In ArcMap 10, data subsets at 11 additional scales were created from base scale for each study site. Scale increments began at the base scale (0.11 m^2) and were increased in size by adding 1/3 of a meter to the length and width of each scale (**Figure 2**). Thus, base scale of 1/3 of a meter had its cells increased in size to length and width of 2/3 meter by 2/3 meter (0.44 m^2) for scale two (**Table 1**). Scale two in turn had 1/3 of a meter added to its cell width and length to create a one meter by one meter cell scale (1 m^2) for scale three (**Table 1**). This process was continued until 12 scale increments were created, including the final scale of 4.0 m × 4.0 m per cell, or 16 m^2 (**Table 1**). The number of cells used when plotted with scale as the x axis follows a power function with the equation $y = 4755.797x^{-1.923}$ (Grayling creek), and $y = 3438.1x^{-1.822}$ (Little Wapiti creek) (**Table 1**).

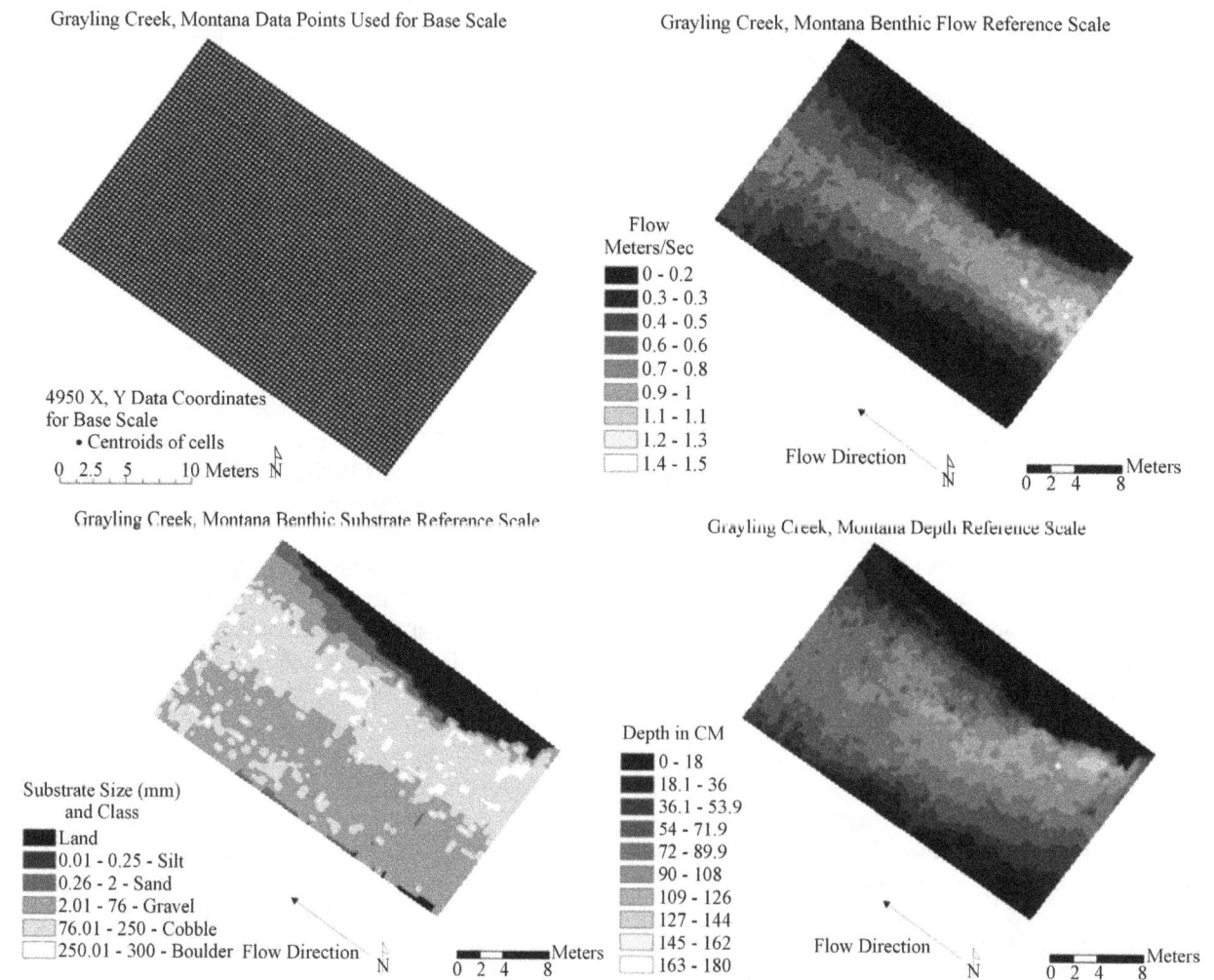

Grayling Creek, Montana Data Points Used for Base Scale

Grayling Creek, Montana Benthic Flow Reference Scale

Grayling Creek, Montana Benthic Substrate Reference Scale

Grayling Creek, Montana Depth Reference Scale

Figure 1. Example of x,y coordinate point grid for Grayling Creek showing 4950 data locations each containing depth, flow, and dominant observed substrate information. Natural neighbor interpolation of the points created the baseline visual map representing reality.

Grids Used to Create Natural Neighbor Interpolations at Different Scales

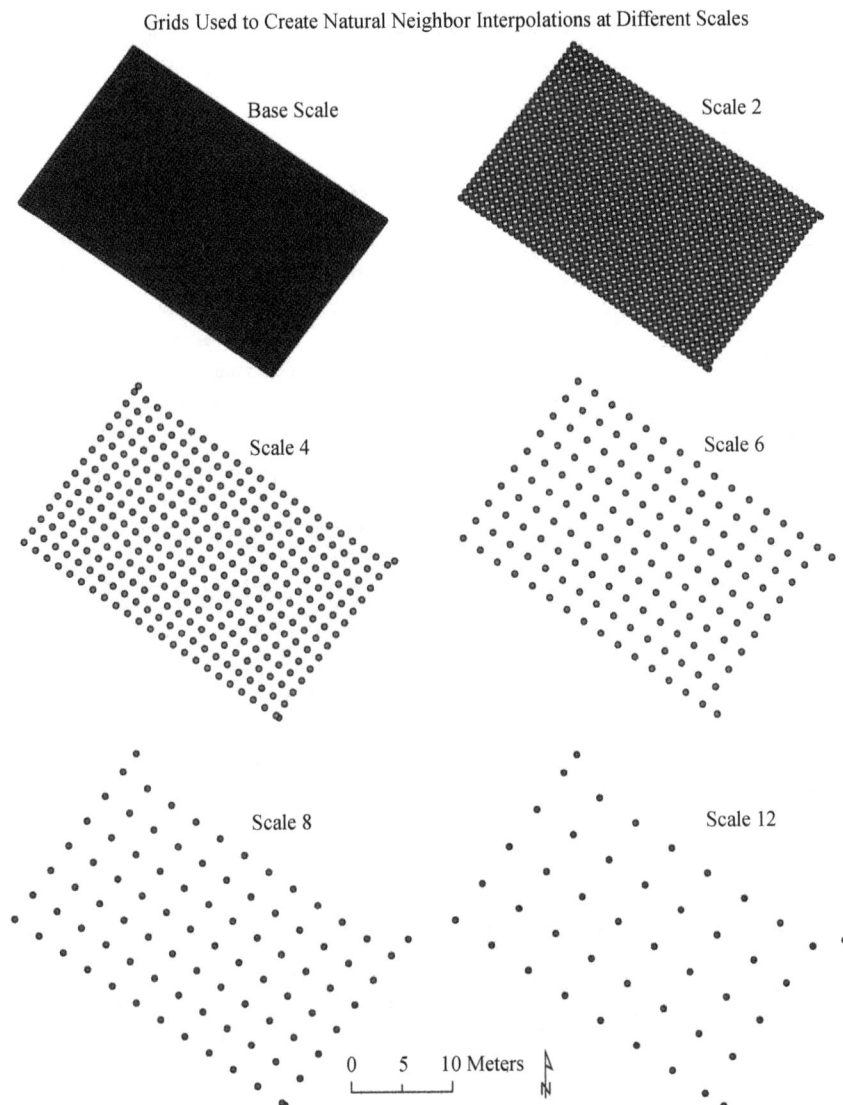

Figure 2. Coordinates of centroids of each cell used to created natural neighbor interpolations of study sites.

Table 1. Data information used for interpolations including raster cell size. The number of cells used when plotted with scale as the x axis follows a power function with the equation $y = 4755.797x^{-1.923}$ (Grayling creek), and $y = 3438.1x^{-1.822}$ (Little Wapiti creek).

Scale	Cell size in m^2	Cell dimensions	Number of cells used to create interpolations (Grayling, Little Wapiti)	Percent of data used compared to base scale Grayling	Percent of data used compared to base scale L. Wapiti
(Base) 1	0.11	0.33×0.33	4950, 3630	-	-
2	0.44	0.66×0.66	1262, 937	25.49%	25.81%
3	1.00	1.00×1.00	572, 409	11.56%	11.27%
4	1.78	1.33×1.33	325, 254	6.57%	7.00%
5	2.78	1.66×1.66	201, 157	4.06%	4.33%
6	4.00	2.0×2.0	152, 117	3.07%	3.22%
7	5.44	2.33×2.33	107, 83	2.16%	2.29%
8	7.11	2.66×2.66	87, 72	1.76%	1.98%
9	9.00	3.0×3.0	70, 55	1.41%	1.52%
10	11.11	3.33×3.33	57, 47	1.15%	1.29%
11	13.44	3.66×3.66	48, 38	0.97%	1.05%
12	16.00	4.0×4.0	43, 33	0.87%	0.91%

Natural neighbor interpolation was run on each scale to create interpolated maps for comparative analysis to the base scale. Natural neighbor maps served as visual and statistical base for scale accuracy comparisons because they have been shown to predict stream habitat data variables depth, and benthic substrate well [19]. When a continuous grid of data is collected at $1/3$ m^2 trend curves indicate natural neighbor interpolations are 100% accurate, thus allowing the base scale interpolation to be used as a digital representation of reality for effective comparison [19]. Although successively fewer points were used to create interpolations two through 12 (**Table 1**), values from the interpolated surface from each site were extracted to the original x,y coordinate points from base scale (3630 for Wapiti and 4950 for Grayling) to allow for exact comparison between base and coarser scales at each site. Extraction was accomplished using the extract values to points tool in ArcMap 10.

Each interpolated dataset from scale two through 12, was then subjected to Ordinary Least Squares (OLS) regression (ArcToolbox, ArcMap 10). Regressions were run with predicted values of depth, flow, and benthic substrate (from interpolated scales 2 - 12) as the dependent variable to explain the expected variable (base scale). In this way, OLS regressions provided comparative r^2 values for interpolations and residuals for each individual x,y coordinate (the original 3630 for Little Wapiti creek and 4950 for Grayling creek). Using OLS in this way provides quantitative, directly proportional (percent) comparison between base scale and scales two through 12 in the form of r^2. Maps of depth, flow velocity, and benthic substrate residuals were created to display positive and negative prediction trends in the form of standard deviation at each x,y location. Mapping residuals is important because it allows for unique examination of regional accuracy of interpolations. Maps were created showing over and under estimation of each coordinate with standard deviation values classes ranging from −2.5 to 2.5. Residual maps were created by performing OLS regression on extracted natural neighbor interpolated values for all scales compared to base scale.

Root mean square error (RMSE) values were then calculated for interpolations at each scale, plotted, and appropriate trend lines applied to all predicted habitat values (depth, flow, benthic substrate) (**Figures 3-7**). Plotting of RMSE values for each scale shows decay of accuracy for each scale effectively. Root mean square error compliments r^2 values because RMSE decreases as proportional r^2 increases. It is important to note that unlike r^2 values from interpolations, r^2 values on RMSE graphs indicate log trend line fit, and not proportional accuracy of interpolations at each scale. Substrate, which contains silt, sand, gravel, cobble, boulder, and land, had RMSE values calculated for all substrate sizes combined,

as well as for each substrate type to allow better understanding of interpolation accuracy (substrate is often discussed in terms of categories, though collected in the form of continuous data).

3. Results

Accuracy of depth, flow, and benthic substrate RMSE values degraded in a logarithmic linear fashion as data scale used to create interpolations became coarser (0.11 m^2 to 16 m^2) (**Figures 3-7**). At both sites, depth and flow interpolations retained accuracy more effectively than for benthic substrate (**Table 2, Figures 3-7**). Grayling creek maintained lower RMSE values than Little Wapiti creek for flow, similar RMSE values for depth, and nearly identical RMSE values at all scales for benthic substrate (**Figures 2-6**). As scale of data (and number of data points) used to create interpolations became coarser, range between maximum and minimum values for variables decreased. An example of decrease in range of values is shown through depth; by the coarsest scale the range of depths produced by interpolations was 0 - 44.9 cm, rather than 0 - 83 cm, a reduction of nearly half. As indicated by r^2 values, interpolation accuracy decreased with use of coarser data, but maintained some integrity even as the amount of data used to create maps decreased nearly 99%, 4950 to 43 for Grayling creek and 3630 to 33 for Little Wapiti creek, from base scale to scale 12 for both sites (**Tables 1** and **2**).

Table 2. Wapiti Creek base scale (0.11 m^2) compared to interpolated values from scales 2, 4, 6, 8, 10, and 12.

Little Wapiti					
Substrate		Depth		Flow	
r^2	Scale	r^2	Scale	r^2	Scale
0.90	2	0.88	2	0.86	2
0.80	4	0.73	4	0.74	4
0.74	6	0.59	6	0.67	6
0.57	8	0.25	8	0.37	8
0.51	10	0.4	10	0.34	10
0.53	12	0.35	12	0.41	12
Grayling					
Substrate		Depth		Flow	
r^2	Scale	r^2	Scale	r^2	Scale
0.86	2	0.95	2	0.95	2
0.78	4	0.91	4	0.94	4
0.69	6	0.86	6	0.87	6
0.67	8	0.87	8	0.9	8
0.56	10	0.78	10	0.78	10
0.54	12	0.81	12	0.78	12

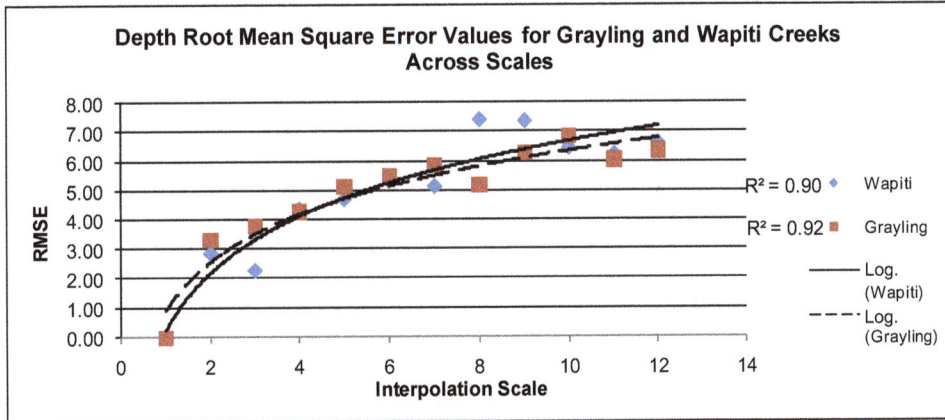

Figure 3. Root mean square error values for natural neighbor predicted maps of depth values on Wapiti Creek. Scales 1 - 12 are found on Table 1 and range from 0.11 to 16 square meter cell size.

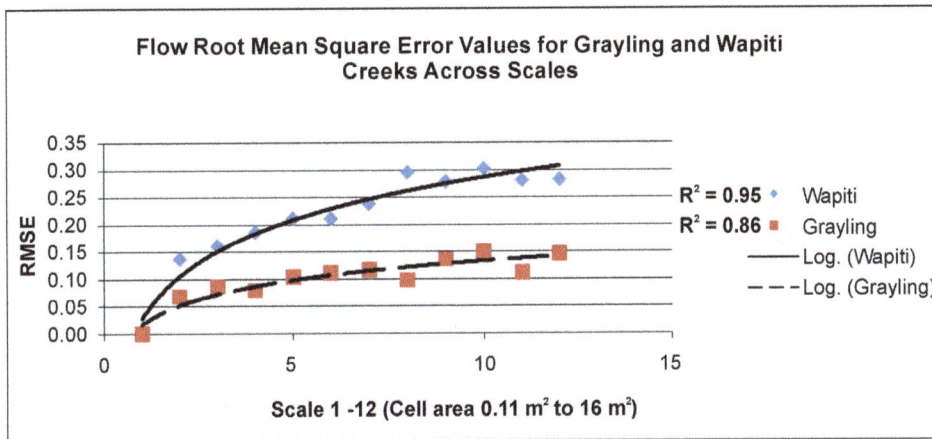

Figure 4. Root mean square error values for natural neighbor predicted maps of flow velocity values on Wapiti Creek. Scales 1 - 12 are found on Table 1 and range from 0.11 at the base scale to 16 square meter cell size used to crease interpolations at scale 12.

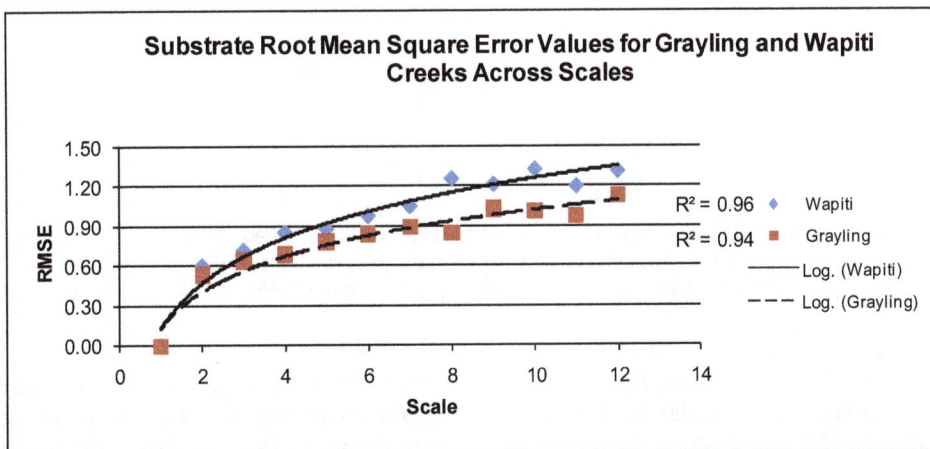

Figure 5. Root mean square error values for predicted maps of all substrate values at Wapiti and Grayling Creeks. Scales 1 - 12 are found on Table 1 and range from 0.11 to 16 square meter cell size.

Interpolated surfaces of Grayling and Wapiti creeks provided visual confirmation of a shrinking range of maximum interpolated values as indicated by RMSE (**Figures 3-7**), r^2 (**Table 2**), and residual standard deviation (**Figures 8** and **9**). Interpolation results grew less spatially complex as scale became coarser for all habitat

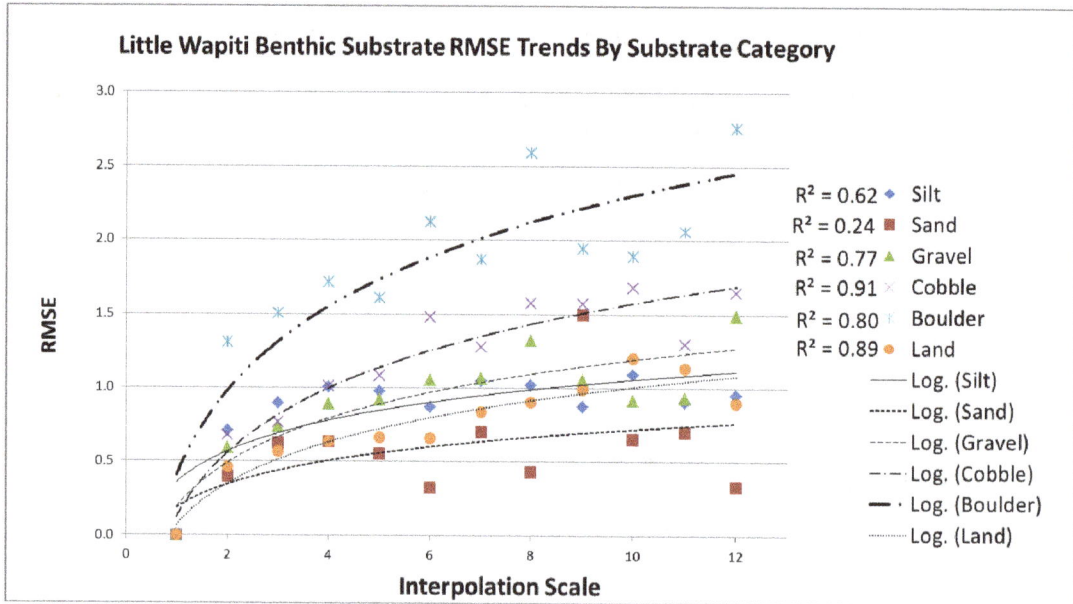

Figure 6. Little Wapiti creek substrate RMSE values with a logarithmic trend line applied showing r² values for each. Increase in RMSE as scale moves away from the baseline reference scale shows a progressive tapering effect of accuracy loss as predictive scale moves away from the baseline. Sand substrate size maintained the smallest RMSE change between scales.

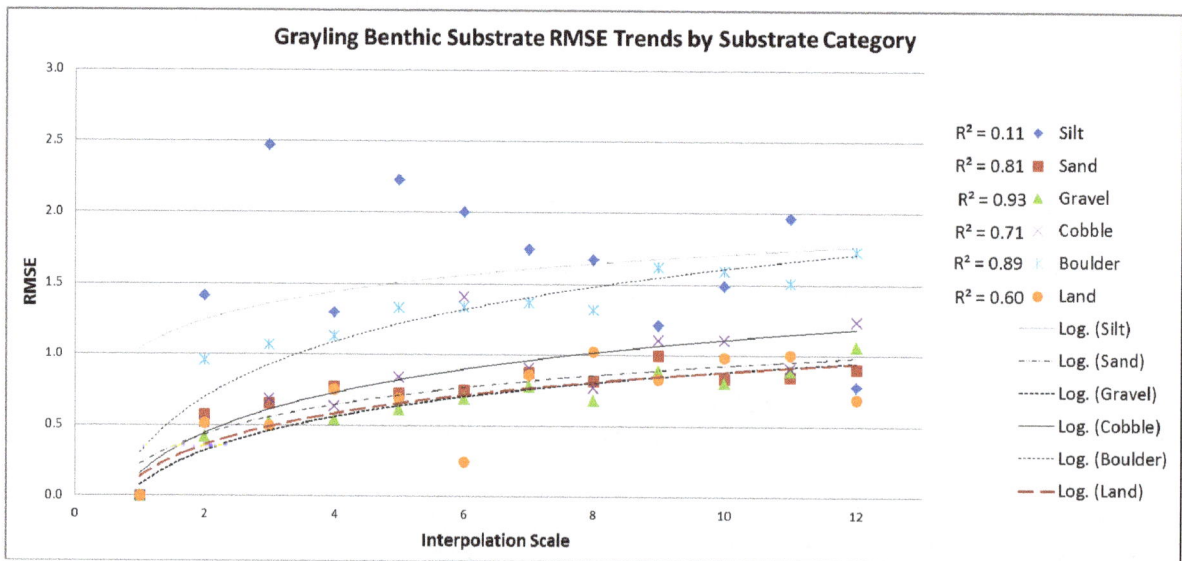

Figure 7. Grayling creek substrate RMSE values with a logarithmic trend line applied showing r² values for each. Increase in RMSE as scale moves away from the baseline reference scale shows a progressive tapering effect of accuracy loss as predictive scale moves away from the baseline. Sand and gravel maintained the smallest RMSE changes between scales.

variables at both sites. Decrease in spatial complexity was due in part to loss of extreme depths, flow variation and atypically located substrate variables located in the sparser data grid used for coarse scale interpolations (**Figure 2**). However, location and shape of deep areas, thalweg, zones of like flow, and substrate depositional areas were generally well maintained even to the terminal scale. Maintenance of spatial integrity is indicated by r² values (percent match) from OLS regressions of each interpolation when compared to base scale (**Table 2**, **Fig-**

ures **8** and **9**).

Ordinary Least Squares regressions demonstrated that all models coarser than base scale tended to underestimate deeper sections of river, and overestimate shallow sections, creating a smoothing effect along both benthic substrate edge boundaries and depth transition zones (**Figures 8** and **9**). This predictive smoothing effect increased in physical area proportional to the original base scale habitat feature as scale decreased. Another way of illustrating this behavior is that residuals from scale two

Wapiti Creek Residual Values at Scale 2 (top row) and Scale 12 (bottom row)

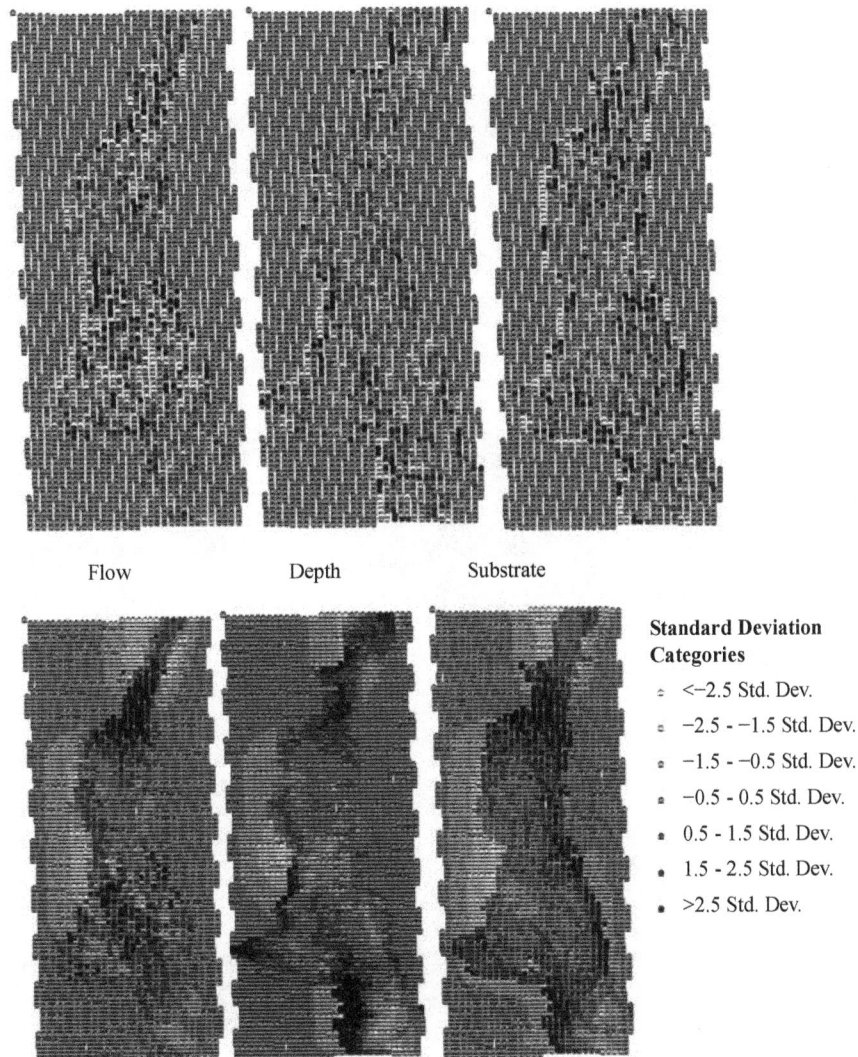

Flow   Depth   Substrate

**Standard Deviation
Categories**

≈ <−2.5 Std. Dev.

≈ −2.5 - −1.5 Std. Dev.

≈ −1.5 - −0.5 Std. Dev.

≈ −0.5 - 0.5 Std. Dev.

• 0.5 - 1.5 Std. Dev.

• 1.5 - 2.5 Std. Dev.

• >2.5 Std. Dev.

Figure 8. Wapiti Creek residual maps showing depth, flow, and substrate standard deviations for each of the 3630 x,y coordinate points at the site. Highly localized regions of standard deviation variation in scale one progress to larger regions of similar standard deviation values as scale becomes smaller, referred to in the text as a smoothing effect. Distribution pattern type of standard deviation types moves from random to clustered.

showed highly localized fluctuation in standard deviation values surrounding habitat zones with high heterogeneity and better lower standard deviation (**Figures 8** and **9**), while interpolations created from coarser scales saw less regional fluctuation and higher overall standard deviation from the base scale (**Figures 8** and **9**). Localized variation in standard deviation has decreased because both maximum range of values and the amount of data used for interpolations had both decreased (**Table 1**). The space between each interpolated data point increased appreciably by the coarsest scale (**Figure 2**), which also contributed to lack of localized variation. Regressions also demonstrated models created from the coarsest scale, using 99% fewer data points and 145 times more coarse

than the original 0.11 m^2, were able to match performance of finer scales for some variables (**Table 2**).

4. Discussion

Our study demonstrated that habitat data collected at a single spatial scale can be successfully used to accurately predict stream habitat at other spatial scales. Our results define the structure of accuracy loss occurring when interpolating coarse resolution with small amounts of fine scale data. As amount of data used to create predictive maps of stream habitat variables departs from the desired resolution, model accuracy decay occurs in a log linear fashion. As accuracy decays, interpolations using less data are able to retain sufficient predictive capability

Grayling Creek Residual Value Changes between Scale 2 (row 1) and 12 (row 2)

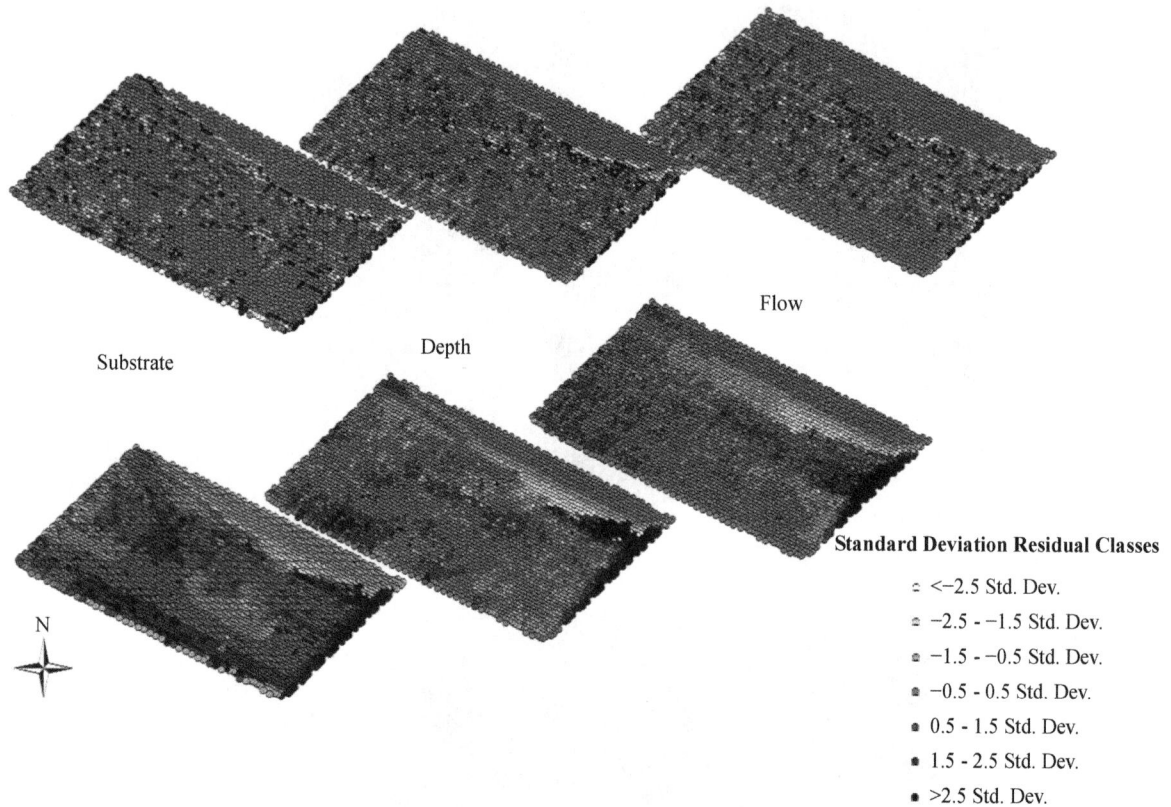

Flow

Substrate   Depth

N

Standard Deviation Residual Classes

○ <−2.5 Std. Dev.

○ −2.5 - −1.5 Std. Dev.

◐ −1.5 - −0.5 Std. Dev.

◕ −0.5 - 0.5 Std. Dev.

● 0.5 - 1.5 Std. Dev.

● 1.5 - 2.5 Std. Dev.

● >2.5 Std. Dev.

Figure 9. Grayling Creek residual maps showing depth, flow, and substrate standard deviations for each of the 3630 x,y co-ordinate points at the site. Abrupt localized changes in standard deviation are more prevalent in scale one. A more gradual change in standard deviation, or smoothing effect, may be seen in scale 12.

required to produce practical (functional, easily interpreted, informative) maps of stream habitat. This is important because adequate accuracy retention between scales affords capability for multi-scale inferences from a single data set. By defining the structure of accuracy loss of natural neighbor interpolations of stream habitat at scales other than the data collection scale through trends in RMSE and OLS regression r^2, we have provided a method for estimating the amount of accuracy lost by interpolating across scales.

By observing the combined results of RMSE trends and residual values as a guide to natural neighbor interpolative inaccuracies, it is possible to see a detailed progression of errors caused by departure from initial scale when interpolating stream habitat variables. Identifying a cause for error propagation is valuable because the source of error in environmental predictive models is not always readily apparent. Maps of interpolations and regression residuals also aid in clarifying the scalability of stream habitat variables by showing specific locations of strong and weak model predictions when moving between scales. This helps quantify what detail is eliminated when using data at a coarse scale, an issue impor-

tant to ecological studies [12,15,36]. The ability to understand accuracy loss when interpolating at coarser scale is better understood, thus increasing the value of a single dataset.

5. Conclusion

This study indicated that the initial scale of collected data and stream size influence the range of scales at which the data set retains usefulness for predictive purposes. For instance Little Wapiti creek showed a drop in predictive accuracy below 60% at the fifth scale removed (from the original) for all habitat variables. The Wapiti site encompassed a series of three pool/riffle zones, while Grayling creek was a single pool riffle interface (three to four times the scale of Little Wapiti). Because of stream size difference in the study, results may have identified presence of a threshold for predictive accuracy purely associated with stream size. This makes ecological sense, in which a larger order stream may have proportionally larger habitat patches which in turn maintains any scale's predictive accuracy at further reaching scales.

Maintaining spatial integrity of site boundaries and habitat transition zones though interpolations at varying scales from a single dataset was shown to be possible in this study. Spatial integrity is important because it allows accurate measurement of area of available habitat. In streams, the amount and distribution of available habitat are closely tied to species occurrence and species diversity, and their examination aids in ecological study at varying spatial scales [5,6,42-45].

An important question with respect to accuracy of stream habitat models, and perhaps all models, is what level of accuracy is acceptable. This paper does not attempt to answer that question, it only helps quantify the level of accuracy possible and identify details and accuracy lost during the collection and analysis process. Acceptable accuracy is often a function of the question being asked and may vary greatly [44]. Though we do not advocate a particular level for acceptable accuracy in this study, the ability to scale a single habitat data set to a scale far removed from the original and still maintain accuracy, is a valuable tool for stream management and assessment purposes.

6. Acknowledgements

We thank Mike Strager for input on the scientific and editing processes. We thank Richard Sheehan for data collection assistance and Nicole Ten Eyck for recording of data. Reference to trade names does not imply endorsement of commercial products by the US government.

REFERENCES

[1] K. Looy, et al., "A Scale-Sensitive Connectivity Analysis to Identify Ecological Networks and Conservation Value in River Networks," Landscape Ecology, Vol. 28, No. 7, 2013, pp. 1239-1249.

[2] D. Ruddell and E. A. Wentz, "Multi-Tasking: Scale in Geography," Geography Compass, Vol. 3, No. 2, 2009, pp. 681-697.

[3] D. L. Urban, "Modeling Ecological Processes across Scales," Ecology, Vol. 86, No. 8, 2005, pp. 1996-2006.

[4] J. T. Petty, et al., "Quantifying Instream Habitat in the Upper Shavers Fork Basin at Multiple Spatial Scales," Proceedings of the Annual Conference of the Southeastern Association of Fish and Wildlife Agencies, 2001.

[5] T. Petty and G. Grossman, "Patch Selection by Mottled Sculpin (Pisces: Cottidae) in a Southern Appalachian Stream," Freshwater Biology, Vol. 35, No. 2, 1996, pp. 261-276.

[6] A. R. Thompson, J. T. Petty and G. D. Grossman, "Multi-Scale Effects of Resource Patchiness on Foraging Behaviour and Habitat Use by Longnose Dace, Rhinichthys Cataractae," Freshwater Biology, Vol. 46, No. 2, 2001, pp.

145-160.

[7] M. Hondzo, et al., "Estimating and Scaling Stream Ecosystem Metabolism along Channels with Heterogeneous Substrate," Ecohydrology, Vol. 6, No. 4, 2013, pp. 679-688.

[8] M. Wheatley and C. Johnson, "Factors Limiting Our Understanding of Ecological Scale," Ecol Complexity, Vol. 6, No. 2, 2009, pp. 150-159.

[9] V. D. Valavanis, et al., "Modelling of Essential Fish Habitat Based on Remote Sensing, Spatial Analysis and GIS," Hydrobiologia, Vol. 612, No. 1, 2008, pp. 5-20.

[10] K. B. Gido, et al., "Fish-Habitat Relations across Spatial Scales in Prairie Streams," In American Fisheries Society Symposium 48, American Fisheries Society, Bethesda, 2006.

[11] J. D. Allan, D. L. Erickson and J. Fay, "The Influence of Catchment Land Use on Stream Integrity across Multiple Spatial Scales," Freshwater Biology, Oxford, Vol. 37, No. 1, 1997, pp. 149-161.

[12] S. A. Levin, "The Problem of Pattern and Scale in Ecology: The Robert H. MacArthur Award Lecture," Ecology, Vol. 73, No. 6, 1992, pp. 1943-1967.

[13] J. A. Wiens, "Spatial Scaling in Ecology," Functional Ecology, Vol. 3, No. 4, 1989, pp. 385-397.

[14] A. L. Sheldon, "Cost and Precision in a Stream Sampling Program," Hydrobiologia, Vol. 111, No. 2, 1984, pp. 147-152.

[15] W. L. Fisher and F. J. Rahel, "Geographic Information Systems Applications in Stream and River Fisheries," In Geographic Information Systems in Fisheries, American Fisheries Society, Bethesda, 2004, pp. 49-84.

[16] C. A. Gotway and L. J. Young, "Combining Incompatible Spatial Data," Journal of the American Statistical Association, Vol. 97, No. 458, 2002, pp. 632-648.

[17] S. D. Cooper, et al., "Implications of Scale for Patterns and Processes in Stream Ecology," Austral Ecology, Vol. 23, No. 1, 1998, pp. 27-40.

[18] S. Gergel, et al., "What Is the Value of a Good Map? An Example Using High Spatial Resolution Imagery to Aid Riparian Restoration," Ecosystems, Vol. 10, No. 5, 2007, pp. 688-702.

[19] K. R. Sheehan and S. A. Welsh, "An Interpolation Method for Stream Habitat Assessments," North American Journal of Fisheries Management, Vol. 29, No. 1, 2009, pp. 1-9.

[20] M. L. Stein, "Interpolation of Spatial Data: Some Theory for Kriging," Springer, 1999.

[21] C. Childs, "Interpolating Surfaces in ArcGIS Spatial Ana-

lyst," In ArcUser, ESRI, California, 2004, pp. 32-35.

[22] J. F. Kratzer, D. B. Hayes and B. E. Thompson, "Methods for Interpolating Stream Width, Depth, and Current Velocity," *Ecological Modelling*, Vol. 196, No. 1-2, 2006, pp. 256-264.

[23] Y. Sun, *et al.*, "Comparison of Interpolation Methods for Depth to Groundwater and Its Temporal and Spatial Variations in the Minqin Oasis of Northwest China," *Environmental Modelling & Software*, Vol. 24, No. 10, 2009, pp. 1163-1170.

[24] V. Merwade, "Effect of Spatial Trends on Interpolation of River Bathymetry," *Journal of Hydrology*, Vol. 371, No. 1, 2009, pp. 169-181.

[25] R. R. Murphy, F. C. Curriero and W. P. Ball, "Comparison of Spatial Interpolation Methods for Water Quality Evaluation in the Chesapeake Bay," *Journal of Environmental Engineering*, Vol. 136, No. 2, 2009, pp. 160-171.

[26] R. J. Bennett, R. P. Haining and D. A. Griffith, "Review Article: The Problem of Missing Data on Spatial Surfaces," *Annals of the Association of American Geographers*, Vol. 74, No. 1, 1984, pp. 138-156.

[27] N. D. Le, W. Sun and J. V. Zidek, "Bayesian Multivariate Spatial Interpolation with Data Missing by Design," *Journal of the Royal Statistical Society, Series B (Methodological)*, Vol. 59, No. 2, 1997, pp. 501-510.

[28] D. J. C. MacKay, "Bayesian Interpolation," *Neural Computation*, Vol. 4, No. 3, 1992, pp. 415-447.

[29] M. Sambridge, J. Braun and H. McQueen, "Geophysical Parametrization and Interpolation of Irregular Data Using Natural Neighbours," *Geophysical Journal International*, Vol. 122, No. 3, 1995, pp. 837-857.

[30] D. Zimmerman, *et al.*, "An Experimental Comparison of Ordinary and Universal Kriging and Inverse Distance Weighting," *Mathematical Geology*, Vol. 31, No. 4, 1999, pp. 375-390.

[31] A. V. Jopling and D. L. Forbes, "Flume Study of Silt Transportation and Deposition," *Geografiska Annaler, Series A, Physical Geography*, Vol. 61, No. 1-2, 1979, pp. 67-85.

[32] I. A. Lunt, J. S. Bridge and R. S. Tye, "A Quantitative, Three-Dimensional Depositional Model of Gravelly Braided Rivers," *Sedimentology*, Vol. 51, No. 3, 2004, pp. 377-414.

[33] B. Purkait, "Patterns of Grain-Size Distribution in Some Point Bars of the Usri River, India," *India Journal of Sedimentary Research*, Vol. 72, No. 3, 2002, pp. 367-375.

[34] G. H. S. Smith and R. I. Ferguson, "The Gravel-Sand Transition along River Channels," *Journal of Sediment Research*, Vol. 65, No. 2a, 1995, pp. 423-430.

[35] B. M. Weigel, *et al.*, "Relative Influence of Variables at Multiple Spatial Scales on Stream Macroinvertebrates in the Northern Lakes and Forest Ecoregion, USA," *Freshwater Biology*, Vol. 48, No. 8, 2003, pp. 1440-1461.

[36] N. B. Kotliar and J. A. Wiens, "Multiple Scales of Patchiness and Patch Structure: A Hierarchical Framework for the Study of Heterogeneity," *Oikos*, Vol. 59, No. 2, 1990, pp. 253-260.

[37] R. H. Waring and S. W. Running, "Forest Ecosystems: Analysis at Multiple Scales," Elsevier, 2010.

[38] A. M. Helton, *et al.*, "Relative Influences of the River Channel, Floodplain Surface, and Alluvial Aquifer on Simulated Hydrologic Residence Time in a Montane River Floodplain," *Geomorphology*, 2012.

[39] E. B. Rastetter, *et al.*, "Aggregating Fine-Scale Ecological Knowledge to Model Coarser-Scale Attributes of Ecosystems," *Ecological Applications*, Vol. 2, No. 1, 1992, pp. 55-70.

[40] K. G. Boykin, *et al.*, "A National Approach for Mapping and Quantifying Habitat-Based Biodiversity Metrics across Multiple Spatial Scales," *Ecological Indicators*, Vol. 33, 2012, pp. 139-147.

[41] M. G. Wolman and A. G. Union, "A Method of Sampling Coarse River-bed Material," *Transactions of the American Geophysical Union*, Vol. 35, 1954, pp. 951-956.

[42] K. O. Winemiller, A. S. Flecker and D. J. Hoeinghaus, "Patch Dynamics and Environmental Heterogeneity in Lotic Ecosystems," *Journal of the North American Benthological Society*, Vol. 29, No. 1, 2010, pp. 84-99.

[43] C. R. Townsend, "The Patch Dynamics Concept of Stream Community Ecology," *Journal of the North American Benthological Society*, Vol. 8, No. 1, 1989, pp. 36-50.

[44] C. M. Pringle, *et al.*, "Patch Dynamics in Lotic Systems: The Stream as a Mosaic," *Journal of the North American Benthological Society*, Vol. 7, No. 4, 1988, pp. 503-524.

[45] I. Hanski and M. E. Gilpin, "Metapopulation Dynamics," *Nature*, Vol. 396, No. 6706, 1998, pp. 41-49.

10

The Accuracy of GIS Tools for Transforming Assumed Total Station Surveys to Real World Coordinates

Ragab Khalil[1,2]
[1]Landscape Architecture Department, KAU, Jeddah, Saudi Arabia
[2]Civil Engineering Department, Assiut University, Assiut, Egypt

ABSTRACT

Most surveying works for mapping or GIS applications are performed with total station. Due to the remote nature of many of the sites surveyed, the surveys are often done in unprojected, local, assumed coordinate systems. However, without the survey data projected in real world coordinates, the range of possible analyses is limited and the value of existing imagery, elevation models, and hydrologic layers cannot be exploited. This requires a transformation from the local assumed to the real world coordinate systems. There are various built-in and add-in tools to perform transformations through GIS programs. This paper studies the effect of using Georeferencing tool, Spatial Adjustment tool (Affine and similarity) and CHaMP tool on the precision and relative accuracy of total station survey. This transformation requires real-world coordinates of at least two control points, which can be collected from different sources. This paper also studies the effect of using geodetic GPS, hand-held GPS, Google Earth (GE) and Bing Basemaps as sources for control points on the precision and relative accuracy of total station survey. These effects have been tested by using 111 points covered area of 60,000 m^2 and the results have shown that the CHaMP tool is the best for preserving the relative accuracy of the transformed points. The Georeferencing and spatial adjustment (similarity) tools give the same results and their accuracy are between 1/1000 and 1/300 depending on the source of control points. The results have also shown that the cornerstone to preserve the precision and relative accuracy of the transformed coordinates is the relative position of the control points despite their source.

Keywords: Control Network; Topographic Surveying; Coordinate Transformation; Spatial Adjustment; Georeferencing; CHaMP; Accuracy; Total Station

1. Introduction

Total station surveys are a widely used method to survey topography [1], with applications ranging from traditional land surveying [2], land form evolution monitoring [3], to land use monitoring [4]. In the geosciences and biological sciences, total stations are now becoming standard tools in monitoring geomorphic change detection of rivers [5-7], streams [8], beaches [9,10] and mass wasting of hill slopes [11,12]. Since many total station surveys are now undertaken in remote and/or undeveloped localities, there is often not an established local control network tied to a projected real world coordinate system [13]. Thus, many of these surveys are done from an unprojected local assumed coordinate system. However, as GIS has become more of an everyday tool for visualization, modeling and analysis of topographic data [3], there is an increasing demand for such surveys to be in real world coordinate system. Transforming total station surveys from unprojected local assumed coordinate system to real world coordinates makes the power of overlaying those data with other datasets (e.g., aerial imagery, vector datasets of roads, political boundaries, etc.) and certain analyses possible. There are various built-in and add-in tools to perform transformations through GIS programs. This paper studies the effect of using Georeferencing tool, Spatial Adjustment tool (Affine and similarity) and CHaMP tool on the precision and relative accuracy of total station survey. This transformation requires real-world coordinates of at least two control points, which can be collected from different sources. This paper also studies the effect of using geodetic GPS, hand-held GPS, Google Earth (GE) and Bing Basemaps as sources for control points on the precision and relative accuracy of total station survey.

2. Transformation Methods

There are numerous transformation methods for transforming between coordinate systems ranging from simple to sophisticated. The choice of appropriate method depends on the specifics of the application and is generally one that should be made by someone with proper training in surveying and geomatics as well as a solid understanding of the source data and how it was collected [13]. In this paper, the transformation tools built in ArcGIS (e.g. spatial adjustment tool and georeferencing tool), and add-in tools such as CHaMP tool were used to transform unprojected total station precise observations into projected real world coordinates. All of these tools use affine transformation. An affine transformation is any transformation that preserves collinearity (*i.e.*, all points lying on a line initially still lie on a line after transformation) and ratios of distances. In general, an affine transformation is a composition of rotations, translations, dilations (scales), and shears (skews) [14]. The transformation functions are based on the comparison of the coordinates of source and destination points, also called control points.

2.1. Spatial Adjustment Tool

Spatial adjustment supports a variety of adjustment methods (Transformation, Rubbersheet and edge snap) and will adjust all editable data sources. The affine and similarity transformations were used in this study as they are the appropriate methods to transform total station surveys to real world coordinates.

2.1.1. Affine Transformation
The Affine transformation can be represented by the following equations (in matrix formation) [15].

$$\begin{bmatrix} x \\ y \end{bmatrix} = \begin{bmatrix} \cos\theta & \sin\theta \\ -\sin\theta & \cos\theta \end{bmatrix} \begin{bmatrix} s_u & s_v \sin\alpha \\ 0 & s_v \cos\alpha \end{bmatrix} \begin{bmatrix} u \\ v \end{bmatrix} + \begin{bmatrix} t_x \\ t_y \end{bmatrix} \quad (1)$$

$$\begin{bmatrix} x \\ y \end{bmatrix} = \begin{bmatrix} s_u \cos\theta & (s_v \sin\alpha \cos\theta + s_v \cos\alpha \sin\theta) \\ -s_u \sin\theta & (-s_v \sin\alpha \sin\theta + s_v \cos\alpha \cos\theta) \end{bmatrix}$$
$$\times \begin{bmatrix} u \\ v \end{bmatrix} + \begin{bmatrix} t_x \\ t_y \end{bmatrix} \quad (2)$$

$$\begin{bmatrix} x \\ y \end{bmatrix} = \begin{bmatrix} A & B \\ D & E \end{bmatrix} \begin{bmatrix} u \\ v \end{bmatrix} + \begin{bmatrix} C \\ F \end{bmatrix} \quad (3)$$

Where:
u, v are coordinates of the input data and
x, y are the transformed coordinates.
A, B, C, D, E, and F are six unknowns; determined by comparing the location of source and destination control points.
Affine transformation can differentially scale the data,

skew it, rotate it, and translate it. As there are six unknowns in the transformation equations, this method requires a minimum of three control points.

2.1.2. Similarity Transformation
The similarity transformation scales, rotates, and translates the data. It will not independently scale the axes, nor will it introduce any skew. It maintains the aspect ratio of the features transformed, which is important if you want to maintain the relative shape of features. The similarity transformation function (in matrix formation) [15] is:

$$\begin{bmatrix} x \\ y \end{bmatrix} = s \begin{bmatrix} \cos\theta & \sin\theta \\ -\sin\theta & \cos\theta \end{bmatrix} \begin{bmatrix} u \\ v \end{bmatrix} + \begin{bmatrix} t_x \\ t_y \end{bmatrix} \quad (4)$$

$$\begin{bmatrix} x \\ y \end{bmatrix} = \begin{bmatrix} s\cos\theta & s\sin\theta \\ -s\sin\theta & s\cos\theta \end{bmatrix} \begin{bmatrix} u \\ v \end{bmatrix} + \begin{bmatrix} t_x \\ t_y \end{bmatrix} \quad (5)$$

$$\begin{bmatrix} x \\ y \end{bmatrix} = \begin{bmatrix} A & B \\ -B & A \end{bmatrix} \begin{bmatrix} u \\ v \end{bmatrix} + \begin{bmatrix} C \\ F \end{bmatrix} \quad (6)$$

In similarity method, the scale is the same in both x and y directions, and a minimum of two control points are required.

2.2. Georeferencing Tool

Georeferencing tool is used to adjust raster data using different polynomial equations and to adjust a CAD dataset using the similarity transformation method. The transformation functions are similar to Equations (4)-(6).

2.3. CHaMP Tool

ChaMP tool was introduced by [13]. It uses a simple affine transformation that just rotates and translates the data. This type of transformation is accurate because the scale is preserved [16]. The equations can be as the followings:

$$\begin{bmatrix} x \\ y \end{bmatrix} = \begin{bmatrix} \cos\theta & \sin\theta \\ -\sin\theta & \cos\theta \end{bmatrix} \begin{bmatrix} u \\ v \end{bmatrix} + \begin{bmatrix} t_x \\ t_y \end{bmatrix} \quad (7)$$

$$\begin{bmatrix} x \\ y \end{bmatrix} = \begin{bmatrix} A & B \\ -B & A \end{bmatrix} \begin{bmatrix} u \\ v \end{bmatrix} + \begin{bmatrix} C \\ F \end{bmatrix} \quad (8)$$

This tool requires two control points but three is essentials as stated by the designers.

3. Control Points

To transform the unprojected total station survey data, coordinates in a projected real world coordinate system for two to three control points which were established and used in the total station survey are required. The rotation is performed about one of these points, where the rotation is computed based on the difference in azimuth between point pairs whose coordinates are known in both systems. The origin shift is computed from the rotation point's coordinates in both systems [16].

There are multiple methods to acquire the real world coordinates of these control points. The most accurate is to use a known pre-existing control network surveyed in real world coordinate system [1,17]. At the same level of accuracy, a geodetic GPS can be used to obtain accurate coordinates for the control points. The problems of these two methods as stated by [13] are that many practitioners often do not have access to geodetic GPS receivers or may not have access to the necessary post-processing software and may work in areas far away of existing control network. Google Earth is a low-cost and readily accessible tool with relatively good spatial accuracy [18, 19]. It offers high resolution imagery from which, it may be possible to derive sufficient quality photo control points if ground features visible in the photo (e.g., fence corner, rockedge, etc.) can be accurately located in the field. The latest versions of ArcGIS offer a high resolution Bing Basemaps which can also be used to drivephoto control points ifground features visible in the photo can beaccurately located in the field. The most common is to use a simple, inexpensive, consumer-grade GPS (e.g., Garmin hand held, Smart-Phone, GPS card in field data collector. The accuracy of such devices is sufficient for purposes of GIS overlay at scales of 1:1000 or coarser [13].

4. Field Data

Topcon total station GPT-7501 was used to collect the coordinates of more than 100 point in a parking and open space area in King Abdulaziz University campus. The total station was first set up on a point with assumed coordinates. Then the total station was oriented with "backsight azimuth" setup which uses bearing to the backsight point, using assumed bearing for the line connected the two points. Once the survey is begun on this assumed coordinate system, all additional station setups and all data, including three control points (which acquired during the survey course), collected in a single unprojected local assumed coordinate system. The collected data was exported to *.txt and once again to *.dxf. The *.txt file was used to generate shapefile while the *.dxf file represents the CAD file. Both files are needed to apply the transformation on.

The projected coordinates of the three control points were collected using four methods 1) RTK GPS with two Topcon GR3 geodetic receivers; one receiver was setup on existing control point at the university main gate while the other receiver was used to acquire the needed points; 2) Garmin handheld GPS; 3) Google Earth at Eye altitude equal to 50 m; and 4) Bing Imagery that is available in ArcGIS as an online basemap layer at scale 1:100.

5. Results and Discussion

To check the accuracy of coordinate transformation tools available in ArcGIS, the relative position of surveyed points were calculated before and after transformation using the different tools and compared to the original positions. The most upper left point was chosen as an origin and the distances from it to all other points were computed using the raw data and data after transforming the coordinates. The difference between distances to the corresponding points were calculated and represented in **Figure 1** for control points acquired using geodetic GPS. **Figures 2-4** represent the errors in relative positions for control points collected using Bing basemap imagery, Google Earth and Hand-held GPS respectively.

From the figures it's clear that there are no errors in

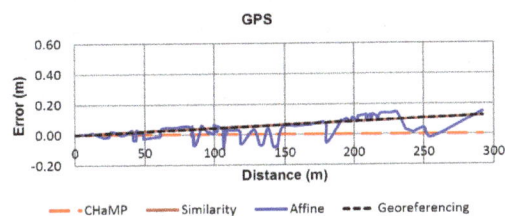

Figure 1. Relative error in point positions using GPS control points.

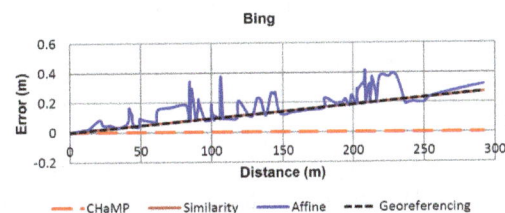

Figure 2. Relative error in point positions using Bing basemap control points.

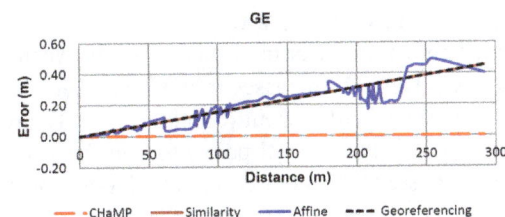

Figure 3. Relative error in point positions using Google Earth control points.

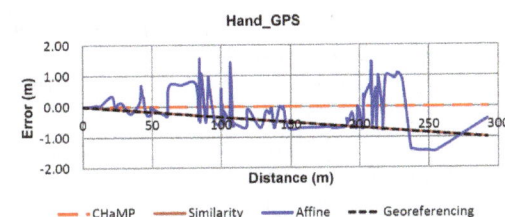

Figure 4. Relative error in point positions using Gamin GPS control points.

coordinates transformed using CHaMP tool regardless the source of the control points. The errors in coordinates resulting using Georeferencing tool and Spatial Adjustment tool (similarity) are the same, and the error increases as the distance from the origin increases. The error at the farthest point was 0.12 m, 0.27 m, 0.45 m and 0.99 m for GPS, Bing basemap, Google Earth and Garmin control points respectively. The error ratios were 1:2400, 1:1100, 1:650 and 1:300 respectively. Errors in coordinates due to using Spatial Adjustment tool (Affine) are undulated and its trend is almost the same as Georeferencing and similarity transformation.

CHaMP tool uses simple mathematical operation of a translation and rotation, so it preserves the relative positional accuracy and precision of the total station survey. The data may be shifted out of its absolute location depending on the absolute accuracy of the control points. A single control point will ultimately be used as the basis for the horizontal translation and datum adjustment, and a bearing based on a second control point will be used to define the rotation [13].

To explain the results of the other tools, let's first have a look on the absolute coordinates of the control points acquired from the different sources which shown in **Table 1** and the deviations in these coordinates related to the GPS points as it is the most accurate which shown in **Table 2**.

From **Table 2**, one can notice that GE points are much closer to GPS points than Bing points, while **Figures 2** and **3** show that errors in the transformed coordinates using GE points are bigger than those resulted when using Bing points. This means that the absolute location of control points does not affect the transformation accuracy.

The distances between control points are shown in **Table 3**, and the deviations from the original (total station) distances are shown in **Table 4**.

From **Table 4**, the error in distance (1-2) between control points (1) and (2) increases from GPS to Bing to GE to Garmin. Spatial Adjustment (Similarity) and Georeferencing tools use two control points to transform the coordinates. These tools preserve the relative geometry of the control points and scale (stretch) the source data to fit the geometry of the controls. This explains increasing the error in the transformed coordinates according to the

increase of that distance error as shown in **Figures 1-4**. These two transformation tools scale the coordinates with the same amount in both X and Y axes, so the error in the transformed coordinates of any point is proportional to the distance to that point from the base control point.

Spatial Adjustment (Affine) tool uses and preserve the relative geometry of three control points, so this tool stretch and skew the source data. The trend of the error in the relative position of the transformed data is due to the error in distance (1-2), while the undulations of error chart is due to the error in distance (1-3) as shown in **Figures 1-4**. The error in distance (1-2) for Bing is smaller than that of GE, so the trend of error in transformed data using Bing control points is smaller than that for GE control points as shown in **Figures 2** and **3**. The distance (1-3) for Bing is greater than that of GE, accordingly the error in relative positions due to using Bing control points is higher than the error due to using GE control points in transformation. The error in relative position of transformed data depending on the point location related to the line connecting the first two control points as shown in **Figure 5**, which show that the maximum error is in the points far away from the mentioned line.

6. Conclusions

From the results and discussion, the followings could be concluded:

1) CHaMP transformation tool uses a simple rotation and translation transformation to preserve the precision and relative accuracy of the total station survey.

2) Georeferencing and Spatial Adjustment similarity transformation tools preserve the location of the two control points used in the transformation, so they stretch the data and introduce errors in point location depending on the relative position of the control points.

3) Spatial Adjustment Affine transformation tool preserves the location of the three control points used in the transformation, so it stretches and warps the data which introduce errors in point location depending on the relative position of the control points and on the point location related to the line connect the first two control points.

4) The cornerstone of the accuracy of build in transformation tools is the relative positions of the control points.

Table 1. The absolute coordinates of the control points.

	GPS		Bing		GE		Garmin	
Point	X	Y	X	Y	X	Y	X	Y
1	524966.50	2376567.22	524966.98	2376568.58	524966.64	2376566.45	524966.00	2376569.00
2	525166.46	2376483.88	525167.01	2376485.69	525166.45	2376483.38	525167.00	2376486.00
3	524966.62	2376460.65	524967.53	2376462.48	524966.62	2376460.09	524969.00	2376464.00

Table 2. The deviations in coordinates from GPS points.

	Bing		GE		Garmin	
Point	D X	D Y	D X	D Y	D X	D Y
1	−0.477	−1.366	−0.139	0.768	0.501	−1.782
2	−0.545	−1.812	0.011	0.497	−0.539	−2.123
3	−0.901	−1.835	0.004	0.558	−2.376	−3.352

Table 3. Distances between control points.

Distance	Total station	GPS	Bing	GE	Garmin
1-2	216.724	216.633	216.524	216.390	217.463
1-3	106.478	106.570	106.102	106.360	105.043

Table 4. The deviations from the original distance.

D Distance	GPS	Bing	GE	Garmin
1-2	0.091	0.200	0.334	−0.739
1-3	−0.092	0.376	0.118	1.436

Figure 5. Location of maximum errors due to affine transformation tool.

7. Acknowledgements

Thanks are due to Mr. Talal Al-Thebety for assistance in performing the field measurements.

REFERENCES

[1] USACE, "Control and Topographic Surveying," Engineering Manual, EM1110-1-1005, US Army Corps of Engineers, Vicksburg, 2007, 498 p.

[2] U. Kizil and L. Tisor, "Evaluation of RTK-GPS and Total Station for Applications in Land Surveying," *Journal of Earth System Science*, Vol. 120, No. 2, 2011, pp. 215-221.

[3] S. N. Lane and J. H. Chandler, "Editorial: The Generation of High Quality Topographic Data for Hydrology and Geomorphology: New Data Sources, New Applications and New Problems," *Earth Surface Processes and Landforms*, Vol. 28, No. 3, 2003, pp. 229-230.

[4] L.-S. Lin, "Application of GPS RTK and Total Station System on Dynamic Monitoring Land Use," *The XXth ISPRS Congress*, Istanbul, July 2004, pp. 12-23.

[5] S. N .Lane, J. H. Chandler and K. S. Richards, "Developments in Monitoring and Modeling Small-Scale River Bed Topography", *Earth Surface Processes and Landforms*, Vol. 19, No. 4, 1994, pp. 349-368.

[6] I. C. Fuller, A. R. G. Large and D. J. Milan, "Quantifying Channel Development and Sediment Transfer Following Chute Cutoff in a Wandering Gravel-Bed River," *Geomorphology*, Vol. 54, No. 3-4, 2003, pp. 307-323.

[7] J. E. Merz, G. B. Pasternack and J. M. Wheaton, "Sediment Budget for Salmonid Spawning Habitat Rehabilitation in a Regulated River," *Geomorphology*, Vol. 76, No. 1-2, 2006, pp. 207-228.

[8] D. M. Walters, D. S. Leigh, M. C. Freeman, B. J. Freeman and C. M. Pringle, "Geomorphology and Fish Assemblages in a Piedmont River Basin, USA," *Fresh-Water Biology*, Vol. 48, No. 11, 2003, pp. 1950-1970.

[9] I. Delgado and G. Lloyd, "A Simple Low Cost Method for One Person Beach Profiling," *Journal of Coastal Research*, Vol. 20, No. 4, 2004, pp. 1246-1253.

[10] P. Baptista, T. R. Cunha, A. Matias, C. Gama, C. Bernardes and O. Ferreira, "New Land-Based Method for Surveying Sandy Shores and Extracting DEMs: The IN-SHORE System," *Environmental Monitoring and Assessment*, Vol. 182, No. 1-4, 2011, pp. 243-257.

[11] J. J. De Sanjose-Blasco, A. D. J. Atkinson-Gordo, F. Salvador-Franch and A. Gomez-Ortiz, "Application of Geomatic Techniques to Monitoring of the Dynamics and to Mapping of the Veleta Rock Glacier (Sierra Nevada, Spain) ," *Zeitschrift Fur Geomorphologie*, Vol. 51, 2007, pp. 79-89.

[12] B. H. Mackey, J. J. Roering and J. A. McKean, "Long-Term Kinematics and Sediment Flux of an Active Earthflow, Eel River, California," *Geology*, Vol. 9, No. 37, 2009, pp. 803-806.

[13] J. M. Wheaton, C. Garrard, K. Whitehead and C. Volk, "A simple, Interactive GIS Tool for Transforming Assumed Total Station Surveys to Real World Coordinates—The CHaMP Transformation Tool," *Computers & Geosciences*, Vol. 42, 2012, pp. 28-36.

[14] E. W. Weisstein, "Affine Transformation," From MathWorld—A Wolfram Web Resource, 1 May 2013.

[15] R. E. Deakin, "Coordinate Transformations in Surveying and Mapping," Geospatial Science, 2004.

[16] W. Sprinsky, "Transformation of Survey Coordinates: Another Look at an Old Problem," *Journal of Surveying Engineering*, Vol. 128, No. 4, 2002, pp. 200-209.

[17] USACE, "Topographic Surveying," USA Rmy Corps of Engineers, Washington DC, 111 p.

[18] D. Kaimaris, O. Georgoulab, P. Patiasb and E. Stylianidis, "Comparative analYsis on the Archaeological Content of Imagery from Google Earth," *Journal of Cultural Heritage*, Vol. 12, No. 3, 2011, pp. 263-269.

[19] N. Q. Chien and S. K. Tan, "Google Earth as a Tool in 2-Dhydrodynamic Modeling," *Computers & Geosciences*, Vol. 37, No. 1, 2011, pp. 38-46.

11

Using GIS Data to Build Informed Virtual Geographic Environments (IVGE)

Mehdi Mekni

Department of Math Science and Technology, University of Minnesota, Crookston, USA

ABSTRACT

In this paper, we propose a novel approach to automatically building Informed Virtual Geographic Environments (IVGE) using data provided by Geographic Information Systems (GIS). The obtained IVGE provides 2D and 3D geographic information for visualization and simulation purposes. Conventional VGE approaches are generally built upon a grid-based representation, raising the well-known problems of the lack of accuracy of the localized data and the difficulty to merge data with multiple semantics. On the contrary, our approach uses a topological model and provides an exact representation of GIS data, allowing an accurate geometrical exploitation. Moreover, our model can merge semantic information, even if spatially overlapping. In addition, the proposed IVGE contains spatial information which can be enhanced thanks to a geometric abstraction method. We illustrate this model with an application which automatically extracts the required data from standard GIS files and allows a user to navigate and retrieve information from the computed IVGE.

Keywords: Geographic Information System (GIS); Informed Virtual Geographic Environments (IVGE); Multi-Agent Geo-Simulation (MAGS)

1. Introduction

Modern geography techniques play an irreplaceable role in exploring temporal-spatial patterns and dynamic processes, and understanding the relationships between geographic features, objects, and actors in the real world. Studying these relationships is fundamental and comprehensive geographical analysis often requires the integration of different disciplines at various scales. By using a variety of different techniques, it is possible to produce virtual reality representations, integrated models, simulations, forecasts and evaluations of the geographic environment. Therefore, the Virtual Geographic Environment (VGE) can be a useful tool for understanding the evolutionary process, temporal-spatial patterns, driving mechanisms, and the direction of succession in the real environment. Current VGE techniques, including three-dimensional (3D) techniques, have extended the potential applications of geographic modeling. The ability of VGE to depict past, present, and future geographic environments has been of particular importance [1].

Traditional methods for extracting and expressing spatial information can be modified to better suit development in the field of geographic modeling [2]. Through using geographic analytical modeling and visualization techniques, VGE can be used to perform geographic analysis, simulate geographic phenomena, represent and predict changes in the geographic environment, and evaluate the influence of human activities on the environment [3]. By using VGE to share ideas, people can cooperate and coordinate their work on geographic objects and phenomena, resulting in more advanced methods of designing and transforming the world (**Figure 1**). VGE will become increasingly important in geography in the future [2].

In this paper, we propose a novel approach to modeling VGE which deals with all these constraints. Our approach provides an exact representation of the geographic environment using vector data and including elevation. This representation is organized as a topological graph, enhanced with data integrating both quantitative (geometry) and qualitative information (zones such as roads and buildings). Moreover, our approach is directed to efficient spatial reasoning like path planning for crowd simulation (**Figure 2**). Hence, we also address the performance issue when exploiting such environments.

Figure 1. 3D city models of East Berlin.

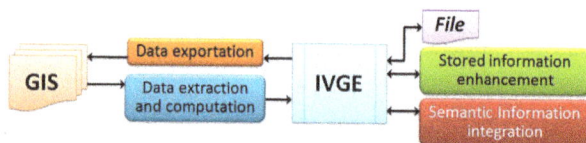

Figure 2. The proposed architecture to generate an IVGE.

We call our model an Informed Virtual Geographic Environment (IVGE) because it is close to standard VGE applications while addressing additional information management issues. IVGE data are directly extracted and computed from standard GIS vector files. It is a one-time automated process based on a Constrained Delaunay Triangulation (CDT) technique [4]. The produced IVGE contains spatial information which can be enhanced thanks to a geometric abstraction method. Finally, the IVGE can either be saved for future use or be exported to a GIS vector file.

The remainder of this chapter finishes with an overview on spatial reasoning and GIS data followed by a discussion of related works on GIS data and the representation of virtual environments. In Section 3, we present our approach to creating an IVGE from GIS data. Section 4 depicts a way to enhance the IVGE information by automatically associating semantics with surfaces based on their boundaries. Next, Section 5 presents two additional information enhancements based on a geometrical abstraction: first, by filtering potential elevation errors related to the input data; second, by qualifying elevation with additional semantics which are associated with specific areas in the environment. Then, Section 6 presents some results obtained by applying our approach to an urban environment. Finally, we provide a summary of this paper and present future work.

2. Spatial Reasoning and GIS Data

Spatial reasoning is a research area with applications in several domains such as crowd simulation and robotics. Reasoning about space does not only require appropriate computation algorithms, but also an efficient description of the spatial environment. Indeed, such a description

must represent the geometrical information which corresponds to geographic features. Moreover, this representation must qualify space by associating semantics to geographic features in order to allow spatial reasoning. A Geographic Information System (GIS) is a system of hardware, software and procedures used to facilitate the management, manipulation, analysis, modeling, representation and display of georeferenced data to solve complex problems regarding planning and management of resources [5]. GISs have emerged in the last decade as an essential tool for urban and resource planning and management [6]. Their capacity to store, retrieve, analyze, model and map large areas with huge volumes of spatial data has led to an extraordinary proliferation of applications [5]. GISs offer two data models to describe a geographic environment: *raster* and *vector* data representations [7].

2.1. Raster Model

The *raster* representation of GIS data is a method for the storage, processing and display of spatial data [8]. Each area is divided into rows and columns, which form a regular grid structure [7]. Each cell must be rectangular in shape, but not necessarily square [6]. Each cell within this matrix contains location coordinates as well as an attribute value [8]. The spatial location of each cell is implicitly contained within the ordering of the matrix [9]. Areas containing the same attribute value are recognized as such, however, raster structures cannot identify the boundaries of such areas as polygons [6]. Raster data are an approximation of the real world where spatial data are expressed as a matrix of cells, with spatial position implicit in the ordering of the pixels [9]. With the raster data model, spatial data are not continuous but divided into discrete units [8]. This makes raster data particularly suitable for certain types of spatial operation, for example overlays or area calculations. Raster structures may lead to increased storage in certain situations, since they store each cell in the matrix regardless of whether it is a feature or simply *empty* space.

2.2. Vector Model

The vector format is defined by the vector representation of its geographic data [10]. Vector storage implies the use of vectors (directional lines) to represent a geographic feature [11]. Vector data is characterized by the use of sequential points or vertices to define a linear segment. Each vertex consists of X and Y coordinates [8]. Vector lines are often referred to as arcs and consist of a string of vertices terminated by a node [6]. A node is defined as a vertex that starts or ends an arc segment. Point features are defined by one coordinate pair, a vertex. Polygonal features are defined by a set of closed coordinate pairs [7]. In vector representation, the storage of the vertices for each feature is important, as well as the connectivity between features, e.g. the sharing of common vertices where features connect [9].

GIS offer two ways to describe an environment: *grid* and *vector* representations. Grids have several drawbacks such as the difficulty to balance accuracy and memory use, making them unusable for the precise exploitation of large scale environments. Vector layers are scalable, but it is difficult to manipulate the stored data in order to simply retrieve pieces of information or to merge data with different semantics. Moreover, while GIS data are stored in a quantitative way which suits to exact calculations, spatial reasoning often needs to manipulate qualitative information. For example, when considering a slope in a landscape, it is simpler and faster to qualify it by an attribute that takes the values light or steep rather than to directly use its angle value with respect to the horizontal plane. This process of qualification, which associates semantics with quantitative intervals of values, greatly simplifies the description and validation of reasoning mechanisms. Moreover, the slope example illustrates another requirement of spatial reasoning which is poorly addressed in the literature: providing a way to handle terrain's elevations. Indeed, a real environment is rarely flat and ignoring this information would greatly decrease the quality of spatial reasoning.

3. Related Work

GIS data are usually available in either raster or vector formats. The raster format subdivides the space into regular square cells, called boxes, associated with space related attributes. This approach generally presents average quantitative data whose precision depends on the scale of the representation. In contrast, the vector format exactly describes geographic information without constraining geometric shapes, and generally associates one qualitative data with each shape. Such data are usually exploited by a VGE in two ways [3]: the grid method and the exact geometric subdivision. The grid method is the direct mapping of the raster format [12], but it can also be applied to the vector format (**Figure 3(a)**). This discrete representation can be used to merge multiple semantic data [13], the locations where to store these data being predefined by the grid cells. The main drawback of the grid method is related to a loss in location accuracy [1], making it difficult to accurately position any information which is not aligned with the subdivision. Another drawback arises when trying to precisely represent large environments using a grid: the number of cells tends to increase dramatically, which makes the environment exploitation very costly [14]. The grid-based method is mainly used for animation purposes [15] and large crowd simulations [16] because of the fast data

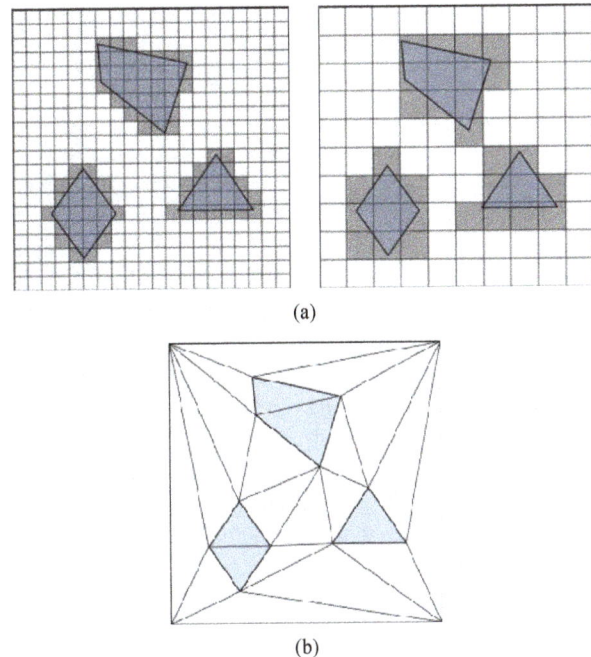

(a)

(b)

Figure 3. Two common cell decomposition techniques: (a) approximate decomposition by grids considering two resolutions. White boxes are free, grey are obstacles; (b) exact decomposition using CDT.

access it provides.

The second method, called exact geometric subdivision, consists in subdividing the environment in convex cells using the vector format as an input. The convex cells can be generated by several algorithms, among which the most popular is the Constrained Delaunay Triangulation (CDT) [17]. This produces triangles while keeping the original geometry segments which are named constraints (**Figure 3(b)**). The first advantage of the exact subdivision method is to preserve the input geometry, allowing accurately manipulating and visualizing the environment at different scales. Another advantage of this approach is that the number of produced cells only depends on the complexity of the input shapes, but not on the environment's size and scale as it is the case with the grid method [18]. The main drawback of this approach is the difficulty to merge multiple semantic data for overlapping shapes. Moreover, this method is generally used to represent planar environments because the CDT can only handle 2D geometries [19]. This method tends to be used for crowd microscopic simulations [2] where motion accuracy is essential.

Both VGE representations, approximate and exact, can be enhanced by an abstraction process. The first goal of an abstraction is to improve the performance of the algorithms based on the environment description, such as path planning, by reducing the number of elements used to describe the environment. The usual abstraction model for grids is mainly geometric: the quadtree groups four boxes of the same kind to create a higher level cell [20]. When considering the exact decomposition, an abstraction is generally based on topological properties rather than on purely geometric ones. Indeed, the exact cell subdivision generates connected triangles which can be manipulated as the nodes of a topological graph. This graph can then be abstracted by grouping the nodes, producing a new graph with fewer nodes [21]. For example, **Figure 4** shows an abstraction which is only based on the nodes' number of connections c: isolated (c = 0), dead-end (c = 1), corridor (c = 2), and crossroad (c = 3). A topological graph can be used for spatial reasoning, like path planning, thanks to traversal algorithms. These algorithms benefit of the abstraction by traversing first the more abstracted graph, and then by refining the computation in the sub-graphs until reaching the graph of the spatial subdivision [22]. This exploitation points out a new need for an abstracted graph which is less addressed in literature: it must contain the minimal information necessary to make a decision. For example, if the width of a path is relevant for a path planning algorithm, this information must be accessible in all the abstracted graphs; if not, the evaluation would be greatly distorted compared to a un-abstracted graph.

Two kinds of information can be stored in the description of an IVGE. Quantitative data are stored as numerical values which are generally used to depict geometric properties (path's width of 2 meters) or statistical values (density of 2.5 persons per square meter). Qualitative data are introduced as identifiers which can be a reference to an external database or a word with an arbitrary semantic, called a label.

Such labels can be used to qualify an area (a road or a building) or to interpret a quantitative value (a narrow passage or a crowded place). An advantage of interpreting quantitative data is to reduce a potentially infinite set of inputs to a discrete set of values, which is particularly useful to condense information in successive abstraction levels to be used for reasoning purposes.

The approach we propose is based on an exact representation whose precision allows realistic applications like crowds' micro-simulations. We will briefly describe how to obtain such a representation since it has already been presented in [16]. We will then show how the obtained topological graph can be improved in two ways. First, by propagating input qualitative information from the arcs of the graph to the nodes, this allows for example to deduce the internal parts of the buildings or of the roads in addition to their outline. Second, we propose a novel approach of information extrapolation using a one-time spatial reasoning process based on a geometric abstraction. This second technique can be used to fix input elevation errors, as well as to create new qualitative data relative to elevation variations. These data are stored as

additional semantics bound to the graph nodes, which can subsequently be used for spatial reasoning.

4. Automated Generation of IVGE

We propose an automated approach to compute the data directly from vector format GIS data. This approach is based on three stages which are detailed in this section (**Figure 5**): *Input data selection*, *spatial decomposition*, and *maps unification*.

The first step of our approach, the inputting data selection is the only one requiring human intervention. The various vector data selected which will be used to build

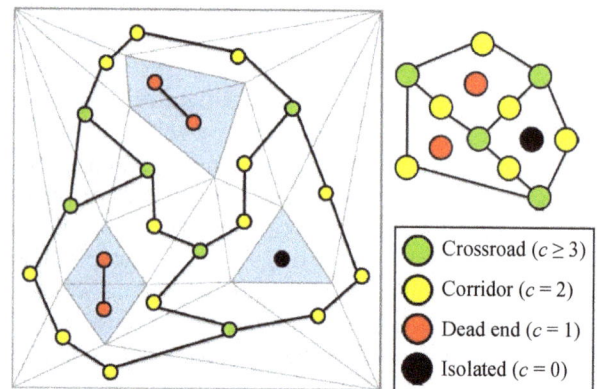

Figure 4. Topologic abstraction of virtual environments.

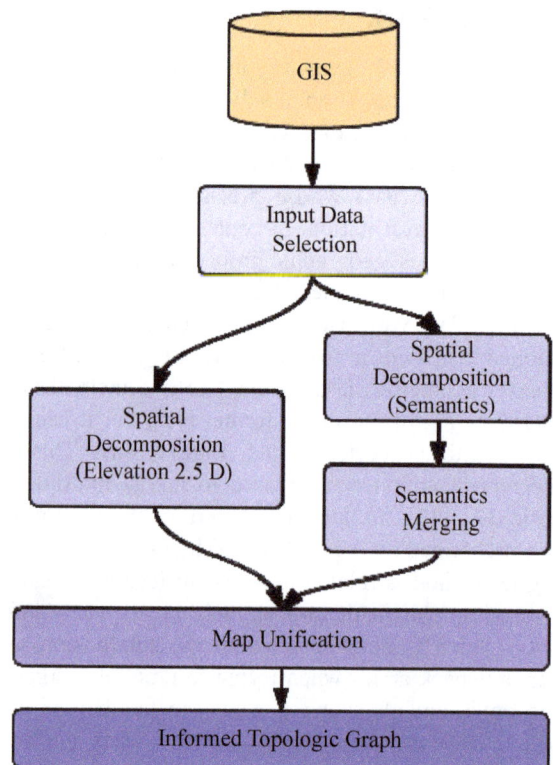

Figure 5. A flow chart illustrating the automated generation of IVGE.

the IVGE. The only restrictions concerning these data are that they need to respect the same scale and be equally geo-referenced. The input data can be organized in two categories. First, elevation layers contain geographical marks indicating absolute terrain elevations. Since we consider the creation of 2.5D, a given coordinate cannot have two different elevations, prohibiting the representation of tunnels for example. Moreover, several elevation layers can be specified, the model being able to merge them automatically. Second, semantic layers are used to qualify various features of the geographic space. As shown in **Figure 4**, each layer indicates the geographic boundaries of a set of features having identical semantics, such as roads and buildings. The boundaries of the features can overlap between two layers, our model is able to merge this information.

The second step of our method, spatial decomposition, consists of obtaining an exact spatial decomposition of the input data in cells. This process uses a Delaunay triangulation and is entirely automatic. It and can be divided into two parts in relation to the previous phase. First, an elevation map is computed, corresponding to the triangulation of the elevation layers. All the elevation points of the layers are injected in a 2D triangulation, the elevation is considered as an additional datum. This process produces an environment subdivision composed of connected triangles. Such a subdivision provides information about coplanar areas: the elevation of any point inside the environment can be deduced thanks to the elevation of the three vertices of the corresponding triangle. Second, a merged semantics map is computed, which corresponds to a Constrained Delaunay Triangulation (CDT) of the semantic layers. Indeed, each segment of a semantic layer is injected as a constraint which keeps track of the original semantic data thanks to an additional datum. Consequently, the resulting map is a merging all input semantics: each constraint represents as many semantics as the number of input layers containing it.

In the third step, the two maps previously obtained are merged. This phase corresponds to the mapping of the 2D merged semantic map on the 2.5D elevation map in order to obtain the final 2.5D elevated merged semantics map. First, a preprocessing is carried out on the merged semantics map in order to preserve the elevation precision inside the unified map. Indeed, all the points of the elevation map are injected in the merged semantics triangulation, and create new triangles. Then, a second process elevates the merged semantics map. The elevation of each merged semantics point P is computed by retrieving the triangle T of the elevation map whose 2D projection contains P. Once T is obtained, the elevation is simply computed by projecting P on the plane defined by T using the Z axis. When P is outside the convex hull of

the elevation map, no triangle can be found and the elevation cannot be directly deduced. In this case, we use the average height of the points of the convex hull which are visible from P.

5. Semantic Information Management

The obtained results of the importation technique emphasize qualified areas by defining the semantics of their boundaries. But these informed boundaries are difficult to exploit when dealing with the semantics associated with a position. For example, it is difficult to check if a position is inside a building. This is why we propose to enhance the information provided by spreading the boundaries' semantics to the cells. Three related processes are necessary, and explained in the following subsections: graph analysis, potential conflicts resolution, and semantics assignment.

5.1. Graph Analysis

The graph analysis is a traversal algorithm which explores the environmental graph while qualifying the cells towards a given semantic. This algorithm is applied to the entire graph one time for each semantic to propagate. While exploring the graph, the algorithm collects three kinds of cells which are stored in three container structures for future use: *Inside*, *Outside* and *Conflict*. Cells are within an area delimited by borders associated to the propagated semantic. *Outside* cells are outside any area defining the propagated semantic. *Conflict* (C) cells are both qualified inside and outside by the algorithm. Non-Conflict (NC) cells are either qualified inside or outside by the algorithm. Three parameters influence the traversal: 1) the semantic *sem* to propagate. 2) a set of *starting cells*, indicating where to start the exploration of the graph; a set is provided instead of a single cell in order to be able to manage disconnected graphs. 3) a boolean value *startin* indicating whether the semantic must be assigned to the starting cells or not. The recursive algorithm is detailed in **Figure 6**.

5.2. Resolution of Conflicts

After each graph traversal, we must deal with the cells that are potentially in conflict. Indeed, these cells must be assigned to either the *Inside* or the *Outside* container in order for the system to continue with the next phase. Cells are in conflict when the shapes of two input features with the same semantic share a segment. Two alternative methods are proposed: 1) a fast assignment where the conflicting cells are arbitrarily transferred to one of the target containers, and 2) a deductive assignment where an algorithm selects the best option based on geometric considerations. The arbitrary assignment is used when the internal details of a shape are not relevant for the target application. The deductive assignment is used when the internal details of a shape are relevant. Both methods are carried out in two steps: 1) a local conflicting graph extraction which is the same for both methods, and 2) a decision step which is specific to each method. The local conflicting graph extraction collects all the cells surrounding a conflicting cell, but only if they are reachable through a border which is not marked by the propagated semantic. Each orange zone in **Figure 6** shows an extracted local conflicting graph. Every time a cell is discovered, it is transferred from the global container *Conflict* to a local container. Then, the algorithm recursively explores the neighbors which are reachable through a border which is not marked by the propagated semantic, transferring them to the local container. At the end, the algorithm obtains a set of local conflict containers, corresponding to the amount of local graphs considered in conflict.

The decision part of the arbitrary assignment only consists in transferring the local conflicting cells to one of the Inside or the *Outside* containers. The decision part of the deducing algorithm is based on geometric considerations. If the local conflicting zone is mainly surrounded by *Outside* cells, then the conflict is resolved as *Inside*, and vice versa. The conflict management algorithm is described in **Figure 7**.

5.3. Semantics Assignment

The last step of the semantic propagation consists in assigning the final semantics to the cells. The process is very simple: each propagated semantic is assigned to all the corresponding *Inside* cells. One can notice that some cells may have multiple semantics when they are present in more than one *Inside* container. Additionally, it is possible to keep track of the Outside cells by assigning them

---

Algorithm 2 Graph Analysis

PARAMETERS: *crt_sem* the propagated semantic label; *start_cells* the starting cells; *start_inside* defining if starting cells are inside or outside.

REQUIRE: *inside_cells* the container of inside cells; *outside_cells* the container of outside cells.

```
1: FOR ALL cell ∈ start_cells DO
2:     IF start_inside THEN
3:         TRAVERSE (cell, inside_cells, outside_cells)
4:     ELSE
5:         TRAVERSE (cell, outside_cells, inside_cells)
6:     END IF
7: END FOR
```

Figure 6. Graph analysis algorithm.

Algorithm 3 Traverse

PARAMETERS: *crt_cell* the current cell; *crt_cont* the current cell container; *oth_cont* the other container.

REQUIRE: *crt_sem* the propagated semantic label; *conflict_cells* the container of conflicting cells.

```
 1: IF   crt_cell ∉ crt_cont ∧ crt_cell ∉ conflict_cells   THEN
 2:     IF   crt_cell ∈ oth_cont   THEN
 3:         oth_cont ← oth_cont \ crt_cell
 4:         conflict_cells ← conflict_cells ∪ {crt_cell}
 5:     ELSE
 6:         crt_cont ← crt_cont ∪ {crt_cell}
 7:         FOR i = 0 to NEIGHBOURS NUMBER (crt_cell) DO
 8:             nxt_border ← BORDER (crt_cell, i)
 9:             nxt_cell ← NEIGHBOURS (crt_cell, i)
10:             IF   crt_sem ∈ SEMANTICS OF (nxt_border) THEN
11:                 TRAVERSE (nxt_cell, oth_cont, crt_cont)
12:             ELSE
13:                 TRAVERSE (nxt_cell, crt_cont, oth_cont)
14:             END IF
15:         END FOR
16:     END IF
17: END IF
```

Figure 7. Semantic propagation and conflict management algorithm.

a negative semantic, as for example in order to know what is road in the environment and what is not.

Finally, an optional process can be performed to remove the borders' semantics of some detected conflicting cells. Indeed, such borders may distort some spatial reasoning algorithms. For example, when considering road borders as obstacles to plan a path, a simulated vehicle would not be able to go through some passageways. After resolution, the semantic of the problematic borders is removed, making them crossable. These problematic borders are the ones which are marked with a propagated semantic and which connect two *Inside* cells. One can note that only the cells previously detected as conflicting need to be tested.

6. Geometric Abstraction of IVGE

Spatial decomposition subdivides the environment into convex cells. Such cells encapsulate various quantitative geometric data which are suitable for accurate computations. Since geographic environments are seldom flat, it is important to consider the terrain's elevation, which is quantitative geometric data. Moreover, while elevation data are stored in a quantitative way which suits exact calculations, spatial reasoning often needs to manipulate qualitative information. Indeed, when considering a slope, it is obviously simpler and faster to qualify it using an attribute such as light and steep rather than using numerical values. However, when dealing with large scale geographic environments, handling the terrain's elevation, including its light variations, may be a complex task. To this end, we propose an abstraction process that uses geometric data to extract the average terrain's elevation information from spatial areas. The objectives of this Geometric Abstraction are threefold [23]. First, it aims to reduce the amount of data used to describe the environment. Second, it helps detect anomalies, deviations, and aberrations in elevation data. Third, the geometric abstraction enhances the environmental description by integrating qualitative information characterizing the terrain's elevations. In this section, we first present the algorithm which computes the geometric abstraction. Then, we describe two processes which use the geometric abstraction: *Filtering elevation anomalies* and *Extracting elevation semantics*.

6.1. Geometric Abstraction Algorithm

The geometric abstraction process gathers cells in groups according to a geometric criterion: we chose the coplanarity of connex cells in order to obtain uniform elevation areas. The algorithm takes advantage of the structure obtained thanks to the IVGE generation process.

The aim of this algorithm is to group cells which verify a geometric criterion in order to build groups of cells.

A cell corresponds to a node in the topological graph. A node represents a triangle generated by the CDT spatial decomposition technique. A cell is characterized by its boundaries, neighboring cells, and surface as well as its normal vector which is a vector perpendicular to its plan **Figure 8(a)**.

A group is a container of adjacent cells. The grouping strategy of this algorithm is based on a coplanarity criterion which is assessed by computing the difference between the normal vectors of two neighboring cells or groups of cells (**Figure 8(a)**). Since a group is basically composed of adjacent cells it is obvious to characterize a group by its boundaries, its neighboring groups, its surface, as well as its normal vector. However, the normal vector of a group must rely on an interpretation of the normal vectors of its composing cells. In order to compute the normal vector of a group, we adopt the area-weight normal vector [24] which takes into account the unit normal vectors of its composing cells as well as their respective surfaces. The area-weight normal vector of a group $\vec{NG}$ is computed using:

$$\vec{NG} = \sum_{c \in G} \left(S_c \cdot \vec{N_c} \right) \Big/ \sum_{c \in G} S_c \qquad (1)$$

where S_c denotes the surface of a cell c and N_c its unit normal vector. Hence, thanks to the area-weight normal vector we are able to compute a normal vector for a group based on the characteristics of its composing cells. The geometric abstraction algorithm uses two input parameters: 1) a set of starting cells which act as access points to the graph structure, and 2) a gradient parameter which corresponds to the maximal allowed difference between cells' inclinations. Indeed, two adjacent cells are considered coplanar and grouped together when the angle between their normal vectors (α in **Figure 8(b)**) is lesser than gradient.

The analysis of the resulting groups helps to identify anomalies in elevation data. Such anomalies need to be fixed in order to build a realistic virtual geographic environment. Furthermore, the average terrain's elevation which characterizes each group is quantitative data described using area-weighted normal vectors. Such quantitative data are too precise to be used by qualitative spatial reasoning. For example, when considering a slope in a landscape, it is simpler and faster to qualify it using a simple attribute that takes the values light and steep rather than to express it using the angle value with respect to the horizontal plane. Hence, a qualification process which aims to associate semantics with quantitative intervals of values characterizing a group's terrain inclinations would greatly simplify spatial reasoning mechanisms. In addition, in order to fix anomalies in elevation data and to qualify the groups' terrain inclinations we propose to apply a technique to improve the geometric

(a)

(b)

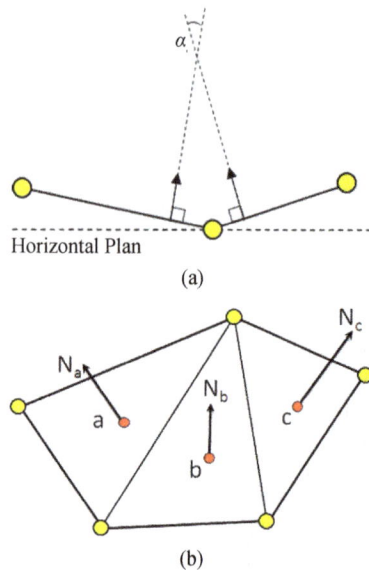

Figure 8. (a) Illustration of two coplanar cells; (b) Unit normal vectors.

(a)

(b)

Figure 9. Profile section of anomalous isolated groups (red color) adjusted to the average elevation of the surrounding ones (yellow color).

abstraction of the VGE. The geometric abstraction allows improving IVGE by filtering the elevation anomalies, qualifying the terrain's elevation using semantics and integrating such semantics in the description of the geographic environment.

A detailed description of the geometric abstraction algorithm as well as GIS data filtering and elevation qualification processes are provided in [23]. The geometric abstraction is built using a graph traversal algorithm. It groups cells based on their area-weighted normal vectors (**Figure 8(b)**). The objectives of the geometric abstraction are threefold. First, it qualifies the terrain's elevation of geographic environments to simplify spatial reasoning mechanisms. Second, it helps identifying and fixing elevation anomalies in initial GIS data. Third, it enriches the description of geographic environments by integrating elevation semantics.

6.2. Filtering Elevation Anomalies

The analysis of the geometric abstraction may reveal some isolated groups which are totally surrounded by a single coherent group. These groups are characterized by a large difference between their respective area-weighted normal vectors. Such isolated groups are often characterized by their small surfaces and can be considered anomalies, deviations, or aberrations in the initial elevation data. The geometric abstraction process helps to identify and allows automatically filtering such anomalies thanks to a two phase process.

First, isolated groups are identified (**Figure 9(a)**). The identification of isolated groups is based on two key parameters: 1) the ratio between the surface of surrounded and surrounding groups, and 2) the difference between

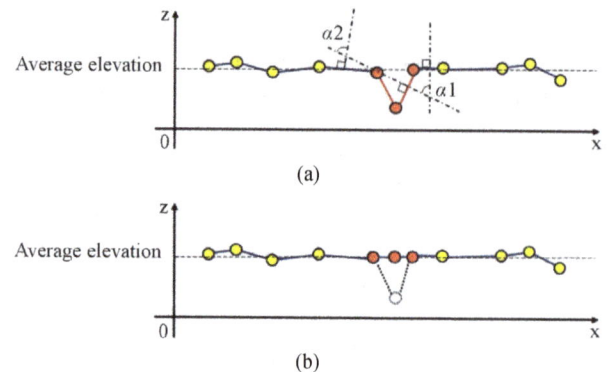

the area-weighted normal vectors of the surrounded and surrounding groups. Second, these isolated groups are elevated at the average level of elevation of the surrounding ones (**Figure 9(b)**). In fact, the lowest and the highest elevations of the isolated group are computed. Then, the elevations of all the vertices of isolated *grp* are updated using the average between the lowest and highest elevation. As a consequence, we obtain more coherent groups in which anomalies of elevation data are corrected.

6.3. Extracting Elevation Semantics

The geometric abstraction algorithm computes quantitative geometric data describing the terrain's inclination. Such data are stored as numerical values which allow to accurately characterize the terrains' elevations.

However, handling and exploiting quantitative data is a complex task as the volume of values may be too large and as a consequence difficult to transcribe and analyze. Therefore, we propose to interpret the quantitative data of the terrain's inclination by qualifying the areas' elevations. Semantic labels, called semantics elevation, are associated to quantitative intervals of values that represent the terrain's elevation. In order to obtain elevation semantics we propose a two steps process taking advantage of the geometric abstraction: 1) a discretization of the angle between the weighted normal vector Ng of a group g and the horizontal plane, 2) an assignment of semantic information to each discrete value which qualifies it. The discretization process can be done in two ways: a customized and automated approach.

The customized approach qualifies the terrain's elevation and requires that the user provides a complete specification of the discretization. Indeed, the user needs to specify a list of angle intervals as well as their associated semantic attributes. The algorithm iterates over the groups obtained by the geometric abstraction. For each group *grp*, it retrieves the terrain inclination value I.

Then, this process checks the interval bounds and determines in which one falls the inclination value I. Finally, the customized discretization extracts the semantic elevation from the selected elevation interval and assigns it to the group *grp*. For example, let us consider the following inclination interval and the associated semantic elevations: f([10; 20]; *gentle slope*); f([20; 25]; *steep slope*). Such a customized specification associates the semantic elevation "gentle slope" to inclination values included in the interval [10; 20] and the semantic elevation "steep slope" to inclination values included in the interval [20; 25]. The automated approach only relies on a list of semantic elevations representing the elevation qualifications. Let N be the number of elements of this list, and T be the total number of groups obtained by the geometric abstraction algorithm. First, the automated discretization orders groups based on their terrain inclination. Then, it iterates over these ordered groups and uniformly associates a new semantic elevation from the semantics set of each T = N processed groups. For example, let us consider the following semantic elevations: gentle; medium and steep. Besides, let us consider an ordered set of groups S which may be composed of six groups of cells.

Let us compare these two discretization approaches. On the one hand, the customized discretization process allows one to freely specify the qualification of the terrain's elevations. However, qualifications resulting from such a flexible approach deeply rely on the correctness of the interval bounds' values. Therefore, the customized discretization method requires having a good knowledge of the terrain characteristics in order to guarantee a valid specification of inclination intervals. On the other hand, the automated discretization process is also able to qualify groups' elevations without the need to specify elevation intervals' bounds. Such a qualification usually produces a visually uniformed semantic assignment. This method also guarantees that all the specified semantic attributes will be assigned to the groups without a prior knowledge of the environment characteristics.

6.4. Enhancing the Geometric Abstraction

The geometric abstraction algorithm produces groups that are built on the basis of their terrain's inclination characteristics. Thanks to the extraction of elevation semantics, the terrain's inclination is qualified using semantic attributes and associated with groups and with their cells. Because of the game of the classification intervals, adjacent groups with different area-weighted normal vectors may obtain the same elevation semantic. In order to improve the results provided by the geometric abstraction, we propose a process that merges adjacent groups which share the same semantic elevation. This process starts by iterating over groups. Then, every time it finds a set of groups sharing an identical semantic ele-

vation, it creates a new group. Next, cells composing the adjacent groups are registered as members of the new group. Finally, the area-weighted normal vector is computed for the new group. Hence, this process guarantees that every group is only surrounded by groups which have different semantic elevations (**Figure 10**). In this section, we proposed a geometric abstraction process as well as three heuristics which take advantage of this process. The geometric abstraction is built using a graph traversal algorithm. It groups cells based on their area-weighted normal vectors. The objectives of the geometric abstraction are threefold. First, it qualifies the terrain's elevation of geographic environments to simplify spatial reasoning mechanisms. Second, it helps identify and fix elevation anomalies in initial GIS data. Third, it enriches the description of geographic environments by integrating elevation semantics.

7. Results

The proposed environment extraction method is used to create an accurate IVGE and provides the advantages of standard GIS visualization techniques including the semantic merging of grids along with the accuracy of vector data layers. Thanks to the automatic extraction method that we propose, our system handles the IVGE construction directly from a specified set of vector format GIS files. The performance of the extraction process is very good and able to process an area such as the center part of Quebec-City, with one elevation map and five semantic layers, in less than five seconds on a standard computer (Intel Core 2 Duo processor 2.13 Ghz, 1 G RAM). The obtained unified map approximately contains 122,000 triangles covering an area of 30 km². Besides, the necessary time to obtain the triangle corresponding to a given coordinate is negligible (less than 10^{-4} seconds). Moreover, the geometric abstraction produces approximately 73,000 groups in 2.8 seconds. Finally, the custom and automated discretization processes are performed respectively in 1.8 and 1.2 seconds using eight semantic elevation labels. The IVGE application provides two visualization modes for the computed data. First, a 3D view (**Figure 10**) allows to freely navigating the virtual environment. We propose an optional mode for this view where the camera is constrained at a given height above the ground, allowing the elevation variations to be followed when navigating. Second, we propose an upper view with orthogonal projection to represent the GIS data as a standard map. In this view the user can scroll and zoom the map (**Figure 10**), and can accurately view any portion of the environment at any scale. Additionally, one can select a position in the environment in order to retrieve the corresponding data (2 in **Figure 10**), such as the underlying triangle geometry, the corresponding height, and the associated semantics, including semantic elevation.

(a)

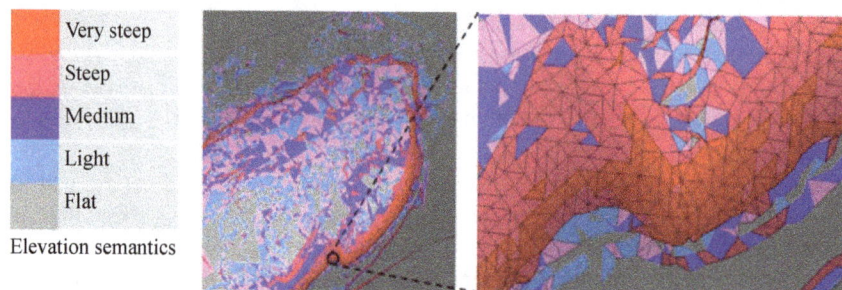

(b)

Figure 10. (a) 2D map visualization of the IVGE: (1) Unified map, (2) Accurate information about the selected position; (b) Results of the geometric abstraction process.

Our approach goes beyond classical grid-based visualization by combining the semantic information and the vector-based representation's accuracy. Indeed, our topological method combines the advantages of grids and vector layers, and avoids their respective drawbacks. Moreover, this data extraction method is fully automated, being able to directly process GIS vector data. In addition, the propagation of the borders' semantics enhances the information provided by the IVGE. As a result, both boundaries and areas are semantically qualified (roads and buildings). Besides, the geometric abstraction algorithm enriches the IVGE description by qualifying the terrain inclination characteristics of cells using semantic elevation. Finally, we have shown the suitability of this method for GIS visualization thanks to an IVGE application which allows two visualization modes: 3D for immersion purpose, and 2D to facilitate geographic data analysis.

8. Conclusion

All of the above-mentioned characteristics allow us to anticipate several applications of this work, mainly thanks to the exploitation of the topological graph. First, we are currently working on the application of the IVGE for crowd simulation. Indeed, this informed environment is particularly well adapted to such a domain, since it allows efficient path planning and navigation algorithms and provides useful spatial data which are needed for agents' behaviors. However, we plan to further improve the environment description by using topological graph abstractions. This will allow us to reduce the complexity of the graph exploration algorithms, and to deduce additional properties such as the reachable areas for a human being with respect to variations of land slopes. Second, we want to append new information to the environment description in order to represent moving and stationary situated elements. These elements could be humans, vehicles, or even street signs, depending on the objectives of the virtual reality application. Our topological graph which describes the geographic environment can be extended by integrating conceptual nodes representing such elements. Thanks to the geometrical accuracy of our approach, it will be relatively easy to add this information at any position.

9. Acknowledgements

This research was supported in part by the Research

Stimulus Dollars Award provided by the University of Minnesota Crookston. The author would like to thank the reviewers for their valuable comments.

REFERENCES

[1]   R. Andersen, J. L. Berrou and A. Gerodimos, "On Some Limitations of Grid-Based (CA) Pedestrian Simulation Models," In *VRlab, EPFL*, 2005.

[2]   M. Kallmann, H. Bieri and D. Thalmann, "Fully Dynamic Constrained Delaunay Triangulations," *Geometric Modelling for Scientific Visualization*, 2003.

[3]   F. Lamarche and S. Donikian, "Crowds of Virtual Humans: A New Approach for Real Time Navigation in Complex and Structured Environments," *Computer Graphics Forum, Eurographics'*04, 2004.

[4]   J. Gong and L. Hui, "Virtual Geographical Environments: Concept, Design, and Applications," 1999.

[5]   D. Demyen and M. Buro, "Efficient Triangulation-Based Pathfinding," 2006.

[6]   M. F. Goodchild, "GIS and Disasters: Planning for Catastrophe," *Computers, Environment and Urban Systems*, Vol. 30, No. 3, 2006, pp. 227-229.

[7]   N. Chrisman, "Exploring Geographical Information Systems," 2nd Edition, John Wiley and Sons Inc., 2001.

[8]   D. Arctur and M. Zeiler, "Designing Geodatabases: Case Studies in GIS Data Modeling," ESRI Press, 2004.

[9]   P. Longley, M. Goodchild, M. David and D. Rhind, "Geographic Information Systems and Science," Wiley and Son. Inc., ESRI Press, 2002.

[10] S. Fortheringham and P. Rogerson, "Spatial Anaysis and GIS's," Taylor & Francis, LTD, 2002.

[11] L. Zhou, G. Lu, Y. Sheng, H. Xu and H. Wang, "A 3D GIS Spatial Data Model Based On Cell Complex," *The International Archives of the Photogrammetry, Remote Sensing and Spatial Information Sciences*, Vol. 37, 2008, pp. 905-908.

[12] M. Mekni, N. Sahli, B. Moulin and H. Haddad, "Using Multi-Agent Geo-Simulation Techniques for the Detection of Risky Areas for Trains," *Simulation*, Vol. 86, No. 12, 2010, pp. 763-775.

[13] J. Zhu, J. Gong, H. Lin, W. Li, J. Zhang and X. Wu, "Spatial Analysis Services in Virtual Geographic Environment Based on Grid Technologies," *MIPPR* 2005: *Geospatial Information, Data Mining, and Applications*, Vol. 6045, No. 1, 2005, pp. 604-615.

[14] M. Mekni and B. Moulin, "Holonic Modelling of Large Scale Geographic Environments," 2007.

[15] F. Tecchia, C. Loscos and Y. Chrysanthou, "Visualizing Crowds in Real-Time," *Computer Graphics Forum*, Vol. 21, No. 4, 2002, pp. 753-765.

[16] M. Mekni, "Crowd Simulation Using Informed Virtual Geographic Environments (IVGE)," 2012.

[17] A. Botea, M. Muller and J. Schaeffer, "Near Optimal Hierarchical Path-Finding," *Journal of Game Development*, Vol. 1, 2004, pp. 7-28.

[18] L. G. Da and S. R. Musse, "Real-Time Generation of Populated Virtual Cities," 2006.

[19] W. Shao and D. Terzopoulos, "Environmental Modeling for Autonomous Virtual Pedestrians," *Digital Human Modeling for Design and Engineering Symposium*, 2005.

[20] S. Paris, S. Donikian and N. Bonvalet, "Environmental Abstraction and Path Planning Techniques for Realistic Crowd Simulation," *Computer Animation and Virtual Worlds*, Vol. 17, No. 3-4, 2006, pp. 325-335.

[21] D. Brown, "Agent-Based Models," *The Earths Changing Land: An Encyclopaedia of Land-Use and Land-Cover Change*, Greenwood Publishing Group, Westport, 2006, pp. 7-13.

[22] M. Mekni, "Abstraction of Informed Virtual Geographic Environments," *Geo-Spatial Information Science*, Vol. 15, No. 1, 2012, pp. 27-36.

[23] S.-G. Chen and J.-Y. Wu, "A Geometric Interpretation of Weighted Normal Vectors and Its Improvements," 2005.

[24] GDAL: Geospatial Data Abstraction Library.

12

Location of Large-Scale Concentrating Solar Power Plants in Northeast Brazil

Verônica Wilma Bezerra Azevedo, Chigueru Tiba
Department of Nuclear Energy, Federal University of Pernambuco, Recife, Brazil

ABSTRACT

Heliothermic electricity generation is gaining popularity in several countries worldwide. In Brazil, this form of energy generation has not yet been explored for large scale projects. However, the country possesses extensive areas with normal and high-intensity direct irradiation and low seasonality factors, particularly in the semi-arid region of the Brazilian Northeast. The region also presents other important features for setting up such plants: proximity to transmission lines, sufficient flatness, non-endangered vegetation, a suitable land use profile low maximum wind speeds, low population density, and more recently, an increase in the demand for local electric energy due to economic growth above the Brazilian average. A Geographic Information System includes a set of specialised resources that allow us to manipulate spatial data, providing quickness and efficiency in the identification of appropriate places for installing solar power plants while also preparing us for future scenarios, with regards to their impacts, costs and benefits. This article presents a study of the optimal location for thermoelectric power plants in the semi-arid region of the Brazilian Northeast on the scale of 1:10,000,000. All provinces with good potential for the implementation of large-scale concentrating solar power plants are identified. Considering that the installed capacity for parabolic cylindrical concentrators in terrains with a steepness of less than 1% is 43.26 MW/km^2 for systems without storage and 30.82 MW/km^2 for systems with 6 hours of storage, the potential of the southeast region of Piauí alone is huge. Even with the lack of information about the urban areas, terrain continuity, and other variables, utilising only 10% of the identified potential area, or 879.7 km^2, would result in an installed capacity of 38.1 - 27.1 GW. This value corresponds to more than 1/3 of the potency of the current Brazilian electric system. If the same calculation is made for the semi-arid region of the Brazilian Northeast, its capacity will be greater than 1000 GW.

Keywords: Solar Energy; Solar Thermoelectric Power Plant; Geographical Information System

1. Introduction

The Solar Electric Generation Systems (SEGS) power plants became the most mature example of thermoelectric solar technology with the usage of parabolic cylindrical concentrators. Nine SEGS power plants were built in three different sites in the Mojave Desert, in California (USA), between 1984 and 1991.

The SEGS are still in commercial operation today, with 354 MW of installed capacity, demonstrating their technical and commercial reliability. Following a halt in the production of this type of system that lasted approximately 15 years, there was a vigorous reintroduction of this technology in the beginning of the last decade. In 2010, the total amount of thermoelectric solar power plants installed worldwide was 941 MW, with a predomi-

nance of parabolic cylindrical concentration technology (95%). The distribution per country was 46.0% for the USA, and 51.3% for Spain. Estimations of thermoelectric solar power plants under construction or publically announced exhibit discrepancies due to the usage of different criteria, such as different periods of consideration and a lack of updates on project modifications in terms of potency. In 2011, 1934 MW were under construction, and 17,539 MW were announced: 9659 MW in the USA, 1080 MW in Spain, and 6800 MW in the rest of the world [1]. Considering these numbers with caution, it is perceivable that the trend of growth, the experience accumulation rate and scale savings in the next years will be spectacular. Thus, according to the learning curve elaborated by [2], it is possible to predict that, in the next five years, the cost of thermoelectric solar energy will be in parity with conventional networks.

The Northeast Region of Brazil has already used nearly all of the large hydraulic resources available for generating electricity, leaving the options of importing energy, exploiting other local renewable sources (wind and solar) or relying on conventional thermoelectric generation with oil fuel or coal or nuclear. The main obstacles to energy importation and conventional thermoelectric and nuclear generation are as follows: 1) the costs of transmission of the Amazon's hydroelectricity, which are much larger than the costs of generation, not to mention the environmental problems that aggregate uncertainties for such ventures; 2) fossil fuels, which exhibit rising prices, environmental problems and "invisible" subsidies (externalities) that are progressively put into question by society; and 3) the risk of catastrophic accidents with regards to nuclear energy.

With the perspective of commercial and technologic maturity for large scale thermoelectric solar energy on the 10-year horizon (2020), it would be appropriate for the Brazilian electric system to follow the evolution of this technology, carry out R&D and assess in detail the potential of the resources available in the Brazilian Northeast. In addition to the renewable aspect and low environmental impact, the implantation of thermoelectric solar power plants is regionally important for the following reasons: 1) They improve the generation "mix" of the Brazilian electric sector and therefore improve the safety of the electric system; 2) Solar resources are more intense in the dry period of the region and thus have a complementary character to the regional hydroelectric system; 3) Solar resources are in phase concurrently with the time period of greatest consumption (summer); 4) Thermoelectric solar energy characteristically has a distributed generation, providing yet another safety factor for the electric system; 5) Thermoelectric solar power plants should be installed in semi-arid regions with low population densities and lands that are not competing for other nobler usages (agriculture, for instance); Finally, 6) such plants would lead to the creation of jobs and income.

The use of GIS in renewable energy began in the 1990s and has gone through considerable development. As a result, various decision support tools have been developed [3]. The pioneering study regarding GIS usage for Concentrated Solar Power (CSP) is that of [4], who analysed Northern Africa and provided a rank of sites with respect to the potentiality and cost of solar thermal electricity for a particular power plant configuration. Recently, this type of study has become widespread: [5] for the Southwest USA, [6] for South Africa, [7] for Oman, [8] for Burkina Faso, [9] for Australia, and [10] for India. In Brazil, this type of study has not yet been conducted; however, country possesses extensive areas with normal and high-intensity direct irradiation and low seasonality

factors, particularly in the semi-arid region of the Brazilian Northeast. Hence, in this study, identification and mapping of the most promising sites in the Northeast of Brazil were conducted so that the country can quickly initiate the process of implanting large scale CSP solar technology.

2. Location of the Solar Power Plants

The renewable energy is a good candidate for rapid and widespread transformation, because renewable power technologies (especially solar) are well-developed and commercially available, and electricity is a versatile energy carrier capable of displacing fossil fuel in other sectors.

Making the best use of renewable resources requires that this "spatiotemporal" complexity be explicitly incorporated into modeling and analysis of alternative energy futures. Toward that end, this paper provides approach for optimized location of the solar thermal power plants. Solar thermal power plants use concentrated solar in order to generate high pressure steam for electricity generation in conventional steam turbines. Thermal storage allows excess heat generated during the day to be stored and utilized later.

Installations are best placed on flat, open terrain free from obstructions, settlements, or dangerous land features. Additionally, solar power plants have a relatively big area demand. The Specific area demand for a parabolic trough power station is ~1 km^2 per 50 MW of installed electric capacity. Typical sites are hot, dry regions like deserts or semi-deserts. Surface waters, forests, settlements, arable and cultivated lands are considered unsuitable for the construction of such plants. Sand deserts are not considered to be criteria for exclusion, but may elevate the cost [4].

The quantity and type of solar radiation are key determinants of overall plant performance. PV technology can utilize all radiation falling on the cell: both the direct sunlight component and diffuse radiation scattered by clouds and aerosols (together, global horizontal irradiance or GHI). CSP utilizes only the direct beam perpendicular to the receiver (direct normal irradiance or DNI).

For CSP configurations, minimizing the cost of production requires optimizing the size of the mirror array ("solar multiple"). For CSP without storage, the optimal solar multiple is ~1.4 for representative sites; In the case of thermal storage, true maximization of profits depends on daily and seasonal variation in the price of electricity since plant operators can somewhat control the timing of sales to the grid. In the absence of extensive electric grids with excess capacity, solar power must be transmitted from a substation near the site of generation to a substation connected to the desired distribution grid [11].

The general procedures used for selecting a site for installing a CSP follow the sequence below:
- Identification of the most promising places;
- Visiting the identified places and their priority;
- Conceptual project;
- Production estimation and sensibility study;
- Final selection.

Identification of the best places for installing large-scale solar electric power plants, whether thermoelectric or photovoltaic requires following the procedures: the first step of the procedure is performed basically with the existing documental information, multiple criteria analysis and a GIS. GIS is a valuable tool for evaluating and developing the usage of renewable energy resources in large regions because it is a tool specially capable of analysing the spatial variability of resources and resolving management and planning problems regarding the installation programs of decentralised systems, which are characterised by great spatial dispersion. The result is a list of potentially attractive sites. The next step is an in-situ visit of the pre-identified places and a reduction and ranking of the previous list.

The next step is setting up solarimetric stations to measure direct solar irradiation, which is performed because solar radiation is the most relevant variable in determining the feasibility of future ventures. If the region is relatively homogeneous from a phytogeographic perspective, each station will be able to cover a radius of 150 km.

3. Methodology

3.1. Equipment and Resources Used

The equipment and resources used for realising this research were 1) a Pentium 4 - 2.80 GHz, 1.0 GB RAM computer; 2) Arc GIS Software version 9× ESRI; 3) vector files of the shape file (shp) type referring to the IL; and 4) images of the SRTM (Shuttle Radar Topography Mission).

3.2. The Northeast Semi-Arid Region

According to [12], the Brazilian semi-arid region is located in parts of the states of the Northeast region and in Minas Gerais and Espírito Santo, occupying an area of 974,752 km^2, where approximately 17,000,000 people live. The semi-aridregion in the Northeast, **Figure 1**, entails parts of the states of Pernambuco, Paraíba, Alagoas, Sergipe, Bahia, Piauí, Ceará and Rio Grande do Norte, occupying 86.48% of the area of all semi-arid regions in Brazil. It is an area where the pluvial regime is irregular, with an annual average that varies between 400 and 800 mm. The predominant climate is hot and dry, with an average annual temperature of 27°C and low thermal amplitudes (of approximately 3°C - 5°C). Its vegetation is deciduous tropical forest (*caatinga*), which develops in a complex mosaic of soils resistant to long drought periods. According to [13], the annual mean daily value of normal direct solar irradiation is approximately 6.0 kWh/m^2.

3.3. Definition of the Spatial Database

A location study of thermoelectric solar power plants for the Northeast semi-arid region required knowledge of the following ILs: soil usage and occupation, solar resources, water supply, terrain topography, connection with the electric network, fuel availability for backup, and access. The ILs indicated for evaluating the viability of implanting solar power plants in the Northeast semi-arid region were selected based on their relevance to the process of defining suitable areas for implantation.

To define the Spatial Database (SD), the ILs were duly evaluated and conformed to avoid cartographic inconsistencies resulting from the conversion between CAD and GIS platforms and to make the ILs compatible in different scales or using documents from distinct Geodesic Reference Systems (GRS). In Brazil, the SIRGAS2000 GRS constitutes the Official System for the Brazilian Geodesic System (SGB) and for the National Cartographic System (SCN). Thus, the ILswere geo-referenced to this GRS. After geo-referencing, a work scale that was compatible for the data presentations as well as their projection system was determined.

3.3.1. Soil Usage and Occupation

In setting up a solar power plant, permissions and restrictions regarding soil usage must be considered. Permissions refer to matters of contractual relations established between land owners and project developers, whereas restrictions refer to the destined utility of the soil.

Regarding restrictions, it is possible to verify that areas protected by legislation (environmental conservation units —integral protection and sustainable usage, indigenous territories, Afro-Brazilian population settlement (quilombola) territory and Atlantic Forest reserves, for instance), urban and urban expansion areas and areas of great agricultural potential are not considered suitable for implanting solar power plant.

In this research, due to the work scale of 1:10,000,000, only restrictions regarding soil usage and occupation were considered. However, it should be highlighted that on larger scales (for example, 1:100,000 or larger) in which terrains can be visualised, questions related to permissions must also be approached. Permissions referring to soil usage are contractual relations established between land owners and project developers.

Urban and urban expansion areas were also not considered on the 1:10,000,000 scale, because in this scale,

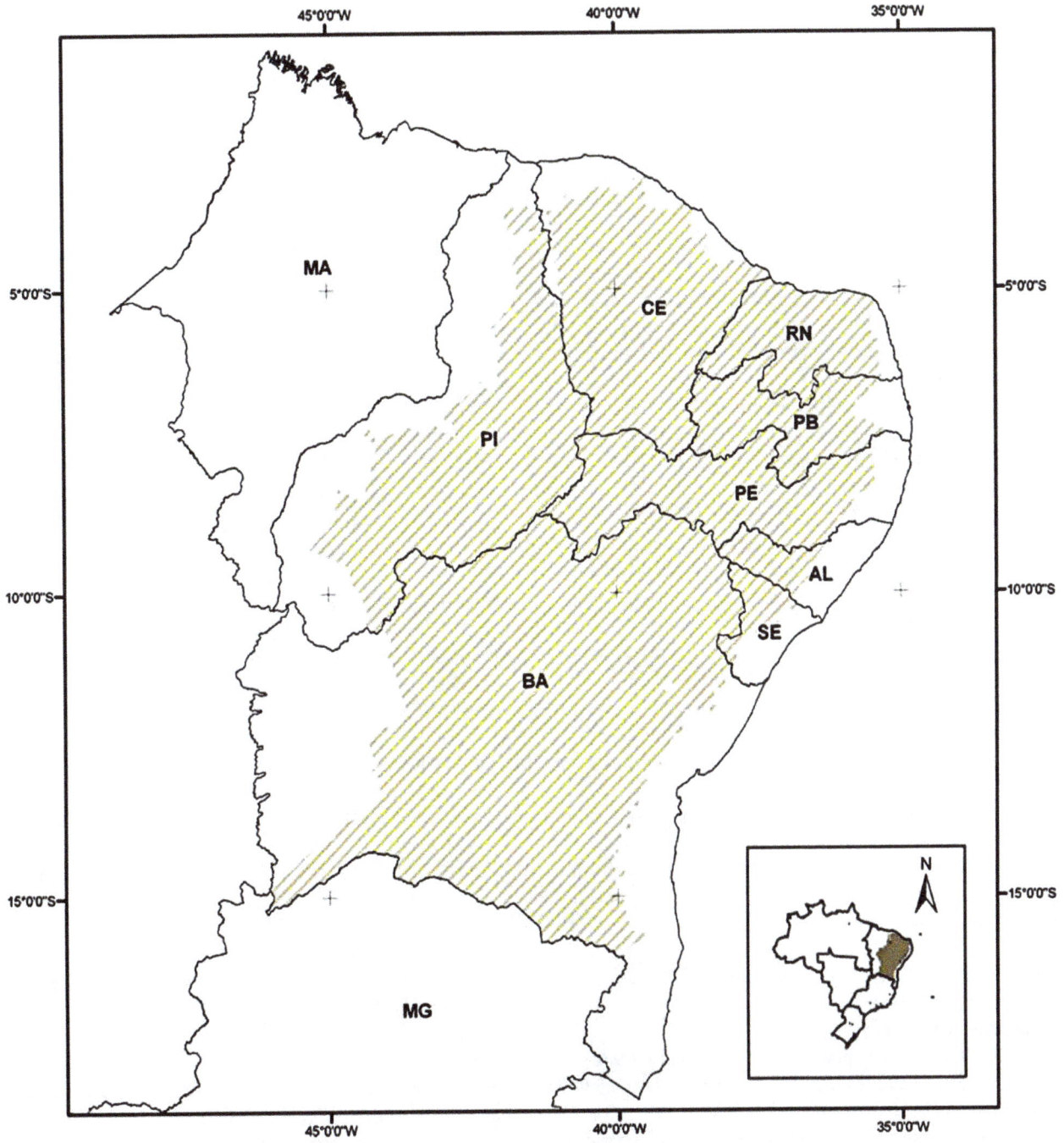

Scale: 1:10,000,000

Geodetic System: SIRGAS2000
Polyconic Projection

Software Spring (Atlas-2008)

Metadata: Instituto Nacional de Pesquisas Espaciais (INPE)

Ministério do Meio Ambiente (MMA)

LEGEND

⬭ States

//// Semi-arid Northeast

Figure 1. Location map of the semi-arid in the Northeast.

representation of these areas is not thematically visible. The quilombola territories were also not considered in this study due to a lack of available information such as spatial data in the Brazilian territory. Nonetheless, in specific studies, for example, on a municipality scale, these region restrictions must be analysed and considered.

3.3.2. Solar Resource

A parabolic cylindrical concentrator practically uses the direct portion of solar irradiation that hits its surface. Therefore, in locating areas suitable for the installation of SEGS solar power plants, knowledge of normal direct solar irradiation is crucial. Normal direct solar irradiation data generally are obtained from measurements made with specific equipment (pyrheliometer) or with mathematical models dealing with global irradiation (direct and diffuse portion). Meteorological satellite data have also been used to provide this information.

The ideal situation for a possible site for a power plant would be the existence of solar resource measurements for a period of at least five years, which would make it possible to perceive annual and seasonal variations. However, very few places have solarimetric stations, and, when they are available, they are insufficient (short temporal series) or low quality; therefore, normal direct irradiation data were used, modelled with satellite images. The resolution of the maps used was of 40 km, and they were modelled in the project known as Solar and Wind Energy Resource Assessment [14].

The values correspond to the annual mean daily values. For the semi-arid region in the Northeast, they were divided into three groups: irradiation from 4.0 to 5.0 kWh/m^2, irradiation from 5.0 to 6.0 kWh/m^2 and irradiation from 6.0 to 7.0 kWh/m^2. For a more detailed approach, ILs must be crossed with seasonal data for solar irradiation (values referring to Summer and Winter).

3.3.3. Water Supply

An SEGS solar power plant (80 MW) operating with an annual capacity factor of 0.27 uses approximately 725,000 m^3 of water [15]. This water quantity is necessary for the cooling towers (approximately 90%), vapour generation in the potency cycle (8%) and cleaning the mirrors (2%). The typical flux for the refrigeration tower is 320 m^3/h. Water must be a proper quality to avoid incrustations and oxidation of the equipment. In the semi-arid region of the Northeast, water availability is a very important and crucial question given the shortness of surface hydraulic sources and that underwater water is normally brackish and has a low flow rate.

3.3.4. Area Availability and Terrain Topography

An 80 MW SEGS power plant requires an area of approximately 2 km^2, of which approximately 500,000 m^2

are for the collector set. The scale factor derived from SEGS experiences shows that it is advantageous to install several adjacent plants. Thus, the minimum required area availability is 8 km^2 for a generation complex of 320 MW. In addition, terrain topography will determine site acceptability according to its impact in the cost relative to terrain preparation. Thus, the site must be as flat as possible, except for a declivity that allows natural terrain drainage.

The data used for determining the declivity were the images from the SRTM (Shuttle Radar Topography Mission) sensor. The images have a spatial resolution of 90 m, are in the Geo TIFF (16 bits) format and are geo-referenced to the Geodesic Reference System (GRS) WGS84, which is compatible with the SIRGAS2000 GRS for the applied work scale.

3.3.5. Connection with the Electric Network

The requirements for connection with the electric network of a solar power plant that uses parabolic cylindrical collectors are similar to those of other vapour plants. Because the construction costs of new transmission lines are, in general, very high and depend on the level of voltage and length of the line, solar power plants must be positioned as closely as possible to transmission lines. According to [16], a power plant that produces 80 MW of potency must have transmission lines of 230 kV potency for transporting the energy.

In the semi-arid region of the Northeast, existing transmission lines are from 230 kV to 500 kV. The 230 kV lines cover all the states of the region, whereas the 500 kV ones cover only the states of Ceará, Piauí, Pernambuco, Alagoas, Bahia and Sergipe.

3.3.6. Fuel Availability for Backup

Backup fuels are necessary for hybrid operation of the power plants. In SEGS plants, natural gas is used as a backup. In the semi-arid region of the Northeast, it is necessary to consider, in addition to natural gas, biodiesel and diesel oil. In addition to fuel availability for backup, the plant's proximity to sources of this fuel is a determining factor. Hence, large distances can make hybrid operation economically unfeasible. In this study, be it due to the insignificance of the implantation of biodiesel power plants or the need for land transportation (because the railroads are precarious) for diesel over considerable distances, it was decided that hybrid operation of the solar power plants would not be considered.

3.3.7. Access

Site access is relevant due to the need to transportlarge-scale and fragile equipment (glass mirror). The criteria for classifying access are the width of highways, the quality of the road surface and the possibility of manoeuvring-

glarge-scale vehicles. Thus, for defining the access issue, a map of the road system of the semi-arid region, with the main federal and state highways that cross the region, was used.

4. Results and Discussion

4.1. GIS Application for the Location Study of Solar Power Plants

Once the Spatial Database (SD) was defined, the ILs were managed in the GIS to provide information that could give support to decisions about the power plant locations. This management was performed in two steps:

First Step: The IL crossings on the 1:10,000,000 scale aiming to provide a pre-location of the most promising areas for installing solar power plants;

Second Step: With this pre-location, more specific and detailed (on scales greater than 1:10,000,000) analyses were conducted for a case study, the state of Piauí.

4.2. Soil Usage and Occupation

Presentation of the result starts with one of the most important crossings between the ILs in this study: an analysis of the available areas for setting up a solar power plant considering only the usage and occupation of the soil in the region.

For Soil Usage and Occupation in the semi-arid region of the Northeast, the following ILs were considered:

- Integral protection conservation units and sustainable usage conservation units, both obtained from the *Instituto Brasileiro do Meio Ambiente e dos Recursos-NaturaisRenováveis* [17];
- Atlantic Forest reserves, obtained from the *Instituto Brasileiro do Meio Ambiente e dos Recursos Naturais Renováveis* [17];
- Indigenous territories, obtained from *Ministério do Meio Ambiente* [18];
- Areas with agricultural potential, obtained from *Empresa Brasileira de Pesquisa Agropecuária* [19].

Urban and urban expansion areas were not considered on the 1:10,000,000 scale given that the representation of such areas would not be thematically visible. Similarly, the *quilombola* territories were not considered in the study due to the lack of information. However, in specific stud -ies (on larger scales, for instance municipalities), urban and *quilombola* areas must be considered in the analysis of soil usage and occupation.

With the ILs numbered above (the first three ILs), an environmental restrictions map was generated for the semi-arid region of the Northeast according to **Figure 2**.

The map in **Figure 3** shows the classification of areas regarding the agricultural potential of their soil. This classification is presented in the spatial data attribute table

and has the following categories: 1) very high; 2) high; 3) medium high; 4) medium; 5) medium low; and 6) low. For the spatial analysis, the objects of this research, the agriculturally suitable areas classified as "very high" and "high," were considered unfit for setting up solar power plants.

Using the maps from **Figures 2** and **3**, **Table 1** was generated, which presents, in quantitative terms, the areas occupied for each information plane regarding the semi-arid region of the Northeast.

Observing **Table 1**, it is possible to see that the areas considered unfit for solar power plants amount to 153123.67 km^2 (18.40% of the area of the semi-arid region of the Northeast). Of this percentage, 9.30% are areas with restricted usage by Brazilian legislation (Conservation Units, Indigenous Territories and Atlantic Forest Reservation). The ILs that integrate the Soil Usage and Occupation group (indicated in **Table 1**) were crossed with the IL of the geographic region of the semi-arid region of the Northeast according to Boolean logic (punctual operations of field algebra). The ILs that integrate the Soil Usage and Occupation group were given a "zero" value in a specific field created in the attributes table.

For the IL that represents the semi-arid region of the Northeast and the areas with agricultural potential between "medium high" and "low," a value of one was attributed in the generated field. In this crossing, the following criteria were used: If the semi-arid region of the Northeast is also one of the areas of the Soil Usage and Occupation group, classify as zero. Otherwise, classify as one. The result of this crossing is the map in **Figure 4**, which shows potentially available areas for inserting solar power plants according to soil usage and occupation. It is observed that there is area availability in all of the states in the semi-arid region of the Northeast, constituteing an enormous area: 694910.33 km^2.

4.3. Direct solar Irradiation in the Northeast Region of Brazil

The IL crossings of the semi-arid region of the Brazilian Northeast with the IL from normal direct solar irradiation hitting the region are represented in **Figure 5** (for annual direct solar irradiation between 4.0 and 5.0 kWh/m^2), **Figure 6** (for annual direct solar irradiation between 5.0 and 6.0 kWh/m^2), and **Figure 7** (for annual direct solar irradiation between 6.0 and 7.0 kWh/m^2).

The annual average of daily direct solar irradiation has significant values in important parts of the Northeast region of Brazil, as can see on **Figure 8**.

The intensity of solar irradiation reaches very high values between 6.0 - 7.0 kWh/m^2 (2190 - 2555 $kWh/(m^2 \cdot year)$ for approximately 1/3 of the semi-arid region (**Table 2**). In the current international literature, the threshold for

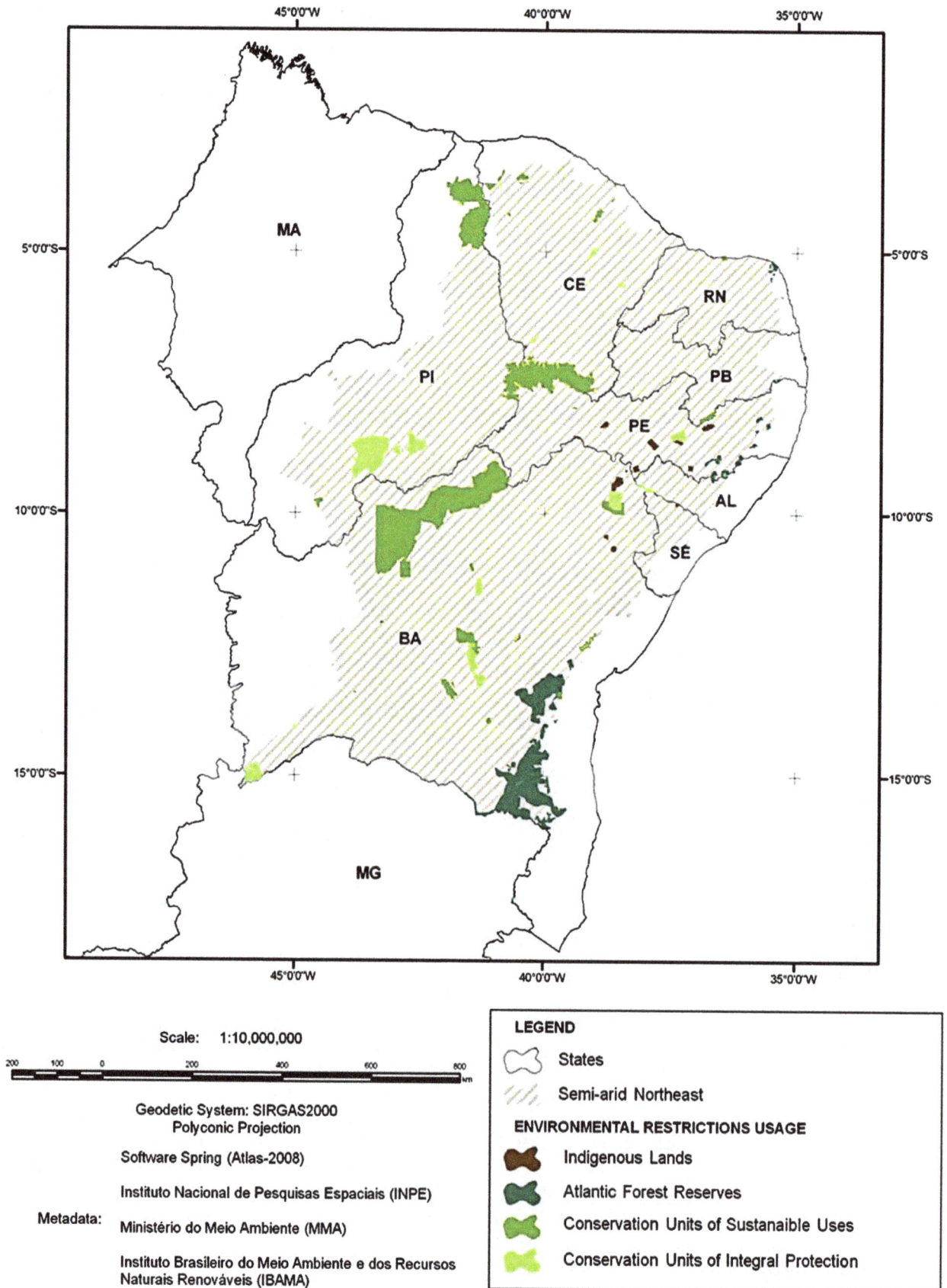

Figure 2. Environmental restrictions usage map for the semi-arid region of the Northeast.

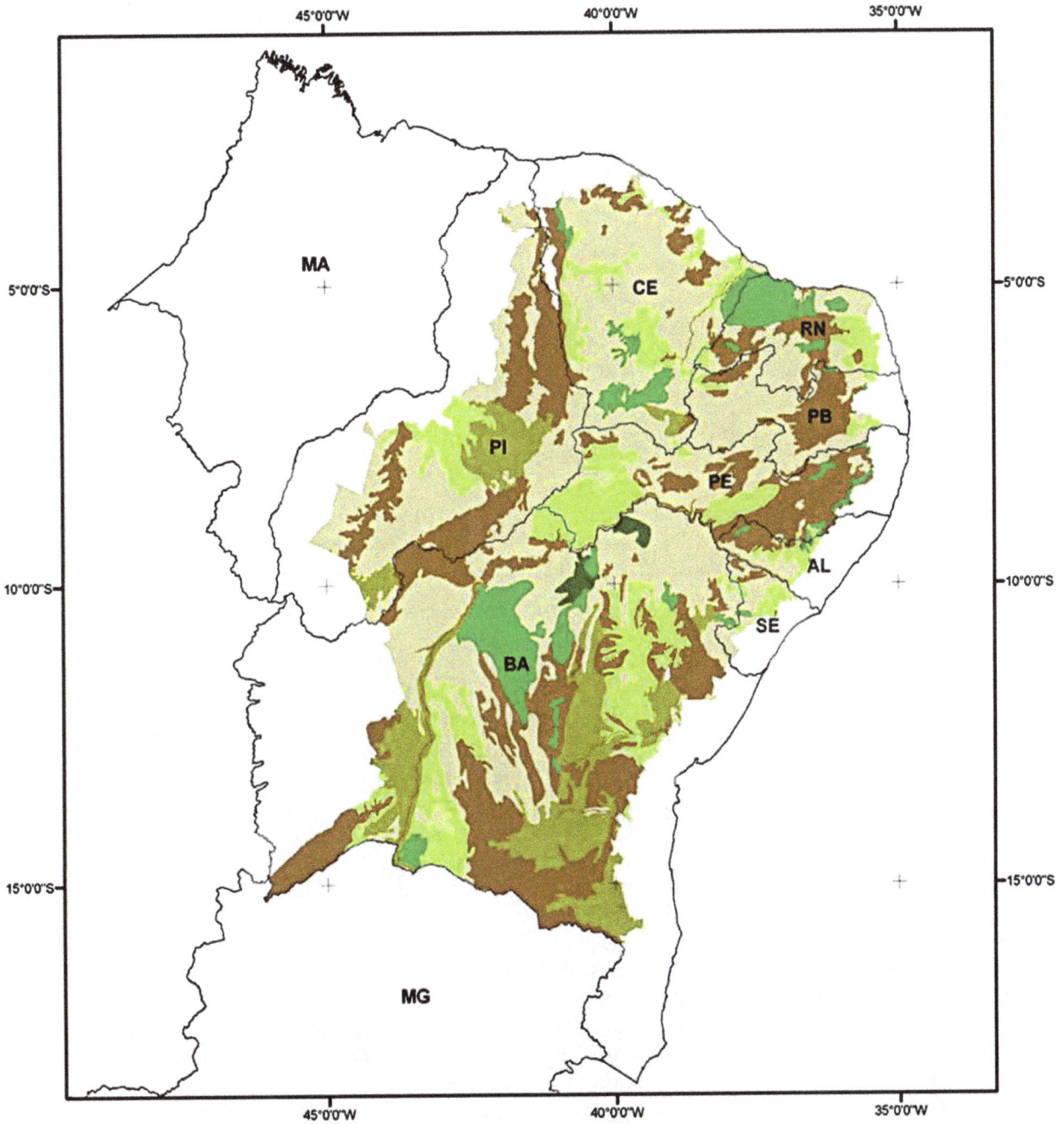

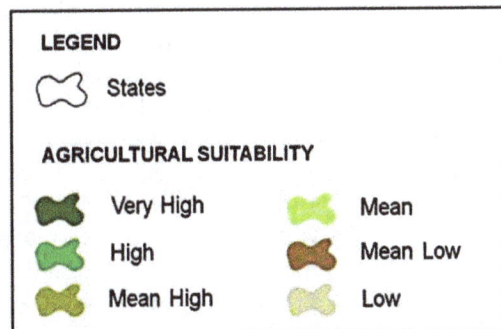

Figure 3. Agricultural potential map for the semi-arid region of the Northeast.

Table 1. Area distribution according to the attributes of the information plane that integrates the soil usage and occupation group.

Information plane	Area (km²)	Percentage in relation to the area of the semi-arid region of the Northeast
Area with Agricultural Potential (High)	71828.48	8.47%
Sustainable Use Conservation Units	45369.82	5.35%
Atlantic Forest	16420.56	2.28%
Integral Protection Conservation Units	12466.10	1.47%
Area with Agricultural Potential (Very High)	5342.61	0.63%
Indigenous Territory	1696.10	0.20%
Quilombola Territory	NC	-
Urban Areas	NC	-
TOTAL	153123.67	18.40%

evaluating sites for future concentration solar centrals is that direct solar irradiation should be greater than 1900 kWh/(m²·year) and preferably greater than 2100 kWh/(m²·year). The discussed region meets those criteria easily. It is also important and interesting to highlight that the maximum seasonality factor in this region is of approximately 1.3, that is, for a solar central to generate constant annual potency, the collectors' field must be scaled 30% larger to meet the demand for the worst period.

4.4. Pre-Location of the Areas Fit for Setting up Thermoelectric Solar Power Plants

Finally, a crossing of the IL of potential areas was performed, on the Soil Usage and Occupation group with normal direct solar irradiation IL, and the most adequate areas in the semi-arid region of the Northeast for installing thermoelectric solar power plants were identified. The criteria used were that the areas suitable for implanttation should have solar irradiation between 6.0 and 7.0 kWh/m² and not occupy any of the areas of the Soil Usage and Occupation group (identified in **Table 2**).

The result is displayed in **Figure 8**, where it can be seen that the large areas that fulfil these criteria are found in Bahia, Piauí, Pernambuco and Paraíba.

After this spatial macro-analysis, smaller spatial regions were detailed on a state level. For this more specific study, other additional criteria were defined, such as terrain declivity, availability and proximity of hydraulic resources, access and transmission lines. The ILs used for the crossings in larger scales (state level) were the same as those used in the 1:10,000,000 scales given that with these data, it is possible to work on a scale of 1:1,000,000, according to the Class A Cartographic Accuracy Standard (CAS).

4.5. Pre-Location of the Areas Fit for the Set up of Thermoelectric Solar Power Plants

The state of Piauí occupies an area of 251529.186 km², divided in 223 municipalities (**Figure 9**). The population estimative in 2009 was 3,145,325 inhabitants, according to [20].

Its declivity is shown in **Figure 10** on the scale of 1:8,000,000. Observing the map, it is possible to verify that the relief is relatively flat with only slight inclinations. Most of the regions display a declivity less than 8%.

A significant portion of the territory of the state is covered by solar irradiation in the interval between 6.0 kWh/m² and 7.0 kWh/m². This region is located in the municipalities in the Southeast region of the state. Buffers in relation to hydraulic resources were generated at distances of 5 km to 20 km. Analogously, a buffer map was made for the access paths that cut through the state. 5 km to 20 km distance buffers were also used. The goal was to identify their proximity to areas with solar irradiation spots.

For transmission lines (230 kV), however, 5 km to 30 km buffers were used. The state is cut by 230 kV and 500 kV lines.

A superposition of the IL was made, as shown in **Figure 11**. It is observed that the region with solar irradiation between 6.0 and 7.0 kWh/m² is cut by several main access roads (highways), hydraulic resources and 230 kV and 500 kV transmission lines. Finally, **Figure 12** is the result of the crossing between **Figure 10** and the declivity map, **Figure 11**. After analysing the results, it was observed that seven municipalities possess suitable conditions for setting up large-scale solar power plants.

This region occupies 8797 km², has a population of 67,362 (a demographic density of 6.7 inhabitants per km²)

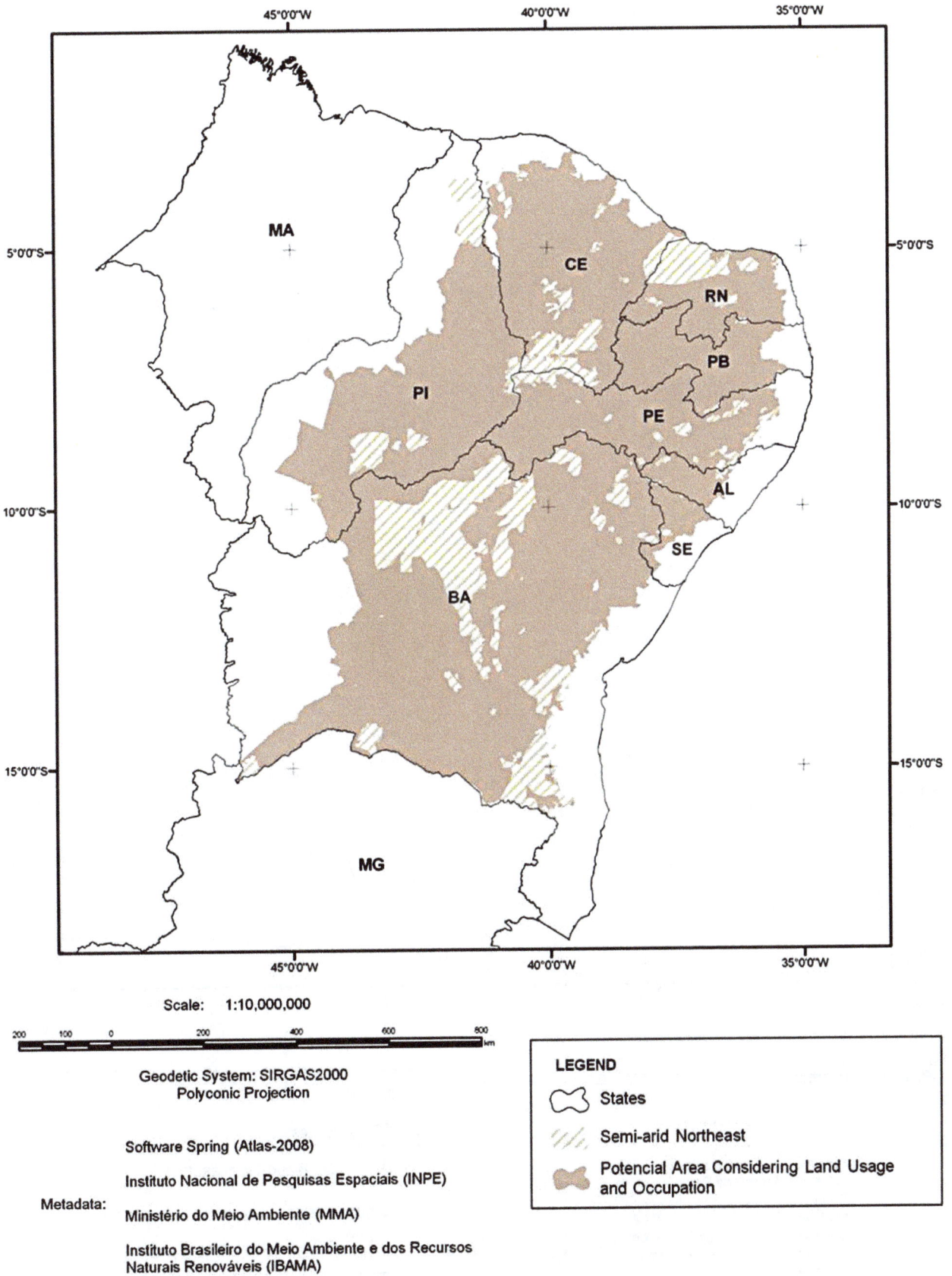

Figure 4. Potentially available areas for inserting solar power plants according to soil usage and occupation in the semi-arid region of the Northeast.

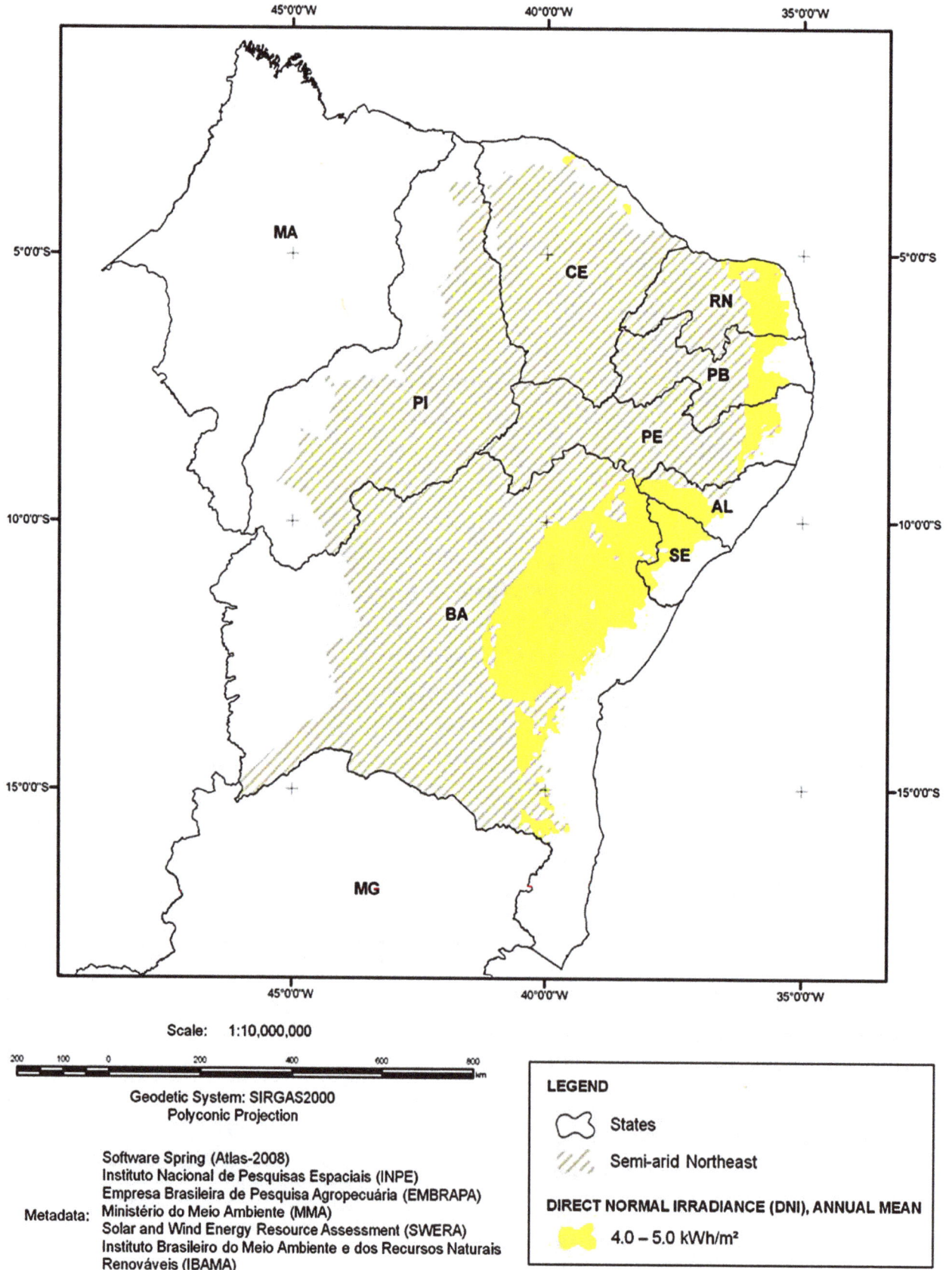

Scale: 1:10,000,000

Geodetic System: SIRGAS2000
Polyconic Projection

Metadata: Software Spring (Atlas-2008)
Instituto Nacional de Pesquisas Espaciais (INPE)
Empresa Brasileira de Pesquisa Agropecuária (EMBRAPA)
Ministério do Meio Ambiente (MMA)
Solar and Wind Energy Resource Assessment (SWERA)
Instituto Brasileiro do Meio Ambiente e dos Recursos Naturais
Renováveis (IBAMA)

LEGEND

⬡ States

▨ Semi-arid Northeast

DIRECT NORMAL IRRADIANCE (DNI), ANNUAL MEAN

🟨 4.0 – 5.0 kWh/m²

Figure 5. Potentially available areas with annual solar irradiation between 4.0 and 5.0 kWh/m².

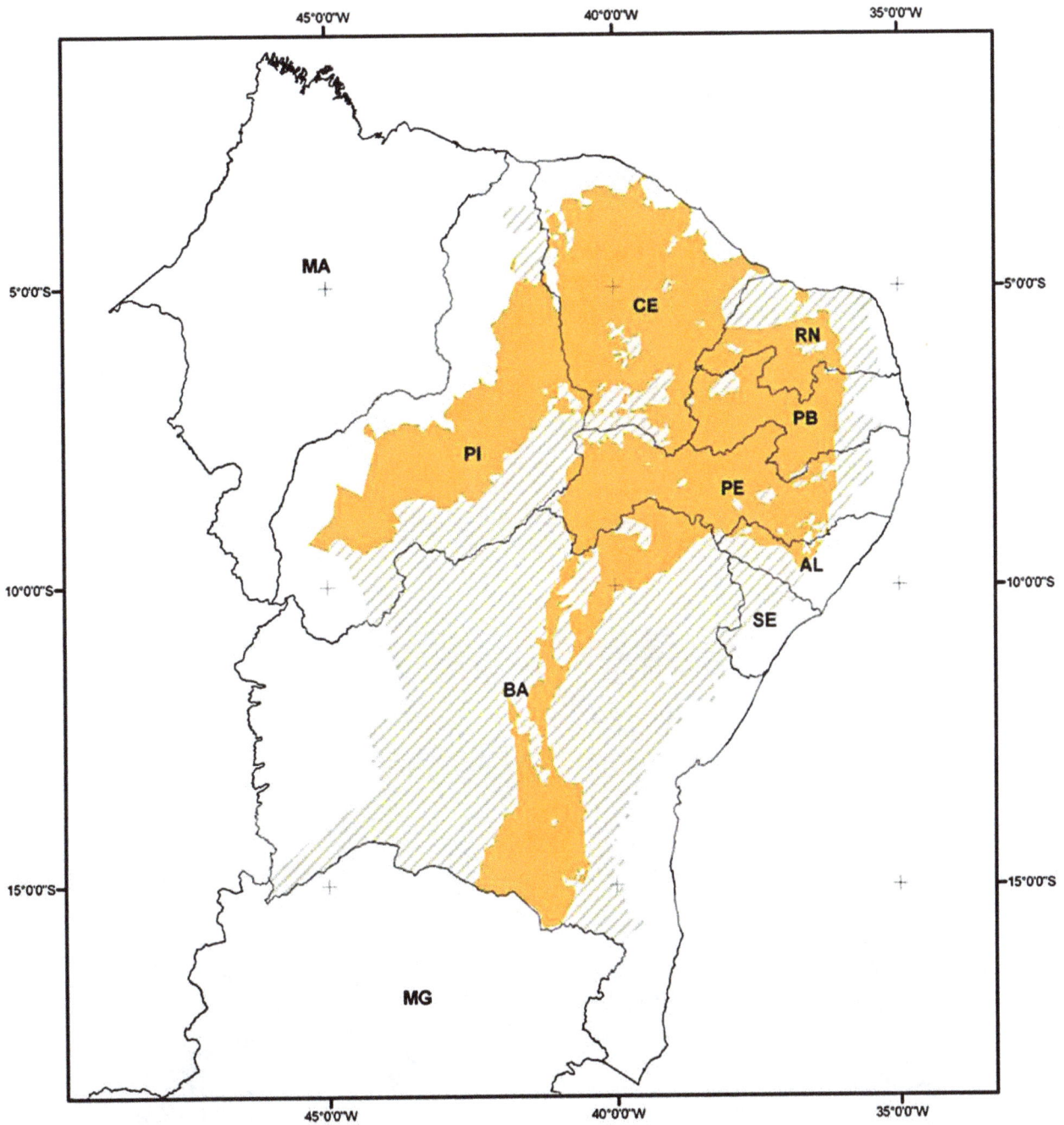

Scale: 1:10,000,000

Geodetic System: SIRGAS2000
Polyconic Projection

Software Spring (Atlas-2008)
Instituto Nacional de Pesquisas Espaciais (INPE)
Empresa Brasileira de Pesquisa Agropecuária (EMBRAPA)
Ministério do Meio Ambiente (MMA)
Metadata: Solar and Wind Energy Resource Assessment (SWERA)
Instituto Brasileiro do Meio Ambiente e dos Recursos Naturais
Renováveis (IBAMA)

LEGEND

States

Semi-arid Northeast

DIRECT NORMAL IRRADIANCE (DNI), ANNUAL MEAN

5.0 – 6.0 kWh/m²

Figure 6. Potentially available areas with annual solar irradiation between 5.0 and 6.0 kWh/m².

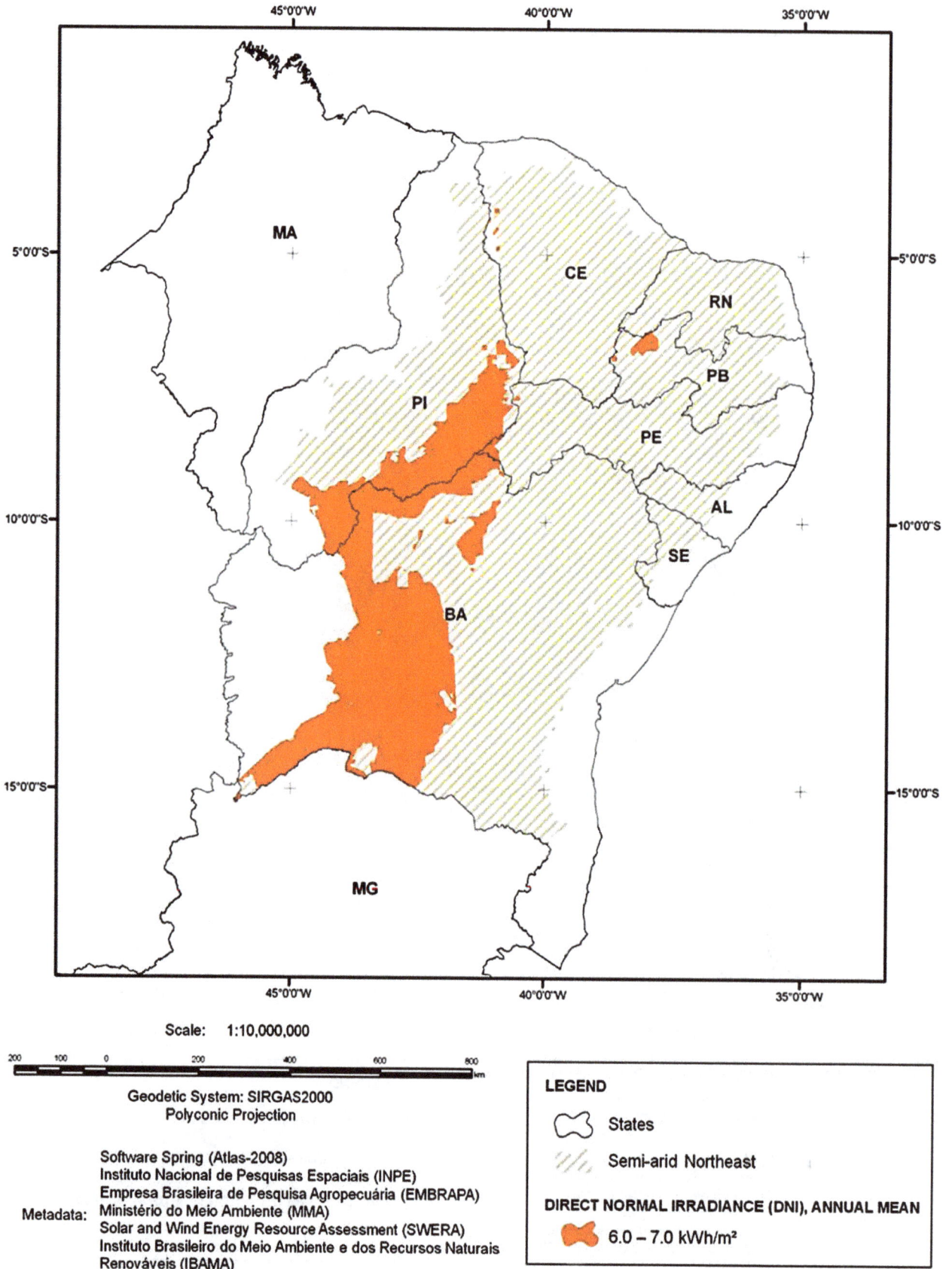

Scale: 1:10,000,000

Geodetic System: SIRGAS2000
Polyconic Projection

Metadata:
Software Spring (Atlas-2008)
Instituto Nacional de Pesquisas Espaciais (INPE)
Empresa Brasileira de Pesquisa Agropecuária (EMBRAPA)
Ministério do Meio Ambiente (MMA)
Solar and Wind Energy Resource Assessment (SWERA)
Instituto Brasileiro do Meio Ambiente e dos Recursos Naturais
Renováveis (IBAMA)

LEGEND

States

Semi-arid Northeast

DIRECT NORMAL IRRADIANCE (DNI), ANNUAL MEAN

6.0 – 7.0 kWh/m²

Figure 7. Potentially available areas with annual radiation between 6.0 and 7.0 kWh/m².

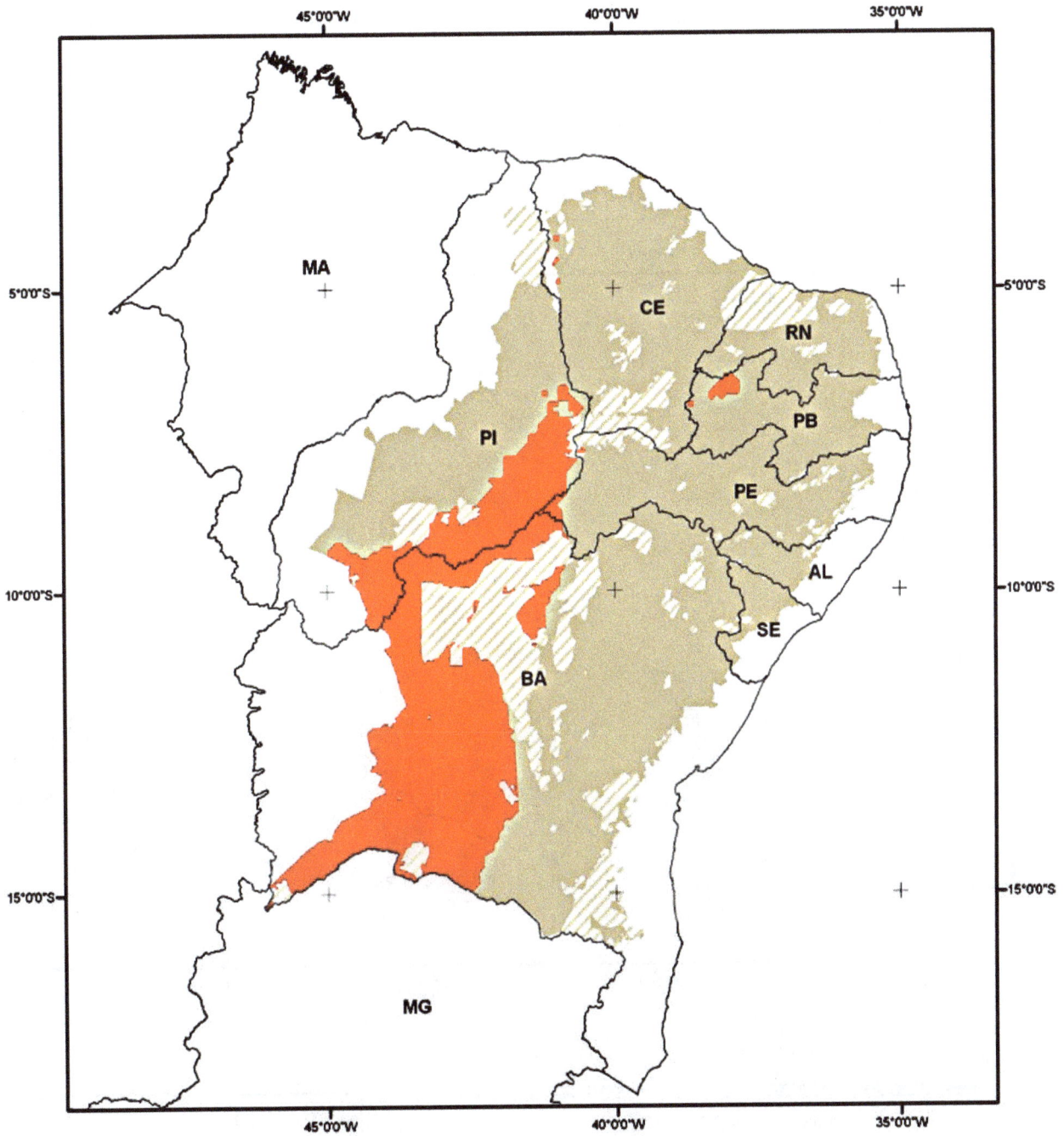

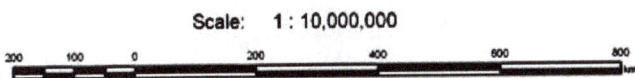

Figure 8. Potentially available areas with annual solar irradiation between 6.0 and 7.0 kWh/m².

Table 2. Potential area availability in the semi-arid region of the Northeast, according to the level of annual direct solar irradiation.

Solar irradiation interval (kWh/m^2)	Area (km^2)	Percentage in relation to the area of the semi-arid region in the Northeast
4.0 - 5.0	120560.00	20%
5.0 - 6.0	282411.80	46.85%
6.0 - 7.0	199828.20	33.15%

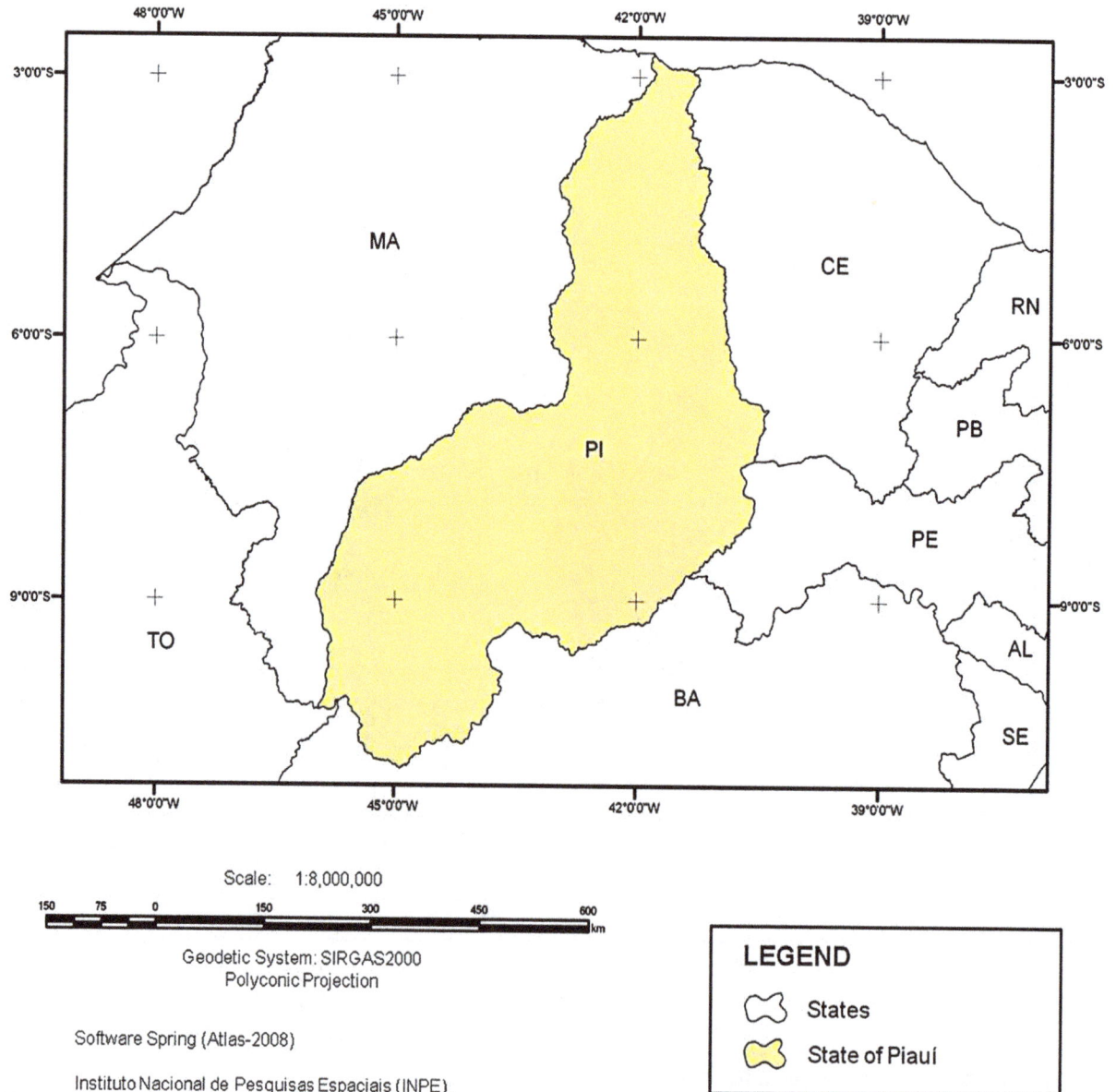

Figure 9. Map location of the Piauí State.

and has an average HDI of 0.555. Its predominant activity is agriculture and farming [20,21]. In addition, it is a region with no environmental reserves or other social restrictions. The low demographic density indicates that there is no need to remove people or excessive competition over soil usage.

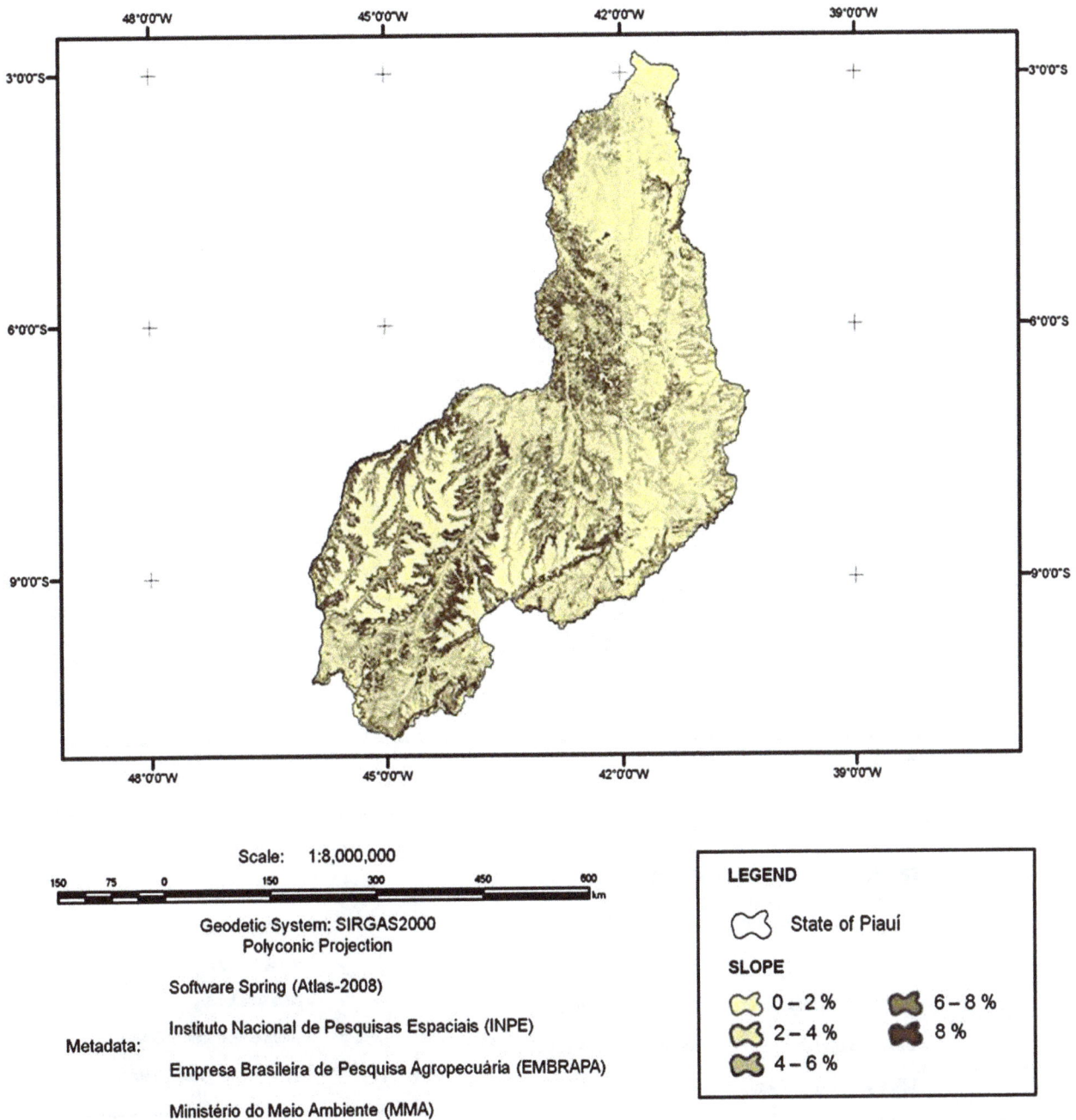

Figure 10. Piauí State declivity.

5. Conclusions

Site evaluation and identification procedures using GIS are innovating decision-making procedures around the world and are contributing decisively to the quick growth in the implantation of CSP technology. Hence, in this study, identification and mapping of the most promising sites in the Northeast of Brazil were conducted so that the country can quickly initiate the process of implanting large-scale CSP solar technology.

The analysis performed in this study consisted of two steps: the first step was a macro approach (small scale—

1:10,000,000—semi-arid region of the Northeast) in which regions that were promising and fit for installation were identified, and the second step was a large-scale approach on the level of the Piauí state.

Very specific regions were outlined so that, in the sequence of this process, they might be visited and prioritised.

According to [22], for California, the installed capacity of parabolic cylindrical concentrators in terrains with a declivity of less than 1% is 43.26 MW/km^2 for systems without storage and 30.82 MW/km^2 for systems with 6

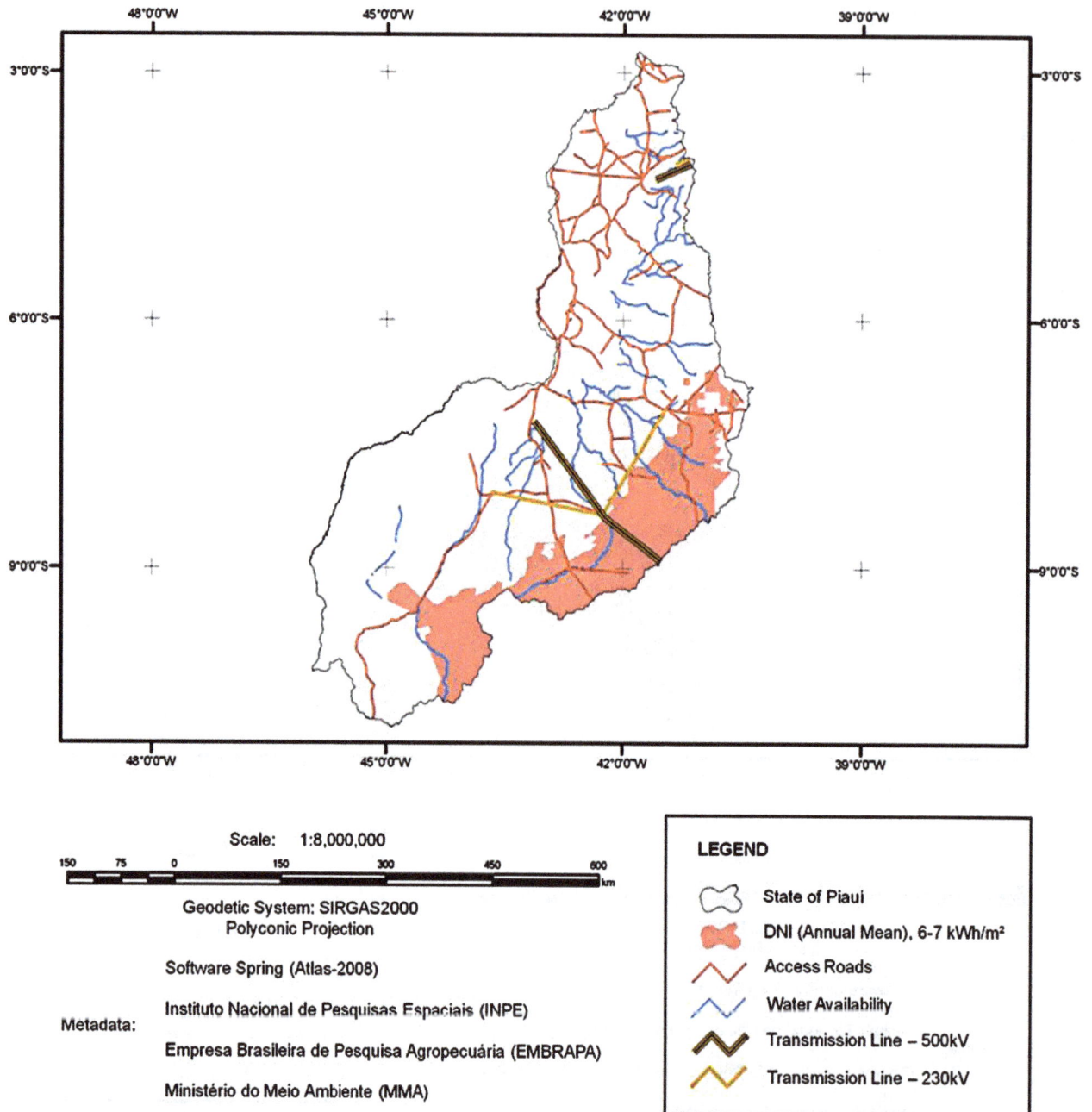

Figure 11. Crossing of the IL of solar irradiation, access paths, hydraulic resources and transmission lines in Piauí State.

hours of storage. For the first step, considering these values are valid for Brazil, the potential for the southeast region of Piauí alone is enormous. Given an absence of information regarding urban areas, terrain continuity, and other variables, we considered that only 10% of the potential area identified above would be adequate for the installation of solar thermoelectric power plants. Still, that area would amount to 879.7 km^2 and had an installed capacity of 38.1 - 27.1 GW. This value corresponds to over 1/3 of the electric potency of the current Brazilian electric system. If the same calculation was

done for the semi-arid region of the Brazilian Northeast, its capacity would be greater than 1000 GW. The considered area corresponds to approximately 3% of the semi-arid region of the Brazilian Northeast.

6. Acknowledgements

We hereby thank the *ConselhoNacional de Pesquisa* (CNPq), *Centrais Elétricas Brasileiras S.A.* (ELETRO-BRAS), *Companhia Hidro Elétrica do São Francisco* (CHESF) and *Coordenação de Aperfeiçoamento de Pessoal de Nível Superior* (CAPES) for their support to the

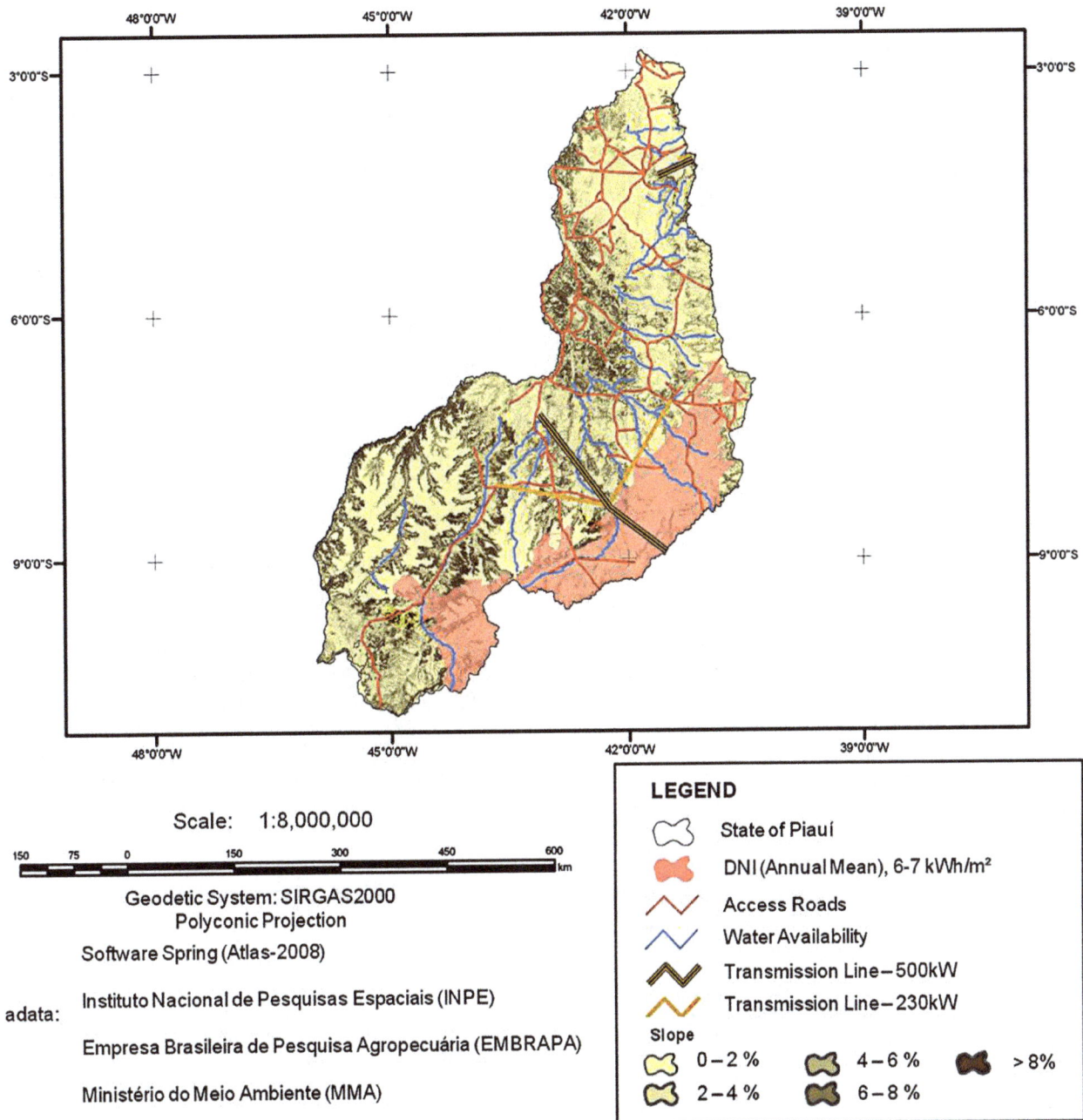

Scale: 1:8,000,000

Geodetic System: SIRGAS2000
Polyconic Projection

Software Spring (Atlas-2008)

adata: Instituto Nacional de Pesquisas Espaciais (INPE)

Empresa Brasileira de Pesquisa Agropecuária (EMBRAPA)

Ministério do Meio Ambiente (MMA)

LEGEND

- State of Piauí
- DNI (Annual Mean), 6-7 kWh/m²
- Access Roads
- Water Availability
- Transmission Line – 500kW
- Transmission Line – 230kW

Slope
- 0 – 2 %
- 2 – 4 %
- 4 – 6 %
- 6 – 8 %
- > 8 %

Figure 12. Crossing of solar irradiation, declivity, access paths, hydraulic resources and transmission lines IL in Piauí State.

solar energy research projects, which have provided the material means and the scientific environment for the execution of this research.

REFERENCES

[1] Wikipedia, "Wikipedia," 2011. http://en.wikipedia.org/wiki/List_of_solar_thermal_power_stations#Operational

[2] Sargent and Lundy LLC Consulting Group, "Assessment of Parabolic Trough and power Tower Solar Technology Cost and Performance Forecasts," DIANE Publishing, 2003.

[3] J. D. Bravo, "Los sistemas de Información Geográfica en la Planificación e Integración de Energías Renovables", Editorial CIEMAT, Madrid, España, 2002.

[4] H. Broesamle, H. Mannstein, C. Schilling and F. Ttieb, "Assessment of Solar Electricity Potentials in North Africa Based on Satellite Data and Geographic Information" *Solar Energy*, Vol. 70, No. 1, 2001, pp. 1-12.

[5] M. Mehos and B. Owens, "An Analysis of Sitting Opportunities for Concentrating Solar Power Plants in the Southwestern United States", World Renewable Energy

Conference VIII, Denver, August 29-September 4, 2004.

[6] T. P. Fluri, "The Potential of Concentrating Solar Power in South Africa", *Energy Policy*, Vol. 37, No. 12, 2009, pp. 5075-5080.

[7] Y. Charabi and A. Gastli, "GIS Assessment of Large CSP Plant in Duqum, Oman", *Renewable and Sustainable Energy Reviews*, Vol. 14, No. 1, 2010, pp. 835-841.

[8] Y. Azoumah, E. W. Ramde, G. Tapsoba and S. Thiam, "Siting Guidelines for Concentrating Solar Power Plants in the Sahel: Case Study of Burkina Faso", *Solar Energy*, Vol. 84, No. 8, 2010, pp. 1545-1553.

[9] J. Clifton and B. Boruff, "Assessing the Potential for Concentrated Solar Power Development in Rural Australia," *Energy Policy*, Vol. 38, No. 9, 2010, pp. 5272-5280.

[10] I. Purohit and H. Purohit, "Techno-Economic Evaluation of Concentrating Solar Power Generation in India", *Energy Policy*, Vol. 38, No. 6, 2010, pp. 3015-3029.

[11] K. Unmel, "Global Prospects for Utility-Scale Power: Toward Spatially Explicit Modeling of Renewable Energy Systems", In: Center for Global Development. 2010.

[12] ASA, "Articulação no Semi-Árido Brasileiro," 2010. http://www.asabrasil.org.br

[13] C. Tiba, N. Fraidenhaich, M. Moskowicz, E. Cavalcanti, F. M. J. Lyra, A. M. B. Nogueira. "Atlas Solarimétrico do Brasil—Banco de Dados Terrestres,".

[14] SWERA, "Solar and Wind Research Assessment," 2009. http://www.swera.unep.net

[15] B. Kelly, "Nexant Parabolic Trough Solar Power Plant Systems Analysis, Task 2: Comparison of Wet and Dry Rankine Cycle Heat Rejection," In: *National Renewable Energy Laboratory* (*NREL*). Assessing the Potencial for Renewable Energy on DOE Legacy Management Lands, 2006.

[16] D. Dahle, D. Elliott, D. Heimiller, M.Mehos, R. Robichaud, M. Schwartz, B. Stafford and A. Walker, "Descriptions of renewable energy Technologies," In: *National Renewable Energy Laboratory* (*NREL*). Assessing the Potencial for Renewable Energy on DOE Legacy Management Lands, 2008.

[17] IBAMA, "Instituto Brasileiro do Meio Ambiente e dos Recursos Naturais Renováveis," 2011. http://www.ibama.gov.br

[18] MMA, "Ministério do Meio Ambiente," 2011. http://www.mma.gov.br

[19] EMBRAPA, "Empresa Brasileira de Pesquisa Agropecuária," 2011. http://www.embrapa.br

[20] IBGE, "Instituto Brasileiro de Geografia e Estatística," 2011. http://www.ibge.gov.br/cidadesat/topwindow.htm?1

[21] PNUD, "Programa das Nações Unidas para o Desenvolvimento,"2011. http://www.pnud.org.br/atlas/ranking/IDH

[22] L. Stoddard, J. "Abiecunas and R. O'Connell, "Economic, Energy, and Environmental Benefits of Concentrating Solar Power in California", In: *National Renewable Energy Laboratory* (*NREL*). Assessing the Potencial for Renewable Energy on DOE Legacy Management Lands, 2006.

13

The Contribution of the Geospatial Information to the Hydrological Modelling of a Watershed with Reservoirs: Case of Low Oum Er Rbiaa Basin (Morocco)

Youness Kharchaf, Hassan Rhinane, Abdelhadi Kaoukaya, Abdelhamid Fadil
Geosciences Laboratory, Faculty of Sciences-Ain Chock, Hassan II University, Casablanca, Morocco

ABSTRACT

Water is undoubtedly the most vital natural resource. Water use management is one of the greatest challenges that face humanity. The demand for water is continuously growing because of the population growth, the intensive urbanization and the development of industrial and agricultural activities. To face the increasing pressure on this vital resource, it is so necessary to set up the adequate instruments to ensure a rational and efficient management of this resource. In this context, the hydrological modeling is largely used as an instrument to assess the functioning of these resources at watershed scale. In addition, the use of spatial models let to depict and simulate the watershed processes at small spatial and heterogeneous scales that reflect the field reality more accurate and more realistic as possible. However, the use of spatial models requires geospatial data that must be gathered at very fine scales. The aim of this study is to highlight the contribution of geospatial data to assess the hydrologic modeling of watershed by using a spatial hydro-agricultural model, notably the SWAT model (Soil and water Assessment Tool). The study area is the Basin of Low Oum Er Rbiaa River which extends from the Al Massira dam to its outlet in the Atlantic Ocean. This watershed includes a set of dams (Daourat, Imfout and Sidi Mâachou) built in waterfall fashion along the river. The objective was to simulate the hydrological functioning of this area that had never been modeled in order to assess the management of these reservoirs used essentially to produce electricity and fresh water. The implementation of the SWAT model required a spatial database that was built from topography, soil, land use and climate data. The calibration and validation of the model was carried out on a daily basis over several years (2001-2010) using The ArcSWAT tool integrated in ArcGIS software and the Parasol optimization method. The calibration of SWAT model was successfully done with 0.6 as value of Nash coefficient used commonly in hydrology to evaluate the model performance. The calibrated model was then used to estimate the hydrological balance sheet of the Low Oum Er Rbiaa to model the intermediate contribution of the three reservoirs situated in the watershed.

Keywords: Modeling; Hydrology; Low Oum Er Rbiaa; Reservoirs; GIS; SWAT; ArcSWAT; Watershed

1. Introduction

Water is a fundamental substance for life preservation. It is also a factor of a great utility in agriculture, energy production, industry, domestic use and in other daily life activities [1]. Taking into account the increasing necessity of the human needs in water, several actions were taken in the Moroccan context to deal with the problems related to water.

To face all these problems, it is essential to set up the adequate instruments for the management, the follow-up and the planning of this vital resource. In this respect, the hydrological modeling is an essential element which allows a better understanding of the hydrological functioning of the studied watershed and the study of the various challenges to which a watershed is exposed: floods, drought, erosion, pollution, climate change, etc. Using a distributed approach based on spatial data is a major asset for the process of watershed modeling since it allows representing and simulating the different components of the model inputs and an accurate and more realistic spatial scale. The use of spatial model is made possible by the spatial information acquisition and the processing techniques, especially remote sensing and GIS. In fact, the global models dealing with watersheds as a single entity are generally contested in their ability to represent the reality of space that is characterized by its complexity, heterogeneity and uniformity.

Hydrological modeling is based on the presentation in mathematical equations of the main components of the water cycle, taking into account the physical and geomorphological characteristics of the study area [2]. The objectives of modeling are multiple: understanding (the functioning of the watershed, the hydrological balance sheet), forecast (floods, drought, management of the irrigation) and simulation of scenarios (climate change, effects of the anthropological arrangements developments) [3].

The presence of natural reservoirs (lakes) or anthropogenic (dams) in the studied area must be taken into account because of the role they play in the flow of water. These reservoirs further complicate the implementation process of the conceptualization of the functioning of watersheds in that all the information concerning the methods of management and operation of these reservoirs: Inputs, outputs, sampling, discharges etc. should be available.

Many hydrological models have been developed and tested worldwide, yet lack of data remains the main obstacle hindering the deployment and widespread use of these models [4]. The emergence of new technologies, acquisition and processing of spatial information notably the GIS and remote sensing constitute a solution and highly promising opportunity that can overcome some of

these barriers by offering new solutions and alternatives for the acquisition and the management of the data particularly on a more interesting spatial scale [5].

In this context, the objective of this study is to propose an approach based on the use of spatial data to simulate the behavior of the watershed low Oum Er Rbiaa, taking into account the existence of the three dams which constitute basin waterfalls. The advantage of this approach lies in having a good spatialized hydrological account of the basin knowing that it has never been modeled due to the lack of gauging stations in the studied area.

The model used is the agro-hydrological model SWAT (Soil and Water Assessment Tool), developed in 1999 by Jeff Arnold for the "USDA—Agriculture Research Service" [6].

2. Materials and Methods

2.1. Description of the Study Area

The basin of the Oum-Er Rbiaa (**Figure 1**) extends over an area of 34,000 km^2, taking the name of one of the largest rivers in the country "Oued Oum Er Rbiaa" with a length of 555 km, located West between longitudes 5°04' and 8°20' north latitude and 31°20' and 33°12'. The source of this river is in the north-western High Atlas limestone, and in the Middle Atlas. It runs across the plain of Tadla and plateau of Phosphates, before skirting the plain of Doukkala. The outlet of this river is in the

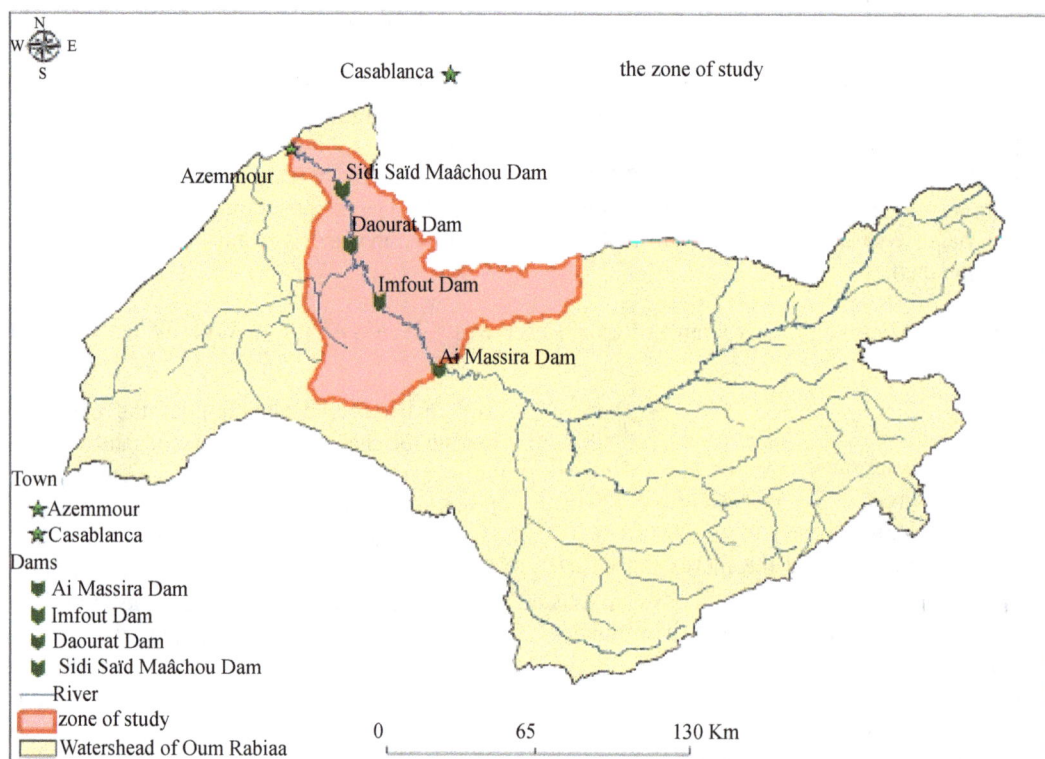

Figure 1. Map of the zone studied.

Atlantic Ocean in the coastal town of Azemmour [7].

This basin is considered a hydraulic reservoir for the country and a set of eight dams have been built on it. Our study concerns specifically the lower area of the Oum Er Rbiaa basin from Al Massira dam to the Atlantic Ocean. This area has three dams Imfout (built in 1944), Daourat (built in 1950), Sidi Maachou (built in 1929). The purpose of these dams is the production of electricity as well as:

- Irrigation of Doukkala and providing drinking water and irrigation for Safi (Imfout dam);
- Providing drinking water and irrigation for Casablanca and Settat (Daourat dam);
- Providing drinking water and irrigation for Casablanca and El Jadida (Sidi Mâachou dam).

The climate of the Oum Er Rbiaa basin is very diverse, ranging from humidity and sub humidity in the mountains to semi-aridity to aridity in the middle and lower parts of the basin. The average rainfall in the Basin is 520 mm with a high spatial and seasonal variation [8].

2.2. Description of the SWAT Model

The Soil and Water Assessment Tool (SWAT) is an agro-hydrological watershed scale model developed by Agricultural Research Services of the USDA. It is a physically based and semi-distributed model that operates on continuous time basis [9].

SWAT allows simulating the major watershed processes as hydrology, sedimentation, nutrients transfer, crop growth, environment and climate change. The aim is to depict the physical functioning of these different components and their interactions as simply and realistically as possible through conceptual equations and the use of available input data so as to make it useful in routine planning and decision making of large catchments management [10].

One of the main goals of SWAT model is to predict the impact of land management practices on water quantity and quality over long periods of time for large complex watersheds that have varying soils, land use and management practices [11]. The model generates a spatial water balance. It controls the various hydrological processes occurring in the basin.

In fact, the SWAT model was used in Indiana for instance to model the movements of pesticides in a basin of 250 km^2. It is used in Germany in the basin Dietzhöle. It has also been adopted in West Africa in the modeling of soil degradation, especially by making scenarios on the impact of climate change and land use [12-14].

2.3. Hydrological Compartments of the Model SWAT

The hydrological shutter is simulated by the model SWAT according to the following Equation (1) [11]:

$$SW_t = SW_0 + \sum_{i=1}^{t} \left(R_{\text{day}} - Q_{\text{surf}} - E_a - W_{\text{seep}} - Q_{gw} \right)_i \quad (1)$$

SW_t = soil water content (mm)
SW_0 = water available to plants (mm)
R_{day} = precipitation (mm)
Q_{surf} = surface runoff (mm).
E_a = evapotranspiration (mm)
W_{seep} = percolation (mm)
Q_{gw} = low flow (mm)
t = time (days)

The **Figure 2** shows the general sequence of processes used by SWAT to model the land phase of the hydrologic cycle [11].

2.4. Creation of the Database

The Implementation of the SWAT model requires the creation of a database containing morphological description (topography), physical description (pedology and land use) and climate (precipitation, temperature, etc.) of the basin. These layers of information must also be associated with a set of attributes describing the properties of each layer.

The representation of these data must be spatial with the highest possible resolution and the one most adapted to the studied area.

Thus, the preparation of the spatial database required a series of tasks carried out primarily through ArcGIS and Erdas Imagine tools and which can be summarized as follows:

- The recovery of digital model representing the relief of the studied area;
- The digitalization of the soil and geological map of the studied area;
- The processing of satellite images through the process of supervised classification for extracting land use map;
- Recovery and structuring in adequate files of data on the climate at the meteorological stations studied;
- The processing of these climate files in order to determine the missing data and complete these files using statistical procedures;
- Collecting and structuring data concerning the inputs and outputs of the three reservoirs: Imfout, Daourat and Sidi Maachou.

All these data have been created or projected to the coordinate system used in this project which is the Lambert Conformal Conic zone one of Morocco.

All these tasks are depicted in **Figure 3** below.

The main sets of data used are briefly explained below:

- **Pedology**

Digitizing the soil map was produced from the soil

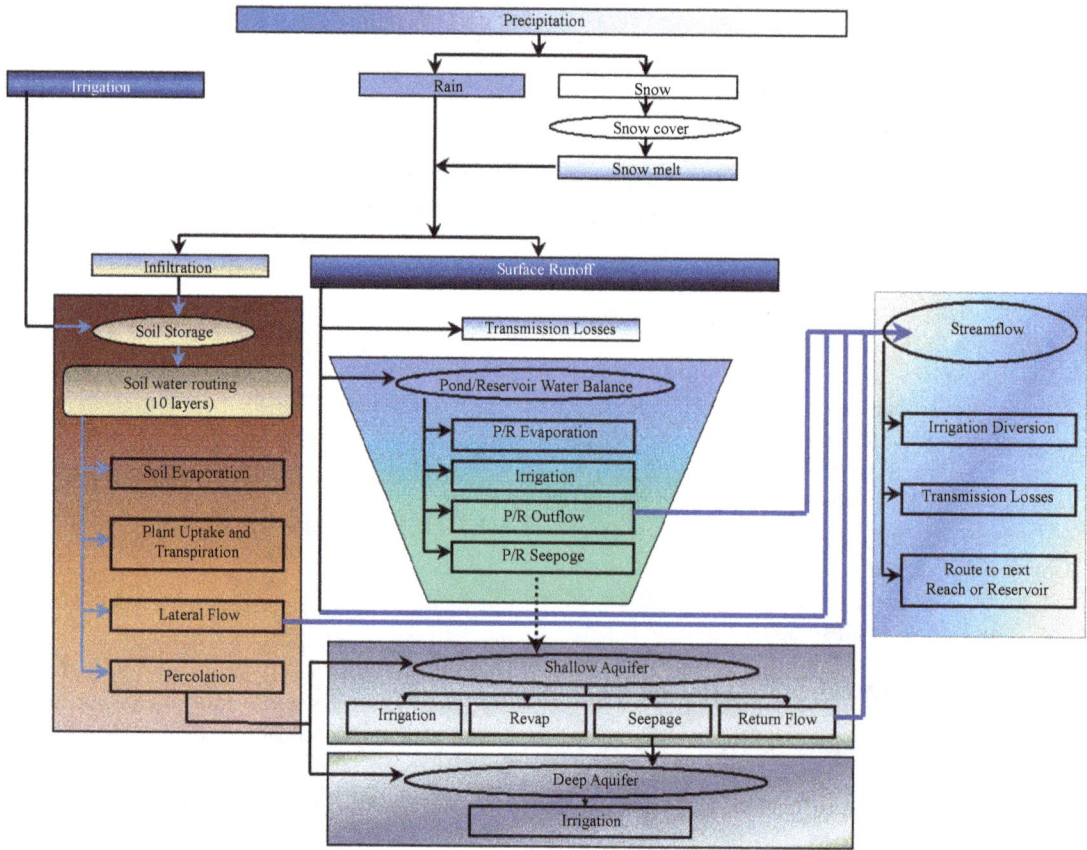

Figure 2. Schematic of pathway available for water movement in SWAT.

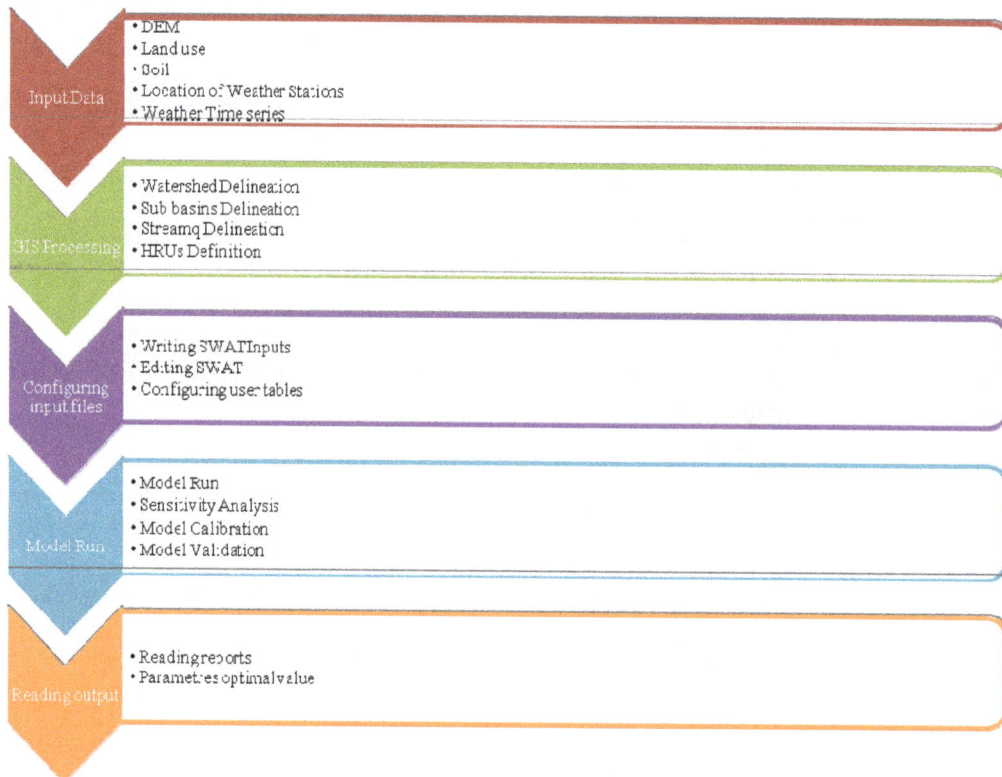

Figure 3. Components and input/output data of SWAT model.

The Contribution of the Geospatial Information to the Hydrological Modelling of a Watershed with Reservoirs:
Case of Low Oum Er Rbiaa Basin (Morocco)

143

map of central Morocco set by the department of the physical environment of the National Institute of Agronomic Research on the scale of 1:500,000 published in 2001. The area consists of several types of soil. the proportion of the area is marked by the dominance of "Calcimagnesian" soil (25%), "Isohumic" soil (17%), the "Mineral Brutes" soil and the "littlel Evolved" soils which are respectively 14% and 13% of the total surface area (**Figure 4(a)**).

- **Land Use**

The land use map (**Figure 4(b)**) was developed from a combination of data sources including:

  ○ The satellite image processing LANDSAT4 through directed classification;

  ○ The data collected and recovered from various agencies: hydraulics agency, agriculture department etc.;

  ○ The type of land use is dominated by the culture Bour which covers 80% of the basin and that spans the entire basin, while the distance covered does not exceed 7% as irrigated crops occupy 5%, and the bare soil 4% of the entire basin. Forest areas are also poorly implemented and only 2% of the total area of the basin.

- **Geology**

The geological map (**Figure 4(c)**) was obtained through the digitization of a set of geological maps (1:250,000) of Morocco. The analysis of this map highlighted certain characteristics of the zone of study as the nature of aquifers and the permeability.

- **Digital Terrain Model (DTM)**

Digital Terrain Model (**Figure 4(d)**) is extracted from the ASTER GDEM satellite with a spatial resolution of 30 m. The DTM is one of the most important data used to run the SWAT model.

Climate and Hydrometry

The study area contains no hydrometric station, no stations of measuring precipitation or temperature. The main station used to this end is located at the dam Al Massira which is managed by the Hydraulic Basin Agency of Oum Er Rbiaa. In addition to this station the Nouasseur station located near the basin was used for the integration of climate information.

For the three reservoirs managed by the National Office of Electricity, the available measures are just the monthly hydric balance sheets of the studied period. These data include the contributions of rivers, outputs, turbined volumes and samples.

The structuring of a series of climate data on a daily scale required some statistical treatments. Indeed, the analysis of temperature data shows some data missing at

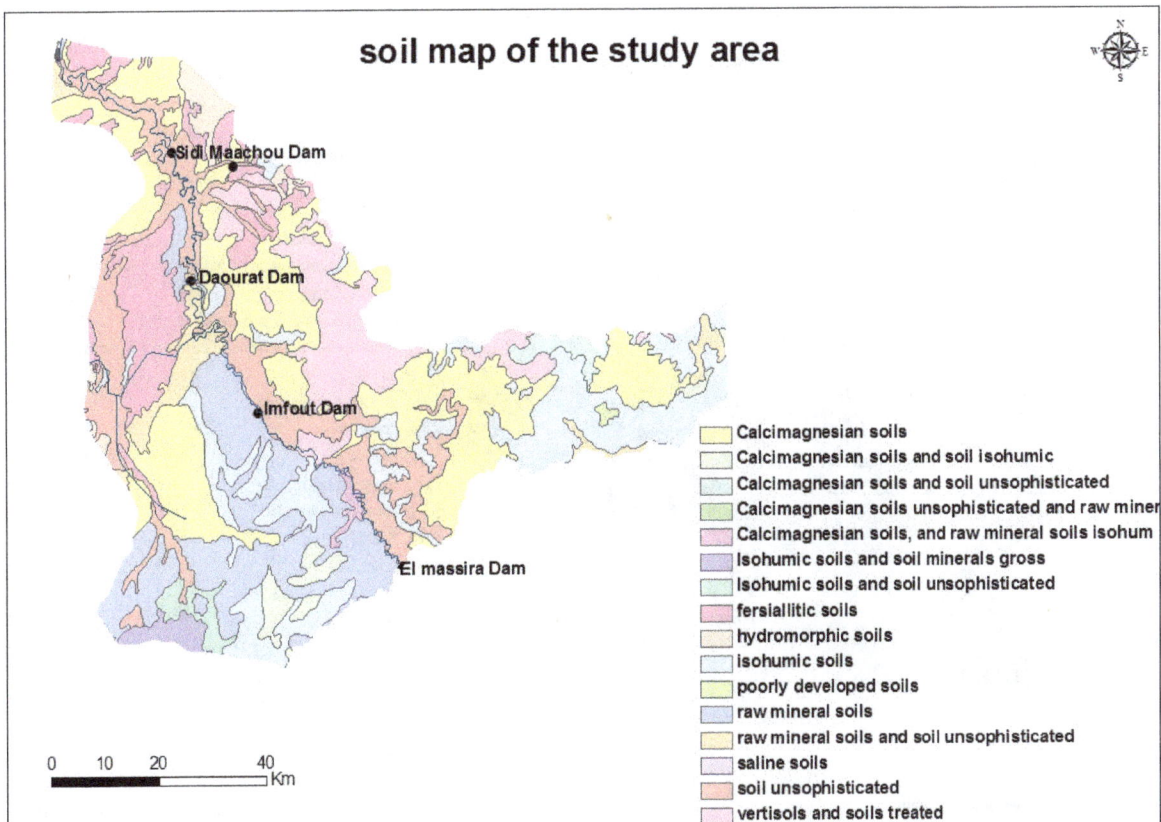

soil map of the study area

Calcimagnesian soils
Calcimagnesian soils and soil isohumic
Calcimagnesian soils and soil unsophisticated
Calcimagnesian soils unsophisticated and raw miner
Calcimagnesian soils, and raw mineral soils isohum
Isohumic soils and soil minerals gross
Isohumic soils and soil unsophisticated
fersiallitic soils
hydromorphic soils
isohumic soils
poorly developed soils
raw mineral soils
raw mineral soils and soil unsophisticated
saline soils
soil unsophisticated
vertisols and soils treated

(a)

Map of land use in the study area

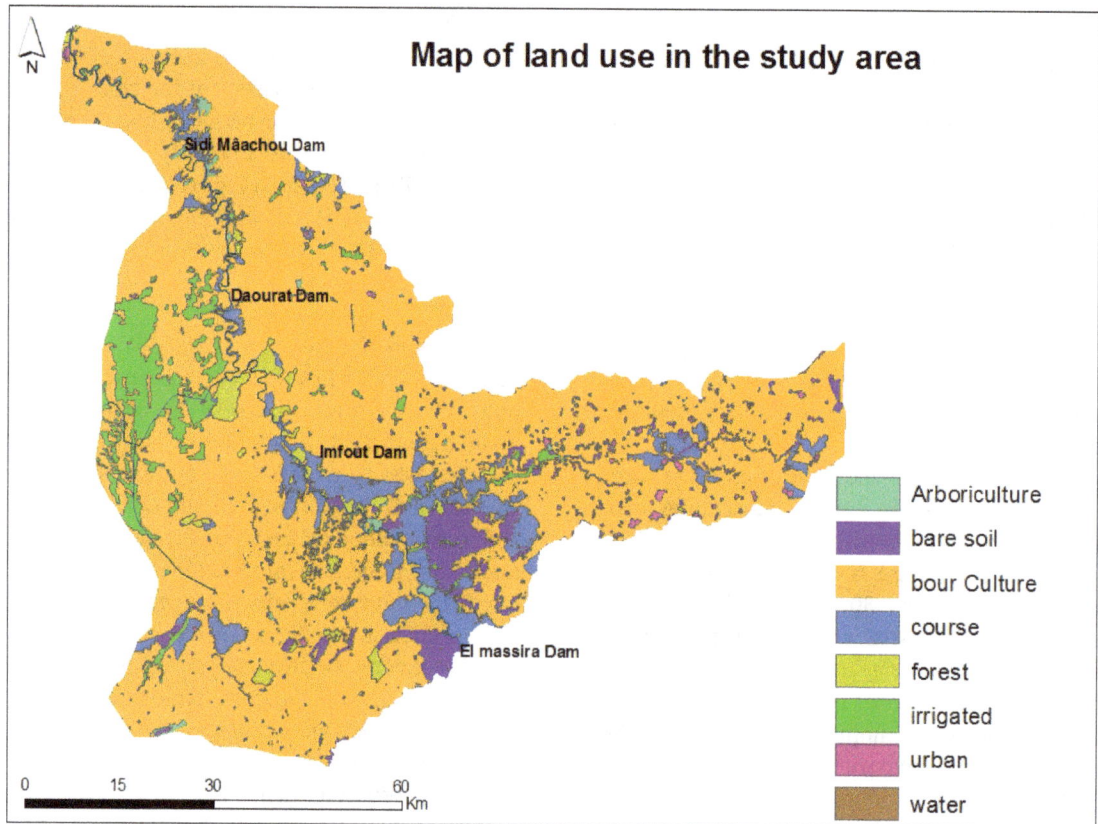

Sidi Mâachou Dam

Daourat Dam

Imfoût Dam

El massira Dam

	Arboriculture
	bare soil
	bour Culture
	course
	forest
	irrigated
	urban
	water

0   15   30   60   Km

(b)

geological map of the study area

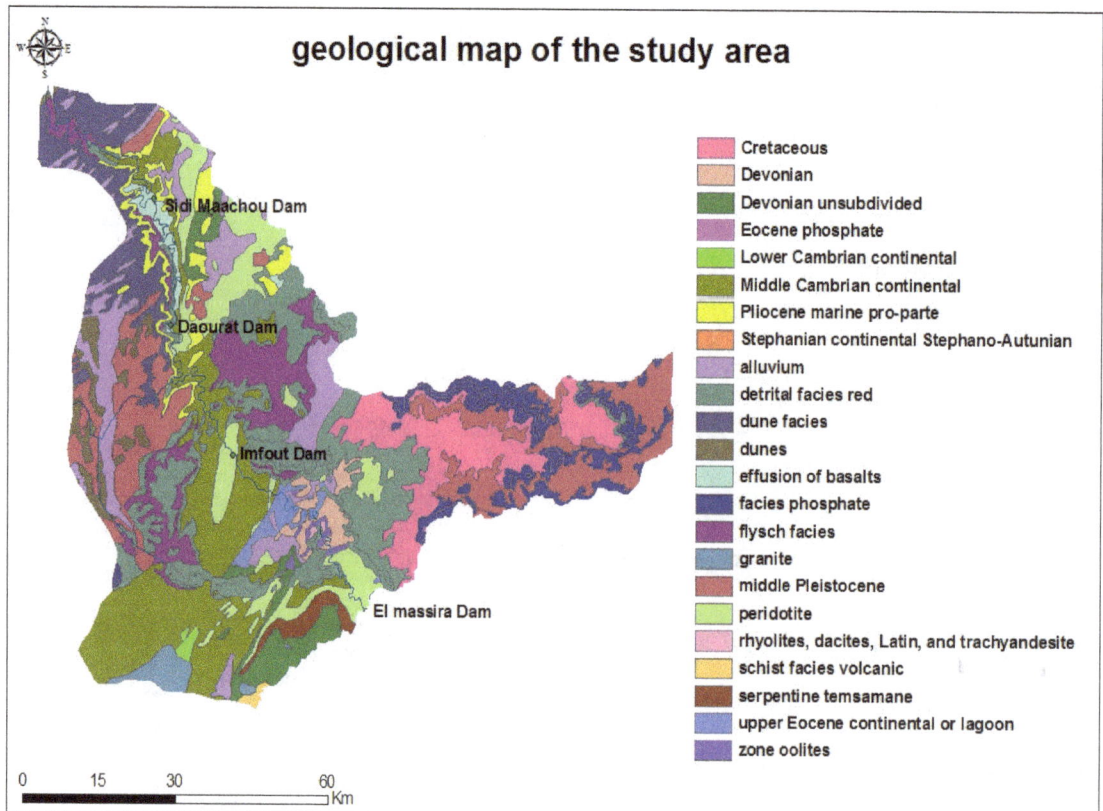

Sidi Maachou Dam

Daourat Dam

Imfout Dam

El massira Dam

	Cretaceous
	Devonian
	Devonian unsubdivided
	Eocene phosphate
	Lower Cambrian continental
	Middle Cambrian continental
	Pliocene marine pro-parte
	Stephanian continental Stephano-Autunian
	alluvium
	detrital facies red
	dune facies
	dunes
	effusion of basalts
	facies phosphate
	flysch facies
	granite
	middle Pleistocene
	peridotite
	rhyolites, dacites, Latin, and trachyandesite
	schist facies volcanic
	serpentine temsamane
	upper Eocene continental or lagoon
	zone oolites

0   15   30   60   Km

(c)

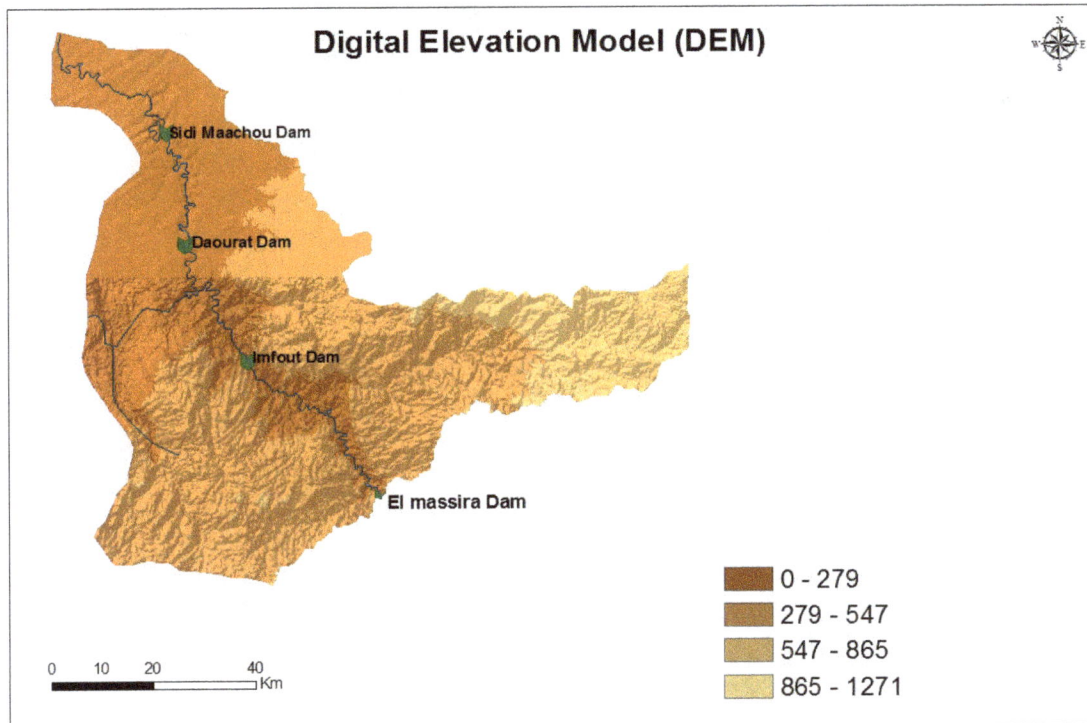

(d)

Figure 4. (a) Soil map; (b) Map of land use; (c) Geological map; (d) Digital elevation model (DEM).

some daily observations. A procedure has been developed to address these missing data by adopting this approach focusing on the following two axes:

- When the number of missing values is less than 4, the missing values are replaced by the average of the observations around.
- When the number of missing values is upper or equals 4, the missing values are replaced by the attested averages for the same periods in similar years for which information is available.

2.5. Configuration Model

After preparing all the data required by the SWAT model, the next step in the modeling process for the Lower Oum Er Rbiaa Basin was to set up the model. Thus, the intersection of temporal data available on climate has led us to choose the period 2001-2010 for the implementation of the model. The configuration of SWAT was made along the following parameters:

- Time input scale: daily
- Time output scale: monthly
- Application period: 10 years (2001-2010)
- Initialization of the model: one year (2001)
- Calibration of the model: 4 years (2002-2005)
- Model Validation: 5 years (2006-2010)

Application of SWAT model requires a number of steps in order to arrive at the desired result. These are

mainly the following steps:
- The introduction of DEM in the model;
- The introduction of the outlet of the study area;
- The generation of drainage;
- Specifying the location of reservoirs, dams in our case;
- The division of the basin into sub-basins, indicating to the model the points to consider as exit points (outfalls secondary sub-basin) and which must include the reservoirs mentioned above. This step helped generate 28 sub-basins constituting the backbone of the study area (**Figure 5**);
- The introduction of layers of soil and land use by filling the necessary attributes.

Following this implementation, the model generates, through the intersection of the DEM, soil, land use and sub-basins, the spatial units of work called "HRU" and which represent a hydrologic response. Each of these units consists of the same class of slope, soil and land use [15]. These units represent the spatial resolution and basic operation of the SWAT model. All processes modeled by SWAT are simulated at the spatial scale of these units.

To avoid generating very small units and a very large number units by combining the four layers mentioned above, a tolerance of 10% was introduced in the model to avoid entities of slope, soil or land use that occupy less than 10% of the surface of the sub-basins while also excluding through this procedure the most sensitive types

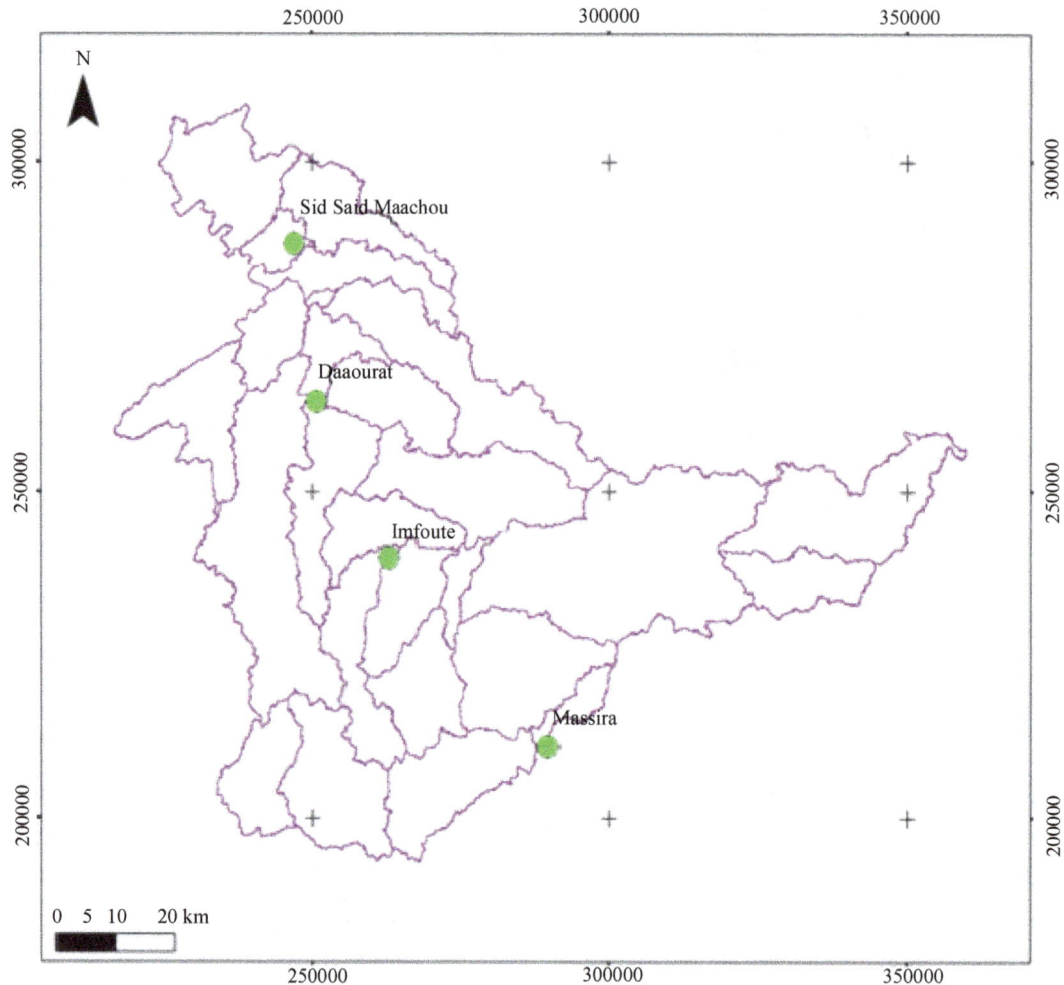

Figure 5. Delineation of sub-basins.

of land use such as urban areas and water surfaces.

Climate data and data from the three dams are subsequently integrated in the appropriate input files of the model.

Once the various input data have been entered, the next step in the process of implementation of the SWAT model is to calibrate and validate the model of the area studied.

The calibration of the model is to find the optimal values of the model parameters which allow better representation of the physical reality of the functioning of the modeled basin [16]. The calibration consists in determining the values which allow the matching of the parameters stimulated by the model and with those observed at a particular point in the basin. The calibration procedure used is the Parasol method [17]. The calibration has been performed at the Imfout dam by comparing monthly water yield of the river at the dam and that simulated by the model. The choice of this reservoir model calibration is dictated by the following considerations:

- Area: the area of the sub-basin associated with this

reservoir is the largest compared to the others.

- Representativity of data: this sub-basin is one for which most climate data is available because it is the dam nearest stations measuring precipitation and temperature.

The calibration consisted in inputting into the model on a daily basis data concerning:

- The year 2001 (365 values) to initialize the parameters of the model;
- The years 2002 to 2005 (4 years * 365 values = 1460 values) for model calibration.

To assess the performance of model calibration, a second phase, called model validation, is implementation. In this phase the values of the model parameters found in the calibration phase are applied to a new period to estimate the difference between the values simulated by the model and those observed in the field [18]. The assessment of this difference is carried out using indicators such as the linear correlation coefficient and the coefficient of Nash. This index provides a more accurate assessment of the efficiency relative to the volumes of the

The Contribution of the Geospatial Information to the Hydrological Modelling of a Watershed with Reservoirs:
Case of Low Oum Er Rbiaa Basin (Morocco)

147

flow and absolute deviations [19]. This is defined by the Formula (2):

$$E\ Nash = 1 - \frac{\sum_{i=1}^{N} (y_{iobs} - y_{sim})}{\sum_{i=1}^{N} (y_{iobs} - \overline{y}_{obs})} \qquad (2)$$

With:

y_{iobs} = mean value of the observed variable

y_{obs} = observed value of the variable

y_{sim} = simulated value of the variable

The Nash criterion varies from $-\infty$ for very bad adjustment to 1. A value close to 1 indicates a strong correlation between observations and simulations [20].

Model validation was conducted over a five year period (2006-2010).

3. Results and Discussion

Results of Calibration and Validation of the Model

Calibration of the model is based on a twenty hydrological parameters. Aiming at reducing the number of iterations and the computing time we selected eleven parameters. The choice of this set of parameters was done based on sensitivity analysis results. SWAT implements the Latin Hypercube method to study how the variation in the input parameters can affect the variation of the output. This study reveals that the most sensitive parameters are CN2, Alpha_BF and soil parameters.

The model requires making twenty thousand iterations to determine the optimum values. This corresponds to duration of treatment of more than 960 hours.

The following table shows the parameters that we have chosen with their meanings and their default values after calibration (**Table 1**).

It appears from **Figures 6** and **7** that the observed flow and simulated flow are well represented. The Nash coefficient was 0.63 for the calibration period and 0.53 for the validation period of the model. The level of this indicator appears satisfactory on the quantity and spatial distribution of data used [21]. This calibration can be further improved by working on a longer and more spatial series of data.

The balance of water generated by the model shows a dominance of evapotranspiration which represents more than 70% of incident rainfall while surface runoff generally does not exceed 25%. The following table summarizes the balance of water for each year of the study period from 2002 to 2010 (**Table 2**).

Table 1. Table of parameters used.

Parameters	Description	Default	Interval of variation	Optimal value
CN2	The curve number	-	−25 à 25	21.8
ALPHA_BF	coefficient of depletion of groundwater	0.048	0 à 1	0.022
CANMX	Index covered up	0	0 à 10	4.24
EPCO	compensation factor evapotranspiration for plant uptake as a function of depth	1	0 à 1	0.85
ESCO	compensation factor soil evaporation according to the depth	0.95	0 à 1	0.58
REVAPMM (mm)	threshold of evaporation from the water table	1	−100 à 100	94.56
SLOPE	the slope	-	−25 à 25	0.87
SOL_AWC	Amount of water available to plants	0.16 - 0.19	−25 à 25	24.56
SOL_K (mm.h-1)	Saturated hydraulic conductivity	22.7 - 24.73	−25 à 25	2.35
SOL_Z (mm)	Layer depth	-	−25 à 25	10.15
SURLAG (jours)	response time of the watershed	4	0 à 10	7.45

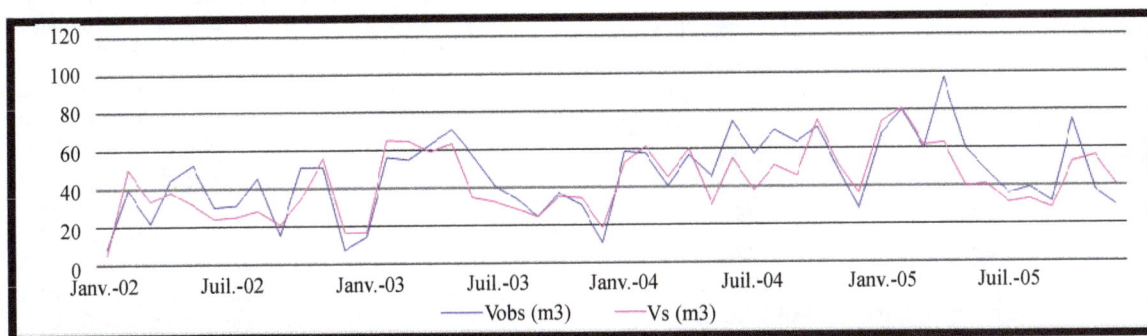

Figure 6. Flow diagram comparing observed and simulated flow for the calibration.

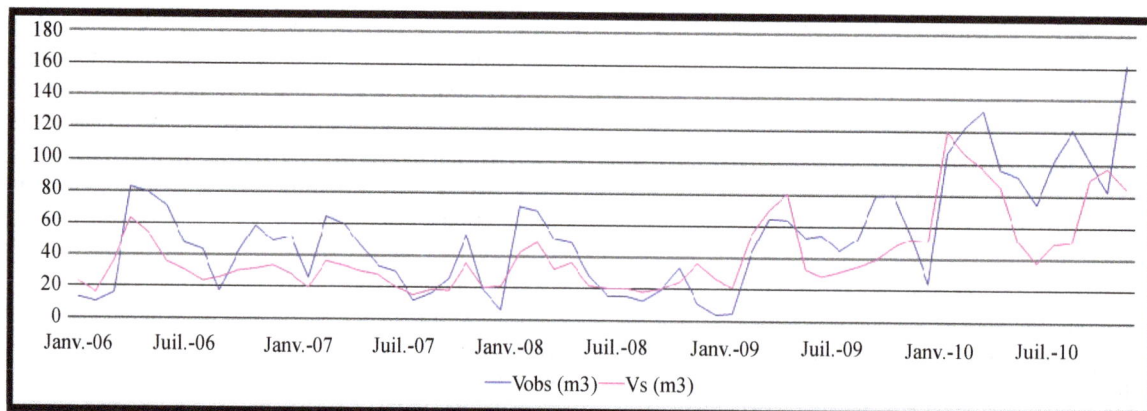

Figure 7. Comparison chart of the observed flow and simulated flow for validation period.

Table 2. Water balance for the period of calibration and validation.

	Years	Runoff (mm)	Evapotranspiration (mm)
Calibration period	2002	72.48	176.11
	2003	86.96	191.91
	2004	93.86	186.96
	2005	36.73	120.57
	2006	72.4	200.83
Validation period	2007	51.83	108.86
	2008	106.74	144.26
	2009	81.89	195.78
	2010	76.38	170.36
	average	75	166

4. Conclusions

In conclusion the first results obtained are satisfactory for both phases: calibration and validation.

The wedging of the model on the studied zone made it possible to determine the optimal values of the various parameters of the SWAT model and to establish the partition of the hydrological balance sheet in the basin of Oum Er Rbiaa. It also made it possible to estimate the importance of different hydrological processes at each sub-basin drained by the three dams. This will provide a better guide for the management methods in these dams, especially as far as the production of electric energy which constitutes currently the main function of these dams is concerned.

Nevertheless, simulations and water balance can be improved by integrating more data, especially climate related data. The calibration of the model also opens new perspectives on the study of the effects of climate change and changes induced by human activities on water resources system in both terms of quantity and quality.

REFERENCES

[1] H. Bouaouda and Y. Timoulali, "The Contribution of Remote Sensing and GIS Hydrogeological Research in the Sahel Region of Central Doukkala (Western Morocco)," Remote Sensing Francophonie: Critical Analysis and Outlook. Éd. AUF, Hachette Diffusion Internationale, Laussane, 2000, pp. 205-213.

[2] B. Héléne, "Variational Data Assimilation for Distributed Hydrological Modeling of Floods with Fast Kinetics," Ph.D. Dissertation, Institut de National Polytecnique, Toulouse, 2008

[3] A. Chaponnière, G. Boulet, A. Chehbouni and M. Aresmouk, "Understanding Hydrological Processes with Scarce Data in a Mountain Environment," Hydrological Processes, Vol. 22, No. 12, 2008, pp. 1908-1921.

[4] J. P. Fortin, R. Moussa, C. Bocquillon and J. P. Villeneuve "Hydrotel, a Distributed Hydrological Model Compatible with Remote Sensing and Geographical Information Systems," Revue des Sciences de l'eau, Vol. 8, No. 1, 1995, pp. 97-124,

[5] R. Srinivasan, T. S. Ramanarayanan, J. G. Arnold, and S.

T. Bednarz, "Large Area Hydrologic Modeling and Assessment. Part 2: Model Application," *Journal of the American Water Resources Association*, Vol. 34, No. 1, 1998, pp. 91-101.

[6] L. Boithias, "Transfer Modeling of Pesticides in Watershed Scale during Vintage," Ph.D. Dissertation, Institut Polytechnique de Toulouse, Toulouse, 2012

[7] H. Somaya, "Study on Wetland Vegetation in Morocco Catalog and Analysis of Plant Biodiversity and Identification of Major Plant Groups," Ph.D. Dissertation, University Mohamed V Rabat, Rabat, 2004, p. 104.

[8] H. Somaya, "Study on Wetland Vegetation in Morocco Catalog and Analysis of Plant Biodiversity and Identification of Major Plant Groups," Ph.D. Dissertation, University Mohamed V Rabat, Rabat, 2004, p. 105.

[9] J. G. Arnold, R. Srinivasan, R. S. Muttiah and J. R. Williams, "Large Area Hydrologic Modelling and Assessment. Part I: Model Development," *Journal of the American Water Resources Association*, Vol. 34, No. 1, 1998, pp. 73-89.

[10] F. L. Ogden, J. Garbrecht, P. A. DeBarry and L. E. Johnson, "GIS and Distributed Watershed Models, II: Modules, Interfaces, and Models," *Journal of Hydraulic Engineering*, Vol. 6, No. 6, 2001, pp. 515-523.

[11] S. L. Neitsch, J. G. Arnold, J. R. Kiniry, J. R. Williams and K. W. King, "Soil and Water Assessment Tool Theoretical Documentation—Version 2005," Soil and Water Research Laboratory, Agricultural Research Service, US Department of Agriculture, Temple, 2005.

[12] B. B. Ashagre, "SWAT to Identify Watershed Management Options: Anjeni Watershed, Blue Nile Basin, Ethiopia," Master's Thesis, Cornell University, New York, 2009.

[13] G. S. Shimelis, R. Srinivasan and B. Dargahi, "Hydrological Modelling in the Lake Tana Basin, Ethiopia Using SWAT Model," *The Open Hydrology Journal*, Vol. 2, No. 1, 2008, pp. 49-62.

[14] C. H. Green, M. D. Tomer, M. Di Luzio and J. G. Arnold,

"Hydrologic Evaluation of the Soil and Water Assessment Tool for a Large Tile-Drained Watershed in IWO," *American Society of Agricultural and Biological Engineers* Vol. 49, No. 2, 2006, pp. 413-422.

[15] A. Sheshukov, K. Douglas-Manking and P. Daggupati, "Evaluating the Effectiveness of Unconfined Livestock BMPs Using SWAT" *Proceeding of the International SWAT Conference*, Colorado, 7-9 August 2009, pp. 204-211

[16] P. W. Gassman, M. R. Reyes, C. H. Green and J. G. Arnold, "The Soil and Water Assessment Tool: Historical Development, Applications, and Future Research Directions," *American Society of Agricultural and Biological Engineers*, Vol. 50, No. 4, 2007, pp. 1211-1250.

[17] A. Van Griensven and T. Meixner, "Methods to Quantify and Identify the Sources of Uncertainty for River Basin Water Quality Models," *Water Science and Technology*, Vol. 53, No. 1, 2006, pp. 51-59.

[18] K. L. White and I. Chaubey, "Sensitivity Analysis, Calibration, and Validations for a Multisite and Multivariable SWAT Model," *Journal of the American Water Resources Association*, Vol. 41, No. 5, 2005, pp. 1077-1089.

[19] P. Reungsang, R. S. Kanwar, M. Jha, P. W. Gassman, K. Ahmad and A. Saleh, "Calibration and Validation of SWAT for the Upper Maquoketa River Watershed," Center for Agricultural and Rural Development, Iowa State University, Ames, 2005.

[20] A. Fadil, H. Rhinane, A. Kaoukaya, Y. Kharchaf and O. Bachir, "Hydrologic Modeling of the Bouregreg Watershed (Morocco) Using GIS and SWAT Model," *Journal of Geographic Information System*, Vol. 3 No. 4, 2011, pp. 279-289.

[21] C. Santhi, J. G. Arnold, J. R. Williams, W. A. Dugas and L. Hauck, "Validation of the SWAT Model on a Large River Basin with Point and Nonpoint Sources,"*Journal of the American Water Resources Association*, Vol. 37, No. 5, 2001, pp. 1169-1188.

Application of LiDAR Data for Hydrologic Assessments of Low-Gradient Coastal Watershed Drainage Characteristics

Devendra Amatya[1], Carl Trettin[1], Sudhanshu Panda[2], Herbert Ssegane[3]
[1]USDA Forest Service, Cordesville, USA
[2]Gainesville State College, Oakwood, USA
[3]University of Georgia, Athens, USA

ABSTRACT

Documenting the recovery of hydrologic functions following perturbations of a landscape/watershed is important to address issues associated with land use change and ecosystem restoration. High resolution LiDAR data for the USDA Forest Service Santee Experimental Forest in coastal South Carolina, USA was used to delineate the remnant historical water management structures within the watersheds supporting bottomland hardwood forests that are typical of the region. Hydrologic functions were altered during the early 1700's agricultural use period for rice cultivation, with changes to detention storage, impoundments, and runoff routing. Since late 1800's, the land was left to revert to forests, without direct intervention. The resultant bottomlands, while typical in terms of vegetative structure and composition, still have altered hydrologic pathways and functions due to the historical land use. Furthermore, an accurate estimate of the watershed drainage area (DA) contributing to stream flow is critical for reliable estimates of peak flow rate, runoff depth and coefficient, as well as water and chemical balance. Peak flow rate, a parameter widely used in design of channels and cross drainage structures, is calculated as a function of the DA and other parameters. However, in contrast with the upland watersheds, currently available topographic maps and digital elevation models (DEMs) used to estimate the DA are not adequate for flat, low-gradient Coastal Plain (LCP) landscape. In this paper we explore a case study of a 3rd order watershed (equivalent to 14-digit hydrologic unit code (HUC)) at headwaters of east branch of Cooper River draining to Charleston Harbor, SC to assess the drainage area and corresponding mean annual runoff coefficient based on various DEMs including LiDAR data. These analyses demonstrate a need for application of LiDAR-based DEMs together with field verification to improve the basis for assessments of hydrology, watershed drainage characteristics, and modeling in the LCP.

Keywords: Santee Experimental Forest; Digital Elevation Models (DEM); Drainage Area; Drainage Network; Low-Gradient Coastal Plain (LCP)

1. Introduction

Watersheds are an organizing framework for the assessment of hydrologic and ecological functions and various impacts of the landscape. Reliable and sustainable water yield from watersheds in the Southeastern Coastal Plain has become an area of concern in recent years because of changing population growth, land use, and potential climate change. To address this concern, there is a need for a reliable understanding of hydrologic processes and water balance of less disturbed, forested watersheds on low-gradient Coastal Plain (LCP) lands [1-7]. This reference water balance could be used to quantify the magnitude and potential change to water balance in the LCP due to the impacts of human and natural disturbances, which is important for economic development and land management practices. Resource data (e.g., topography, hydrography, land use/land cover, soils) characterizing the watersheds are the basis for those assessments, and its resolution may affect results and interpretations. Monitoring and modeling approaches are often used to understand the processes and quantify the runoff, water balance, and pollutant loads [8]. However, there are challenges in accurately quantifying the water and chemical balance of these LCP systems primarily due to the difficulty in accurately estimating the watershed drainage area used in depth-based runoff.

The drainage area can be estimated with a reasonable accuracy for high-gradient upland watersheds with regularly available US Geological Survey (USGS) topographic quadrangle maps (Scale 1" = 1 mi with 20-ft contour intervals) or digital elevation models (DEMs). Most of the analyses in the study were conducted using USGS topographic-quad map based DEMs. The DEM is a digital cartographic/geographic dataset of elevations in xyz coordinates. The terrain elevations for ground positions are sampled at regularly spaced horizontal intervals. DEMs are derived from hypsographic data (contour lines) and/or photogrammetric methods using USGS 7.5-minute, 15-minute, 2-arc-second (30- by 60-minute), and 1-degree (1:250,000-scale) topographic quadrangle maps." http://tahoe.usgs.gov/DEM.html. The 30 m DEM developed from hypsographic data (contour lines) and/or photogrammetric methods using the 7.5' USGS topographic quadrangle map is generally the hydrologists' choice for watershed drainage area delineation [9,10]. However, more accurate topographic maps may be needed to obtain an accurate estimate of the drainage area in the LCP where the land slope is very flat with a few contours. For example, recently [11] found huge discrepancies in the elevation data of Interstate (I-95) obtained by 10 m DEM created from 10 m contours developed by hand digitization of USGS 1:24,000 scale topographic maps compared to the DEM developed from the 2010 Light Detection and Ranging (LiDAR) data. The older 10 m DEM and the latest LiDAR-DEM provided an average difference of 0.9 m on randomly selected locations on the I-95. They converted the 10 m DEM for the study area (Camden County, GA, a southern coastal county) using the algorithm developed with differentiation values of both. The study area, USDA Forest Service Santee Experimental Forest (SEF), used in this study is of similar topographic nature. Similarly, [12] reported that the USGS topographic maps are considered insufficiently accurate in their topographical representation of watershed boundaries, slopes, and upslope contributing areas to meaningfully apply detailed process-based soil erosion assessment tools at the field scale. However, the authors also concluded that DEMs based on USGS 10-ft contour lines from publicly available data can be as good as the most accurate datasets obtained from real-time kinematic differential GPS (RTK-DGPS) in estimating average annual off-site runoff (−18.3% error) and sediment yield using the WEPP model within a 30-ha upland watershed. Most recently, [13] reported the effects of uncertainty in estimating elevations from various DEM types on erosion rates. Similarly, in another study by [14], the authors argued that it is often difficult to quantify soil loss due to gully erosion because

the footprint of individual gullies is too small to be captured by most generally available DEMs, such as the USGS National Elevation Dataset. James et al. [15] also noted that the standard DEMs generally lack the spatial and temporal resolution to perform change detection at the local gully scale. Depending on data sources, methods and procedures used to generate field DEMs, the DEM estimates contain errors [16]. Thus LiDAR data are being increasingly used to derive information on elevation [17]. However, only a very few or no studies have been conducted so far on the effects of errors in DEMs in estimating drainage area of watersheds in low-gradient coastal landscapes (LCP).

DEMs created using elevations from the USGS topographic quad maps may often be altered by construction of roads and cross-drainage structures, more so in flat LCP. In these flat lands such a road bed may serve as a boundary of the watershed that may not have been reflected in the old USGS topographic map based DEMs [18]. If not field verified for those road infrastructures and reconditioned in the DEM, the actual estimated drainage area as well as drainage network may well be different from that developed using the available DEMs. This clearly suggests a necessity for field investigation of road network and other flow control structures in addition to having accurate high resolution DEM for hydrologic analysis and modeling, especially that involves drainage area calculation, in coastal plains.

In this paper we illustrate how the recognition of historical land use features, including water management structures, as a result of high resolution spatial data, affects our understanding of hydrologic processes and pathways. We provide examples of hydrologic models to illustrate how the resolution of the resource data (e.g., soils, vegetation, land use, topography, and hydrography) used as model inputs and the model design may affect interpretations. Most process based models require some form of calibration and/or validation prior to their applications [8,19-21]; that calibration process typically involves modifying parameters or coefficients for specific processes to achieve reasonable model performance with respect to the output of interest. The assumption is that reasonable agreement between the simulated and measured data (e.g., stream discharge) reflects an accurate representation of the processes within the watershed. However, seemingly accurate predictions of stream flow at the watershed outlet may be achieved by complementary errors from internal processes resulting in inaccurate predictions of in-stream flows, water table depths, and soil moisture within the watershed [19].

Floodplains in the Coastal Plain of the southeastern United States were the principal agricultural zone during the early colonial era (e.g., late 1600's and early 1700's). In South Carolina, the freshwater floodplains were used for rice cultivation. The development of the land included water management features like reservoirs, impoundments, diversion and distribution channels, diked fields, and lateral ditches and collector canals [22]. Those man-made features remain on the landscape, however they are not apparent in the classical resource data used for hydrologic assessments and modeling [9,23]. Thus the current USGS topographic survey information is of insufficient resolution to demark these important land features affecting hydrologic pathways and functions as well as to delineate the LCP watersheds.

The objectives of this paper were: 1) to analyze LiDAR data and summarize field observations of the stream channel network and other various drainage and legacy water management features on the Santee Experimental Forest, South Carolina and 2) to summarize the effects of various topographic maps and DEM resolutions used in estimating drainage area on average annual runoff coefficient for a forested watershed (Turkey Creek) in the Atlantic Coastal Plain. Furthermore, the effects of roads and road culverts on estimated actual effective drainage area using the available DEM are also discussed in the

manuscript.

2. Materials and Methods

2.1. Site Description

2.1.1. Santee Experimental Forest

This work was conducted on the US Forest Service's Santee Experimental Forest (SEF) in South Carolina. The SEF is representative of the lower Coastal Plain landscape (LCP), characterized by low relief, mixed hardwood-pine flatwoods, and bottomland hardwood floodplains. The SEF was part of the Cypress Baroney that was conveyed by King Charles in 1697; the land was subsequently divided into three plantations in 1707, which is when the agricultural development began. The floodplains of first, second and third order streams were developed into rice fields during the early 1700's period. The present topographic, hydrography, soils and vegetative information for the forest convey a uniform, low-relief landscape (**Figure 1**). These are the typical spatial data that are used for hydrologic modeling [1,8,21].

LiDAR data for the SEF were obtained using Airborne

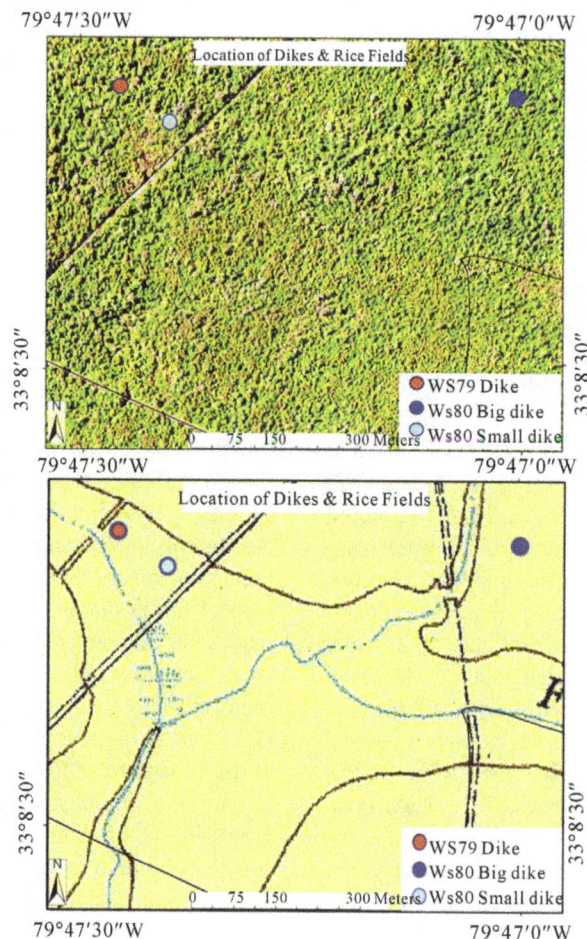

Figure 1. The aerial photograph, and USGS topographic map (1:24,000), of a section of the Huger Creek, Santee Experimental Forest.

Laser Terrain Mapping technique in 2007 by [24]. The raw LiDAR data were collected with a horizontal resolution of 0.1 m and a vertical accuracy of 0.07 - 0.15 m. The bare-earth return data were processed in ArcGIS to smooth the DEM and map potential stream channels using the ArcHydro extension (flow direction, length).

2.1.2. Turkey Creek Watershed

Most of the Turkey Creek watershed is in the Francis Marion National Forest on the coastal plain of South Carolina (**Figure 2**) with a small downstream portion including the gauging station within the Santee Experimental Forest. The US Forest Service established a stream gauging station in Turkey Creek in 1964 and monitored the watershed until 1984 only. Nevertheless,

researchers recognized the importance of stream gauging and other hydro-meteorological data from a forested coastal watershed as a reference in a rapidly changing coastal environment. As a result, in 2004, the US Forest Service, in cooperation with the College of Charleston and the USGS, reinstalled a real-time streamflow and rain gauging station, (http://waterdata.usgs.gov/sc/nwis/uv?site_no=02172035) approximately 800 m upstream of the historic gauging station [25].

The study watershed is located within the USGS quadrangle maps of Huger (NE), Bethera (SE), Shulerville (SW and SE), and Ocean Bay (NW and NE) with the approximate coordinate ranges of 610,400 to 628,600 easting and 3,658,500 to 3,670,500 northing [1].

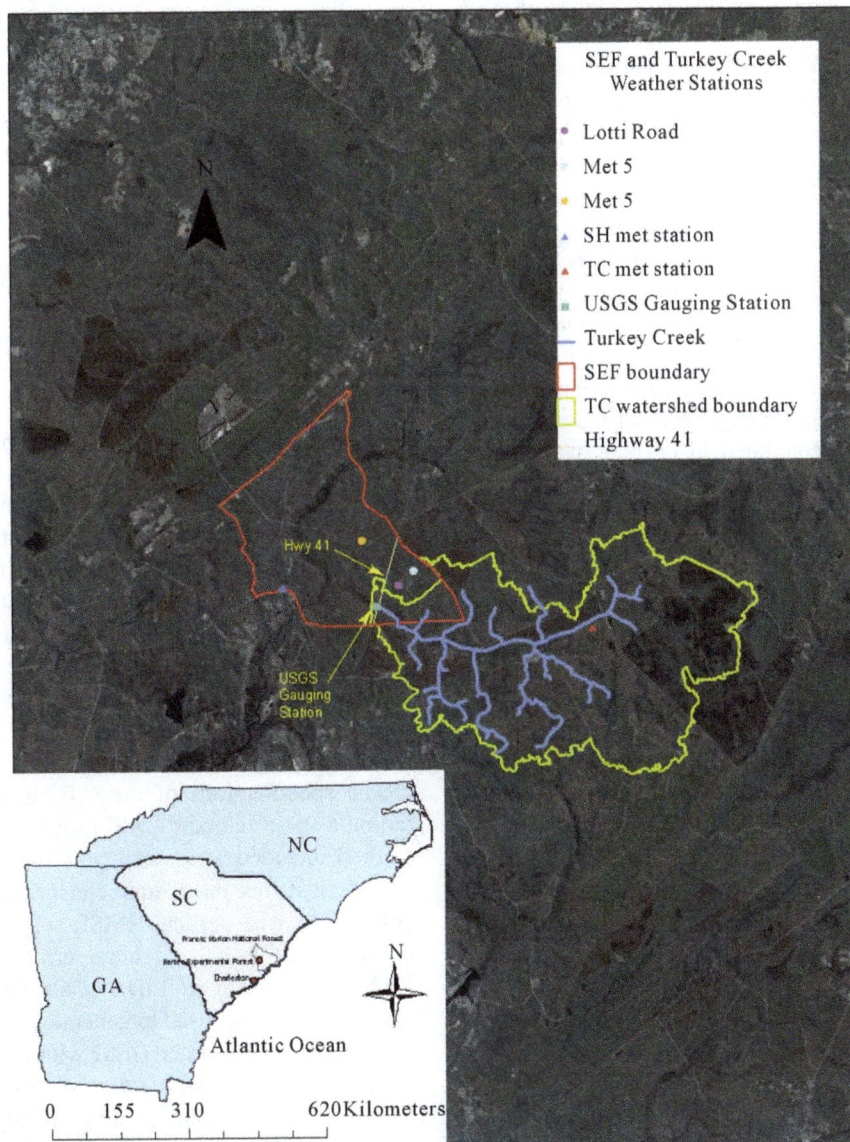

Figure 2. Location of the Turkey Creek watershed (green boundary) adjacent to the Santee Experimental Forest (red boundary) in coastal South Carolina (SC), also shown are the monitoring stations in and around the watershed.

Located within a 12-digit hydrologic unit code (HUC 030502010301) of the Catawba-Santee basin [26] at the headwaters of East Cooper River, a major tributary of the Cooper River draining to the Charleston harbor system, Turkey Creek (WS 78) is typical of other watersheds in the south Atlantic Coastal Plain, where rapid urban development is taking place. Technically, Turkey Creek is a 6th level hydrologic unit that qualifies only as a sub-watershed although we refer to it as a "watershed" in this paper. The topographic elevation of the watershed varies from about 2.0 m at the stream gauging station to 14 m above mean sea level [1].

2.2. Evolution of Drainage Area

The estimation of drainage area of the Turkey Creek watershed changed as more accurate maps and associated DEMs were available in the course of this 48-year (1964-2011) period (**Table 1**). When the watershed was established in 1964, the drainage area boundary was approximated using the then available USGS topographic map with a scale of 1 inch = 2 miles. Later on in early 1969 the 1 inch = 1 mile scale USGS map was used to obtain a new boundary and watershed area. Although flow data on the watershed were continuously collected until 1984, no analysis or publication was done with these historic data, except for internal station reports. After discontinuation of flow measurements in 1984, it was not reestablished until late 2004. Accordingly, there were not any updates in drainage area of the watershed although new DEMs continued to become available in the late 80's and 90's.

The first literature search on recent DEMs in late 2004, when the Turkey Creek watershed was reestablished, resulted in a 1999 publication for 14-digit hydrologic unit code (HUC) development for South Carolina [26] which used the 1:24,000-scale 7.5-minute series topographic maps as the source maps and the base maps from 1:100,000-scale Digital Line Graphs; however, the data

Table 1. Drainage areas and calculated average annual runoff coefficients (ROC) for Turkey Creek watershed based on map or DEM types used during 1964-2011 period.

Time	Map/DEM Type	Delineation Method	Drainage Area (ha)	ROC
1964	1" = 2 mi Topo	Manual	3240	0.38
1969	1" = 1 mi Topo	Manual	4575	0.27
2004	30 m DEM	ArcHydro	4920	0.25
2005	14-digit HUC	ArcInfo	5880	0.21
2008	10 m DEM	AV/SWAT	7260	0.17
2010	Partial LiDAR	ArcSWAT	6510	0.19
2011	Full LiDAR	ArcSWAT	5240	0.24

were published at a scale of 1:500,000. In the 2005 Hydrologic Unit Code (HUC) map for South Carolina [26], Turkey Creek area was listed as 6685 ha (16,508 ac) to its confluence with Nicholson Creek, which is further downstream from the current gauging station. The source maps for the basin delineations using the HUCs are 1:24,000-scale 7.5-minute series topographic maps, and 1:24,000-scale digital raster graphics. Same map was used by the USGS to delineate the area at the current gauging station on Highway 41N near Huger. Similarly, DEMs with 30 m horizontal resolution and 1 m vertical resolution were also available from SC Department of Natural Resources in 2004. Later in 2006, the USGS Enhanced 1:24,000 true 10 m horizontal 1 m vertical DEM was obtained from the USGS [28]. The cross-drainage structures on forest roads in and around the Turkey Creek watershed were also surveyed using 3-D Delorme topographic quad maps of 2002 [29] with a scale of 1:25,000 and a 1988 Forest Service Francis Marion National Forest map with a scale of 1:126,720 that showed perennial streams, bridges, and road names. The field equipment included a GPS unit-Garmin GPS V personal navigator (with an accuracy of 6 - 9 m), a digital camera, a 50 m measuring tape, and a compass during December 2006-April 2007 period. The details of the field survey results of cross-drainage structures (44 culverts) are given by [28].

The LiDAR data obtained in 2007 for Santee Experimental Forest [30] also contained a small part of the downstream portion of the Turkey Creek watershed adjacent to the Forest (**Figure 2**). This was available in very high resolution DEMs (at a 2 m point spacing or better, and gridded with a 1 m resolution and a vertical accuracy of 0.07 - 0.15 m). The estimate of drainage area was updated using DEMs covering that smaller area with LiDAR data. Finally, by August 2011, similar high resolution LIDAR data (draft only) for the Berkeley County, SC containing the whole Turkey Creek watershed was obtained from the SC Department of Natural Resources (SCDNR) [31].

The specifications of the LiDAR elevation data acquired as point cloud ("LAS") for the Turkey Creek watershed included a nominal point spacing of 1.5 meter and vertical root mean square error (RMSE) of 18.4 cm (**Figure 3**). The vertical RMSE suggests that there is a possibility of 18.4 cm average error in elevation in the DEM created from the LiDAR data. Quick Terrain Modeler® ×64 [32] was used to generate the bare earth ground and water surface raster DEM with a cell size of 1.5 m. Although the LiDAR data resolution was 0.3 m, the DEM of 1.5 m resolution was developed to reduce the file size and thus promote faster analysis. The generated DEM was then exported as an ERDAS IMAGINE image [33] and projected to State Plane NAD-83-South Carolina

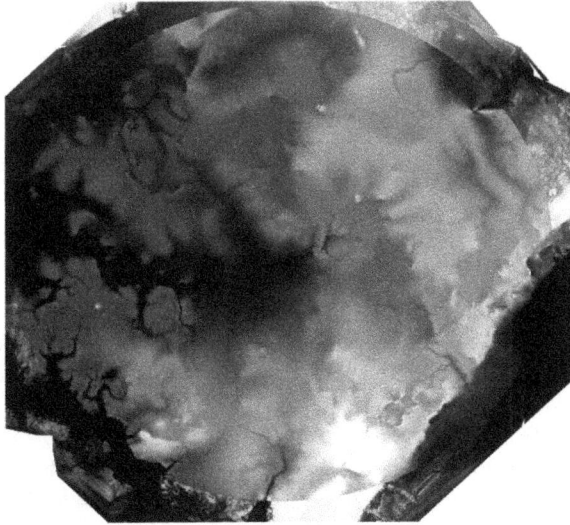

Figure 3. LiDAR-based DEM of 1.5 m × 1.5 m horizontal and 0.18 m vertical resolution.

FIPS 3900 feet, using ESRI® ArcGIS 9.3.1 [34].

2.3. LiDAR-DEM Reconditioning for Culverts

The DEM was further pre-processed to minimize generation of discontinuous streams due to presence of bridges and culverts. Additional challenges of working with low relief and very low slope watersheds such as Turkey Creek are such that wide areas can have a slope of near zero and the area is covered by a network of raised and compressed logging and forest service road beds. These effectively act as runoff barriers or miniature dams. To address some of these challenges, a comprehensive culvert survey of the study area was carried out in 2010 and 2011 using a combination of identifiers to predict the likely locations of culverts.

First, the National Hydrography Dataset (NHD; [35]) was used to locate where confirmed water vectors intersect the roads. National Agricultural Imagery program (NAIP) orthoimagery specific to our study area was used to identify where denser brighter green vegetation common to riparian zones and wetlands (areas where water accumulates for extended periods of time) approach roads. Lastly, the elevation data was used to identify where roads span across depressions where water can collect. These points were plotted at these locations and the data was taken to the field as coordinate lists and maps. Each location was driven to and searched. If a culvert was confirmed, its GPS coordinates and orientation were noted. Any additional culverts that were identified in the field were also recorded. Ideally, each feature would include location, orientation, width, length, and depth. However, the three later parameters were limited by the raster resolution and thus all culverts were represented by three-pixel wide (4.5 m) V-channels.

The DEM reconditioning was done using the Agree Streams capability of the ArcHydro extension for ArcGIS. The process lowers the pixel values that coincide with the chosen polyline shapefile by a pair of defined depth values for a given width in pixels. **Figure 4** shows where a culvert line was drawn along Forest Route 167 at a GPS point where the culvert was identified in a 120 m wide depression area just 15 m from the NHD location of Kutz Creek in the watershed. The channels for bridges had previously been inserted by the agency that collected and processed the LiDAR.

Total of 138 culverts were found by examination of bare earth LiDAR DEM, followed by field visits that were made in July 2010 (18 more culverts) and August 2011 (76 additional culverts) as shown in **Figure 5**.

These findings are consistent with those by [36] who showed that fine-scale LiDAR-derived maps can significantly improve field survey-based inventories of landslides with a subdued morphology in hilly regions. This is also similar to the observations of [14,15] who hypothesized that the ability of LiDAR data to map gullies and channels in a forested landscape should improve channel-network maps and topological models.

A reference stream layer containing the culvert locations and their sizes (lengthwise on the stream segments)

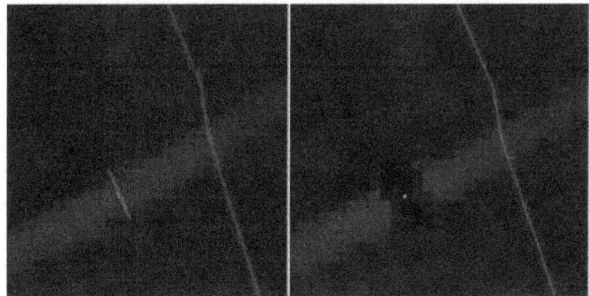

Figure 4. Before and after making channels at culvert locations using ArcHydro agree streams.

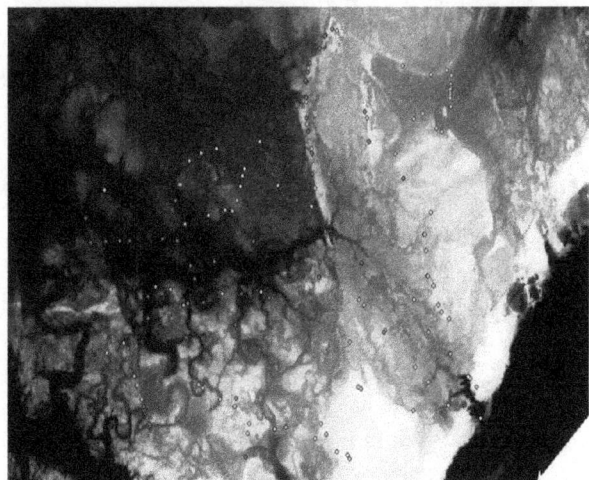

Figure 5. Location of culverts identified in LiDAR-DEM.

was used for the DEM reconditioning. The depth of culvert was used to recondition the DEM on the locations of the culverts through the ArcHydro DEM reconditioning function. Thus, the elevation at the culvert locations become accurate (lower than the elevations obtained from LiDAR data based DEM as LiDAR data cannot locate the culvert openings because they are buried below the roads).

The new DEM obtained after reconditioning for all culvert elevations was used to generate a new Turkey Creek watershed boundary and subwatersheds with ArcSWAT [20,37]. The automated watershed delineation processes of programs such as ArcSWAT [37] or BASINS model [38] use the Deterministic 8-neighbor (D8) algorithm for flow direction [39] based on the concept of steepest slope. Some of the documented weaknesses of D8-algorithm include tendency to generate parallel flow paths in flat areas, high sensitivity to inherent DEM errors, and inaccurate flow direction on convex slopes [40-42]. Other algorithms such as multiple flow direction (MFD), Deterministic-Infinity (D∞), Random-Eight Neighbor (Rho8), and Digital Elevation Model networks (DEMON) have been developed to address the above challenges [43]. This study does not address the effect of the algorithm used to automatically delineate watershed drainage area from DEM, but highlights the effects of DEM resolution using ArcSWAT which implements the D8 algorithm [39].

2.4. Hydrologic Analysis with Changed Watershed Boundary and Area

The historic stream flow data obtained by using the measured stage data and the stage-discharge relationship established in the early period [44] were archived in Forest Service data base [25]. No studies, however, were conducted or published using these flow data from this watershed for this period. The daily stream flow data was processed to obtain volume of water in cubic meters discharged in each year by multiplying by the drainage area with the given DEM method and conversion factors. The calculated mean annual outflow volume of 16.5 million cubic meters for the 13-year period was then divided by the calculated drainage area of the watershed obtained by each of the DEM methods since mid-1960's to obtain the depth-based outflow. The 50-year (1951- 2000) average annual rainfall of 1370 mm obtained from the data collected at a nearby weather station at Santee Experimental Forest (SEF) headquarters was used to calculate the average annual runoff coefficient (ROC). The ROC is the ratio of average annual depth-based outflow (mm) and average annual rainfall (mm). The ROC was calculated to assess its uncertainty, for that matter the uncertainty of the overall average annual water balance of the watershed,

due to uncertainty in drainage area obtained from each DEM type. The ROC calculated using the drainage area derived from the DEM based on the recent high resolution Li-DAR data was considered as a reference for compareson with the ROCs obtained using historic results.

3. Results and Discussion

3.1. Santee Experimental Forest

3.1.1. Detection of Historic Land Use Features

The LiDAR data were effective in delineating the drainage and agricultural water management features associated with the rice cultivation in the floodplain (**Figure 6**). The features range in size from dikes and dams (0.2 - 1.6 m height) to ditches (0.2 - 0.3 m depth). It's important to note the prominence of these features on the watershed, and to realize that their occurrence is within a watershed that has a total relief of less than 4 m. Within the context of this landscape, these dikes and ditches are major topographic features. The only reflection of these historical agricultural water management features on the current USGS topographic map (scale 1:24,000) are the major impoundment structures (see **Figure 1(b)**), but only a few of those existing structures are denoted.

In a similar application, [15] used LiDAR data to map gullies and headwater streams under a forest canopy in South Carolina, and found that LiDAR data provided robust detection of small gullies and channels, except where they are narrow or parallel and closely spaced. They reported that the ability of LiDAR data to map gullies and channels in a forested landscape should improve channel network maps and topological models.

3.1.2. Effects of Historical Water Management Features on Watershed Hydrology

The historical water management features may have been

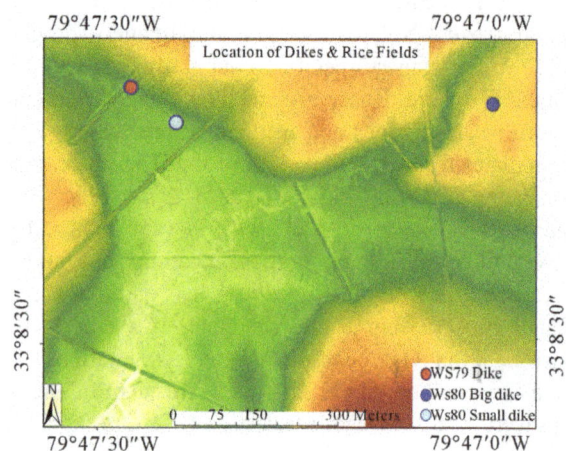

Figure 6. Depiction of surface topography derived from LiDAR data for a section of Huger Creek, Santee Experimental Forest. The location of dikes and ditches are noted.

affecting current watershed hydrology in several ways. Diversion ditches are affecting upland runoff processes including overland flow paths. These ditches were constructed to shunt water from reservoirs to fields located in the floodplain; hence they run perpendicular to the slope. The ditches, with the associated spoil bank, serve to interrupt surface runoff and to channel the runoff at points where water control structures existed (**Figure 7**).

The presence of these features is a major contradiction to the assumptions of hill slope runoff from the traditional resource data (topographic map). The effects of the collection and rechannelization are evident by drainage rivulets into the floodplain. The net effect of these ditches is to interrupt hillslope flow path, pool runoff, and redirect it through small channels. It is also likely that subsurface runoff is also affected. This may also ultimately alter travel time and time to peak of flooding at the watershed outlet.

The old field ditches and banks also affect runoff within the floodplain; these are major topographic features that will affect transport and routing, especially during flood stages. During non-flood periods, if the old ditches are not hydraulically linked to the stream, they may function as detention storage areas affecting infiltration positively and stream flow negatively.

3.1.3. Detection of Stream Network
The high resolution LiDAR data also proved useful in delineating the stream location. The USGS topographic maps convey a rather straight or direct-flowing stream; in contrast the stream generated with the LiDAR data illustrates a meandering channel (**Figure 8**).

As a result, the difference in stream channel configurations like the length and sinuosity among the two data sources is pronounced; for the stream reaches denoted in

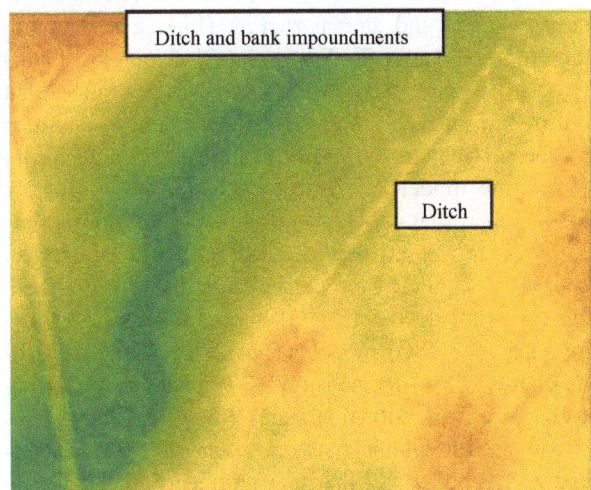

Figure 7. LIDAR image showing a stream diversion ditch running parallel to the present channel. The ditch and associated berm interrupt surface runoff.

Figure 8, the total channel length from the USGS map is 1853 m and it is 2962 m from the LiDAR data (**Table 2**).

Sinuosity, a ratio of channel (thalweg) distance and downvalley distance, describes whether a channel is straight or meandering. The sinuosity values range from 1.0 for straight channels to as high as 4.0 for highly intricate meandering streams [45]. The calculated sinuosity values were also different for these reaches when calculated with the USGS and LiDAR stream data (**Table 2**). Both the length as well as sinuosity was not even detected for reaches R6 to R8 when the USGS map was used.

None of the stream reaches would be considered meandering with a sinuosity ratio of 1.5 or greater) when calculated from the USGS topographic map. In contrast two of the stream reaches (R2 and R3) actually meander, based on the sinuosity values > 1.5 calculated from the LiDAR data (**Table 2**). The 61% increase in channel length and recognition of sinuosity in some reaches has important ramifications when considering peak discharge, time to peak, routing, in-stream processes and pollutant export from these watersheds.

3.1.4. Changes in Hydrologic Functions
Water management structures that were devised for rice cultivation in the floodplain that began 300 years ago are affecting contemporary surface water hydrology and stream channel hydraulics. As a result, hydrologic and hydraulic functions of the watershed have been potentially altered from conditions that were presumed to exist in these now-forested watersheds (**Table 3**).

The changes are associated with alterations to hill slope runoff including its pathways, structures within the floodplain changing depressional storage, and increased channel length and flow routing which results in longer

Table 2. Stream length and sinuosity for segments identified in Figure 8.

Stream Reach	Length		Sinuosity	
	USGS (m)	LiDAR (m)	USGS	LiDAR
R1	203.9	210.3	1	1.3
R2	432.3	692.6	1.1	1.6
R3	339.8	495.8	1.2	1.7
R4	438.2	562.7	1.1	1.2
R5	439.2	468.2	1	1.1
R6	N/A	130.3	N/A	1
R7	N/A	123	N/A	1
R8	N/A	279.2	N/A	1.4
Total	1853.4	2962.1		

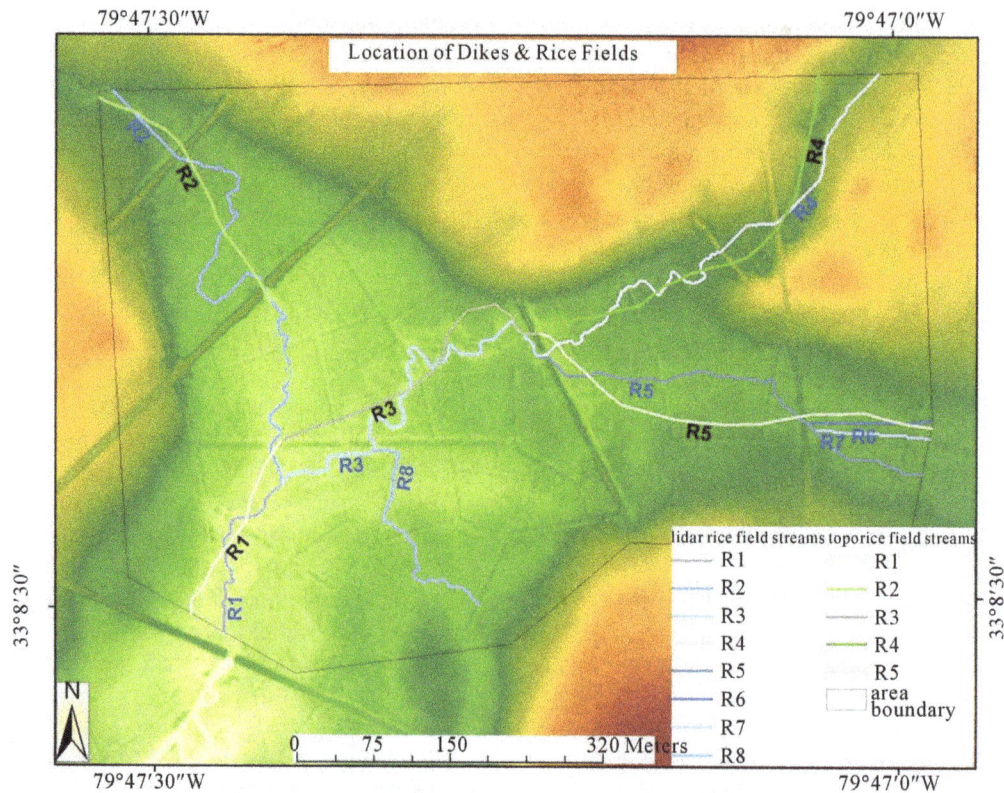

Figure 8. Overlay of the USGS topographic map (green) and LiDAR derived stream channel (blue).

Table 3. Effects of historical agricultural water management systems on hydrologic functions in floodplains of the Santee Experimental Forest.

Function	Rationale for Altered Functionality
Surface Storage	Interruptions in overland runoff may retard the runoff rate and increase infiltration and ET.
Runoff Routing	Interruptions in overland runoff effectively pool runoff and channelize the flow into the riparian zone.
Stream Routing	Development of a meandering stream system following agricultural abandonment has resulted in longer flow path than represented on topographic maps.
Flood Storage	Flood storage is likely increased with the presence of the dikes within the floodplain.
Water Table Depth	Longer surface water retention due to structures increase the water table elevation and soil moisture.

time to peak and reduced peak runoff rate. While active water management structures increase surface depressional storage enhancing the wetland hydrologic functions (e.g., water table elevations and soil moisture), it is uncertain how these relic structures affect depressional storage since the control structures are not functional.

3.1.5. Implications for Modeling
When modeling hydrology on the SEF watershed, the

landscape is represented by the readily available resource data (e.g., **Figure 1**). During model calibration, parameters and coefficients may be modified to achieve reasonable simulations, as compared to measured stream discharge. As an example, a common parameter to adjust for peak flow rates during calibration is the depressional storage, a parameter that is very difficult to measure directly [23]. It is evident that adjustments to depressional storage could mask or compensate for the effects of the actual channel and stream routing (**Figure 8**) and hill slope runoff (**Figure 7**). For example, depressional storage is a key parameter in the DRAINMOD model that controls the surface runoff rate after the soil is saturated and the surface storage is filled [46-48]. The effect is to modify the model behavior to achieve more accurate output, but if that calibration does not reflect actual hydrological processes, then the end results do not reflect accurately simulated processes within the watershed. Recently, [23] developed a GIS-based depressional storage capacity (DSC) model using USGS DEM data for one of the SEF watersheds (WS-77) and estimated 1 cm of effective depressional storage. When that storage factor was used to simulate stream discharge for the watershed using both DRAINMOD and its watershed-scale version, DRAINWAT [23,49], higher simulated peak flow rates were obtained for both the models for the 2003-2007 simulation period; that effect is likely due to an underestimation of the surface storage parameter for this watershed, which could result from not recognizing the historical water management structures not reflected in current DEMs.

3.2. Turkey Creek Watershed

3.2.1. Evolution of Drainage Areas

The watershed drainage area delineated in 1964 using the USGS topographic map with 1 inch = 2 miles scale is shown in **Figure 9**. Based on this the preliminary drainage area at that time was reported as 3240 ha (8000 ac) [44]. Some of the boundary followed the existing road. This drainage area resulted in the average annual ROC of 0.38 (**Table 1**). Such a value is generally observed for mountainous upland conditions [50] and is considered high for predominantly forested low-gradient watersheds in the humid coastal region [5,50].

The second estimated drainage area of the Turkey creek watershed in early 1970s was based on the USGS topographic map of 1 inch = 1 mile scale (**Figure 10, Table 1**).

The area of the watershed using this map was estimated at 4575 ha (11,300 acres) [51], which was about 41% higher than the initial estimate of 3240 ha. As a result, the calculated average annual ROC for this DEM was 0.27 only (**Table 1**), which is in the range of values obtained for similar coastal forested watersheds [52].

Later when the watershed gauging station was revitalized in late 2004 for conducting multi-collaborative hydrologic studies on this and adjacent 1st and 2nd order watersheds, a need of more reliable drainage area estimate was perceived. Accordingly, the new DEMs with 30 m horizontal and 1 m vertical resolution available in 2004 from SCDNR were used to delineate the watershed using the ArcView GIS software tools (**Figure 11**).

The new watershed boundary shown in **Figure 11** yielded the drainage area approximately at 4920 ha (12,150 acres) (**Table 1**). This was an increase of 7.5% compared to the estimate and a 52% increase from the initial (1964) estimate. Thus the drainage area continued to increase. Accordingly, the calculated average annual ROC continued to decrease to 0.25 (**Table 1**), which is also in the range of published data for this type of coastal forested watersheds [50,52,53]. Amatya and Radecki-Pawlik [25] used these data to compare the streamflow dynamics of this watershed with the two adjacent 1st and 2nd order watersheds. No field verification was done.

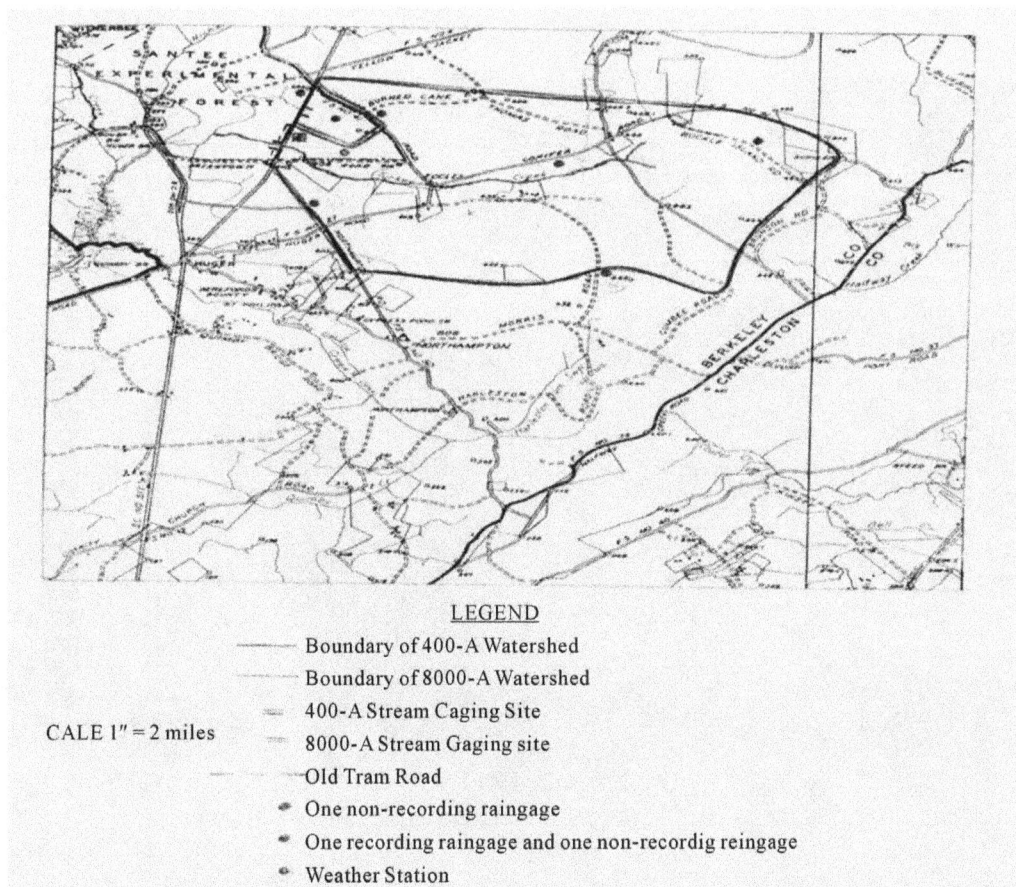

LEGEND

——— Boundary of 400-A Watershed

------- Boundary of 8000-A Watershed

= 400-A Stream Caging Site

= 8000-A Stream Gaging site

CALE 1″ = 2 miles

---- Old Tram Road

• One non-recording raingage

• One recording raingage and one non-recordig reingage

• Weather Station

Figure 9. Watershed boundary using 1" = 2 mile USGS topographic map.

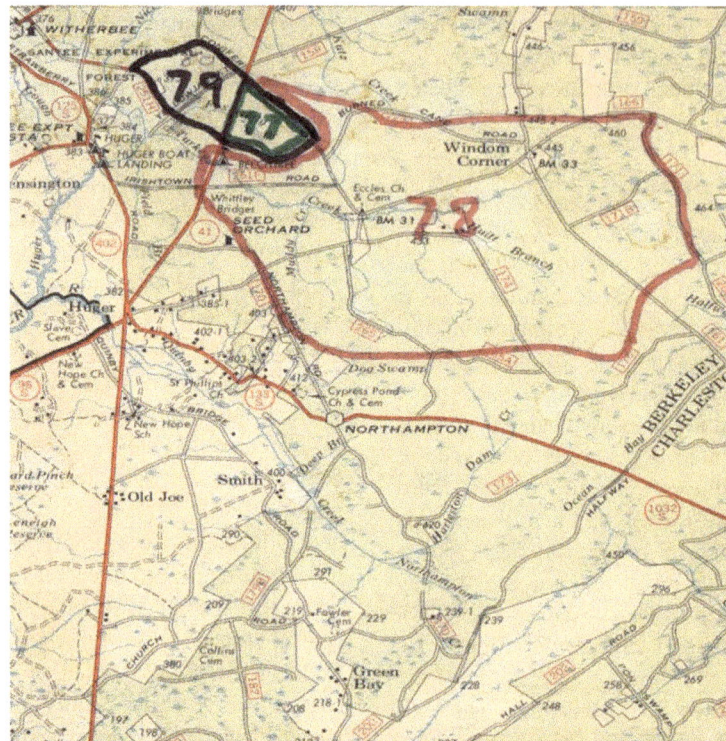

Figure 10. Watershed boundary using 1" = 1 mile USGS topographic map.

Watershed 78: Turkey Creek Topography

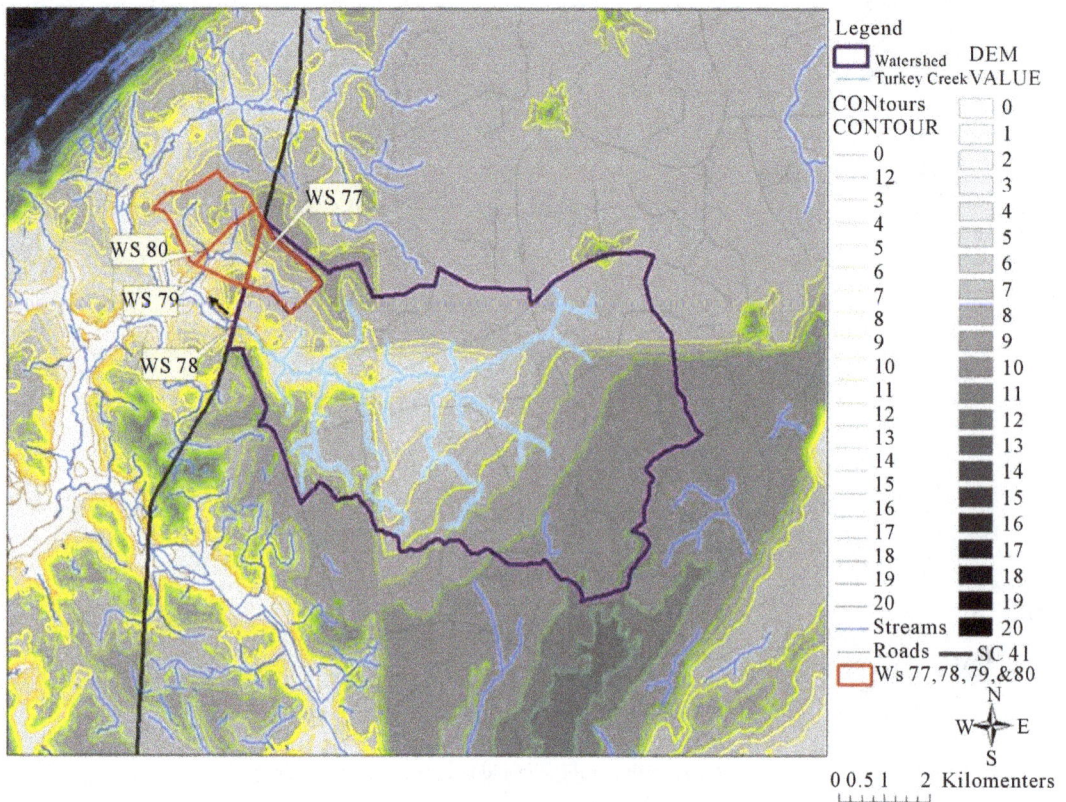

Figure 11. Watershed boundary using 30 m horizontal and 1 m vertical resolution DEM (SCDNR, 2004).

Later in 2005 USGS obtained an estimate of 5880 ha (14,520 acres) as the drainage area at the outlet of the current gauging station (**Table 1**, **Figure 12**, green color) using the SC 14-digit HUC with the 1:24,000-scale 7.5-minute series topographic maps as the source maps and 1:100,000-scale Digital Line Graphs as base maps [27]. This was still a 19.5% increase from the previous SC-DNR2004 based DEM result. No field checking for road boundaries as well as cross drainage structures like culverts was done for this estimate although these may have dramatic effects on the reconditioned DEMs and the derived drainage area. Accordingly, the average annual ROC was found to be 0.21 for this area (**Table 1**), which is also in the range of the published data for the coastal forested watersheds [50,52,53].

In 2006, new DEMs obtained from the USGS Enhanced 1:24,000 true 10 m horizontal 1 m vertical DEM became available to update the watershed boundary. In this case the SWAT interface in ArcView [37] was used to delineate the watershed as shown in **Figure 13**. The delineation also considered the drainage pathways due to forest roads and 44 culverts surveyed during the 2006-2007 period [28]. The ArcView SWAT delineation yielded the highest drainage area of 7260 ha (**Table 1**). This was 124% higher than the initial estimate of 3240 ha and 47.6% higher than the estimate using the 30 m × 30 m DEM. This was possibly due to errors in the 10 m enhanced DEM as reported recently by [11].

This new drainage area resulted in the lowest average annual ROC value of 0.17 (**Table 1**). Ongoing studies on the Turkey Creek watershed at that time used this drainage area to calculate the new depth-based outflow in estimating field water balance [2], assessing the rainfall-runoff storm event dynamics [5], as well as validating a SWAT model [1,28].

Later in 2010 using partial LiDAR dataset for the

Figure 12. Watershed boundary using SC 14-digit HUC (green) overlaid on SCDNR2005 boundary (red).

small lower part of the Turkey Creek watershed, which is within the Santee Experimental Forest, a new drainage area of 6,510 ha was calculated, which was 10.3% lower than the ArcView/SWAT computed area of 7260 ha and 10.7% higher than the SC 14-digit HUC generated area of 5980 ha. The average annual ROC was 0.19 based on this area of 6510 ha (**Table 1**) and is lower compared to similar coastal watersheds.

Finally, in 2011 the most updated high resolution 1.5 m × 1.5 m and 0.18 m vertical resolution DEM based on the complete LiDAR data for the watershed was used to redefine the boundary and corresponding area using the ArcSWAT model as shown in blue color in **Figure 14**. The area calculated by this method as a reference was 5240 ha with a corresponding calculated average annual ROC of 0.24 (**Table 1**), which was very close to the similar other forested watersheds in the coastal plain [50,52,53]. The area of only 5240 ha estimated using the Li-DAR-based DEM is about 6.5% higher than the previously estimated area of 4920 ha by SCDNR2004 method, indicating that the 30 m × 30 m horizontal and 1 m vertical resolution DEM was well within the water balance errors [18]. The USGS defined drainage area of 5880 ha based on 14-digit HUC was 12.2% higher than this Li-DAR-based DEM as a reference.

Interestingly, if the areas that were drained by the road culverts (based on DEM reconditioning and field verification) outside of the watershed boundary are not considered (or included as if there were no culverts) the Li-DAR-based DEM (brown color) also provides exactly the same drainage area (5880 ha) as the one obtained by using the 2011-USGS 14-digit HUC data. This indicates that for areas without culverts the DEM based on the 14-digit HUC may be as accurate as the LiDAR based DEM for these LCP watersheds. However, further analysis of more LCP watersheds is needed to definitively confirm this observation.

Data in **Figure 15** shows the uncertainty in average annual runoff coefficients as a result of variability in the corresponding estimated drainage areas of the Turkey Creek watershed for various DEM types used since 1960's. Clearly, a highest ROC of 0.38 was obtained for the lowest area estimated using very first initial USGS topographic map of 1" = 2 mi scale in 1960's. The lowest ROC of 0.17 was obtained for the largest estimated area of 7260 ha by using ArcSWAT delineation with the 1:24,000-scale true 10 m horizontal and 1 m vertical resolution DEMs. We believe both of these DEMs produced larger errors in estimating average annual ROCs, as a result of errors in drainage areas.

The drainage area of 4920 ha calculated using 30 m DEM from SCDNR in 2004 was the closest, with only about 6% underestimate, to the area of 5240 ha obtained by the LiDAR data in 2011. However, the historic 1969

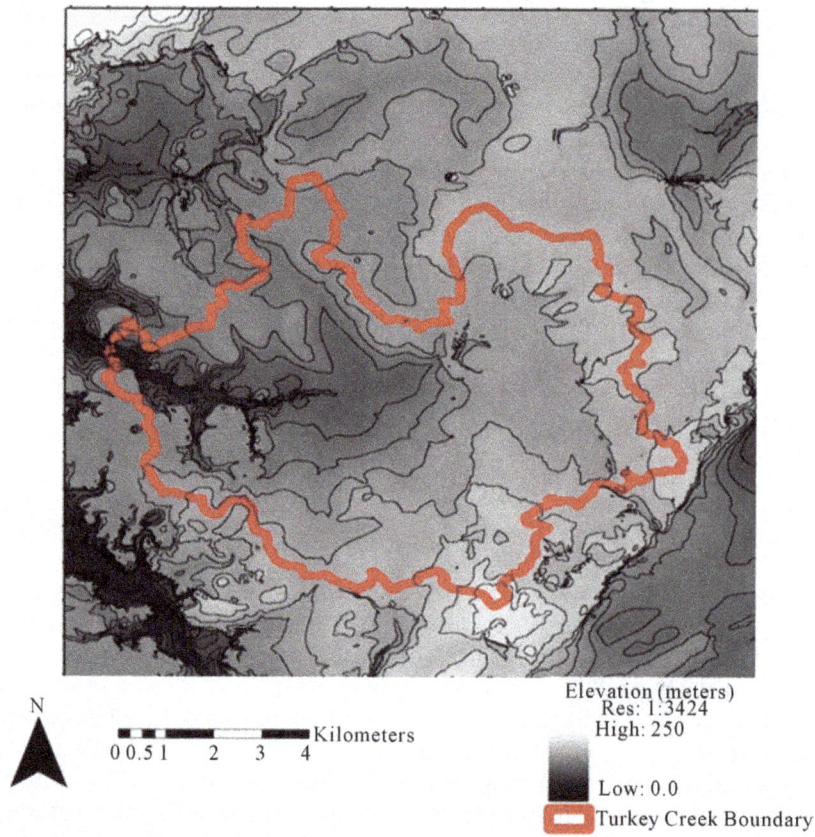

Figure 13. Watershed boundary using 2005 USGS enhanced 1:24,000 scale 10 m × 10 m and 1 m vertical DEM.

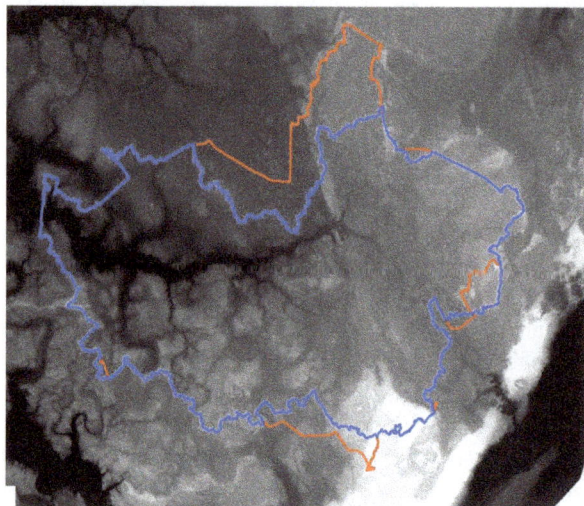

Figure 14. Watershed boundary using 1.5 m horizontal and 0.18 m vertical resolution DEM from the 2011 LiDAR data (Blue color with and brown color without accounting for culverts).

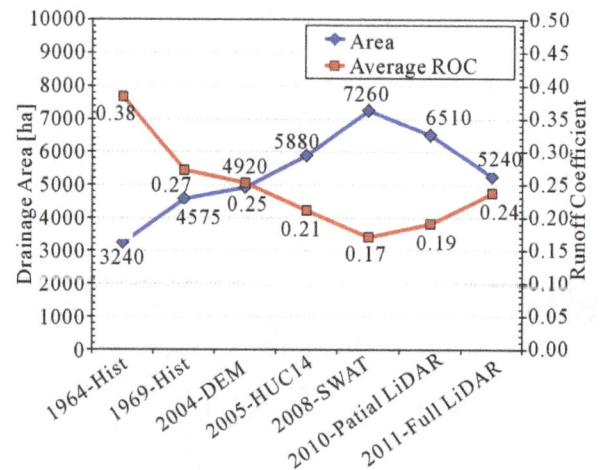

Figure 15. Estimated drainage areas of the Turkey Creek watershed using various DEM types and the corresponding average annual runoff coefficients.

1" = 1 mi topographic map based area of 4575 ha was 12.6% lower than the LiDAR-based estimate. These results are consistent with those by the NOAA's Office of Hydrologic Development [54] who reported a reduction of errors in basin area estimates by more than half by

using 30 m DEM compared to the 400 m DEMs for mild to moderate sloped watersheds with less than 78 km^2 area in the states of Oklahoma and Kansas.

The effects of uncertainty in drainage areas obtained by various DEM types can be propagated in many hydrologic studies including water and nutrient balances, spatially distributed modeling (SWAT, MIKESHE, HSPF,

DRAINWAT, WEPP, etc.) and engineering designs for water resources structures involving Rational Method (Equation (1)), USGS regional flood discharge formula (Equation (2)) [55], peak discharge estimates (Equation (3)) [56], pollutant export coefficients, estimating flow rates using empirical approaches for ungauged basins (Equation (4)) [57], designing the best management practices, and nutrient loading estimates (Equation (5)) [58,59] generally used in Total Maximum Daily Load (TMDL) developments.

$$Q_{peak} = C \times I \times A \qquad (1)$$

$$Q_{10} = 157 A^{0.59} \qquad (2)$$

$$Q_{peak} = 1.097 A^{0.81} \qquad (3)$$

$$Q_p = a10^b A^c X_1^d X_2^e \qquad (4)$$

$$L_{io} = L_{ec} \times A \qquad (5)$$

where Q_{peak} is the peak discharge, C is a runoff coefficient, I is rainfall intensity, A is field or drainage area, Q_{10} is peak discharge with a return period of 10 years, Q_p is long-term daily flow exceeded p% of the time, [a, b, c, d, and e] are regional coefficients, X_1 and X_2 are watershed descriptors, L_{io} is a nutrient loading at a field edge, and L_{ec} is the average export coefficient for a given land management practice (soil, crop/vegetation, and water management).

While the importance of high resolution LiDAR-based DEM for delineation of accurate drainage areas was demonstrated, it may have other applications also as shown by [30] for the Santee Experimental Forest. The authors reported that improvements in hydrologic functions and pathways used in hydrologic models and assessments can be achieved by using these high resolution DEMs that are capable of identifying the remnants of legacy water management structures left from rice planting in 1700's in the SC Lower Coastal Plain, which otherwise were not identified using the regular and enhanced DEMs. Similarly, [23] used DEM-based approach including LiDAR data to quantify the surface storage parameter widely used in hydrologic models for predicting the peak flow rates. Lang and McCarty [17] demonstrated the ability of LiDAR intensity data for forest inundation mapping on forested wetlands in eastern shore of Maryland. Eeckhaut et al. [36] concluded that LiDAR data enabled experts to find ten new landslides and to correct the boundaries of 11 of the 77 landslides mapped during the field survey. Our work here used recent high density LiDAR data with field verification to identify more culverts in the stream tributary network than possible using the NHD dataset. Our findings are consistent with recent findings of [60], who reported that stream datasets derived from semi-automated and auto-mated interpretation of LiDAR derived DEMs were considerably more accurate than NHD high resolution and Plus datasets. These authors found that LiDAR derived datasets significantly increased percent area and total number of wetlands that were considered connected.

4. Summary and Conclusions

There is a tremendous need to accurately represent environmental processes on the landscape. Questions involving climate change, land use effects, urbanization, etc., require a thorough understanding of the processes regulated by hydrology because the consequential thresholds are usually small. While models are the principal tools for conducting assessments, representations from spatially distributed, physically-based models are only as effective as the mathematical representation of the processes and the accuracy and resolution of the supporting data. We have shown that historical land use features may affect contemporary watershed hydrologic and transport processes, to illustrate that the modeling process (e.g., calibration) may compensate for inherent features in the landscape. Adoption of higher resolution data, whenever available, will challenge and ultimately improve our understanding of hydrologic and pollutant cycling processes and hence model applications [10,12-15,17,60]. With the improvement in computer processing speed, sensor technology, and advanced models, high resolution data are being analyzed on a faster scale and providing better and efficient results. In areas where water resources are critical and existing data relatively poor (e.g., coastal plain), acquisition of high resolution topographic data will greatly enhance our ability to assess hydrologic functions including water, nutrient and carbon balances.

DEMs based on high resolution topographic data such as LiDAR with field verification for cross drainage structures and roadbeds of the study watershed should be used for estimating more reliable boundary/drainage areas often used in hydrologic studies in the flat, low-gradient coastal plain landscapes. The area delineated by using the 14-digit HUC may be as accurate as the one obtained by the LiDAR-based DEM for these LCP watersheds as long there are no roads or culverts on the landscape. The drainage area estimated using the classical 30 m horizontal and 1 m vertical resolution DEM was found to be much more accurate than the USGS enhanced 1:24,000 true 10 m horizontal 1 m vertical DEM when compared to the LiDAR-based DEM in this study. Although the high resolution LiDAR-based DEM was considered as the most accurate for use as a reference in this study, we should still acknowledge some uncertainties and errors in the LiDAR data processing also as various software used to process them have some limitations and potential errors (e.g., errors due to inherent structure of algorithms for automatic watershed delineation).

5. Acknowledgements

The authors would like to acknowledge the USDA Forest Service Southern Research Station Civil Rights Committee Summer Student Hire Program for supporting a part of this study using the LiDAR data and field verification works by Jose Martin, the former student at Gainesville State College. We also would like to thank Dr. Jim Scurry at SC Department of Natural Resources for providing the much needed LiDAR data (draft), Beth Haley, former student of College of Charleston for field survey of culverts and processing the 2005 enhanced DEM data, Lisa Wilson, former student of College of Charleston for processing the SCDNR2005 data, Andy Harrison, Hydrologic Technician at Forest Service in Cordesville, SC for various levels of support in field/data, and TetraTech for digitizing the historic streamflow data used in this study. Bill Hansen, Forest Hydrologist, at Francis Marion & Sumter National Forests office, SC are also acknowledged for their advice and suggestions on DEM data.

REFERENCES

[1]   D. M. Amatya and M. K. Jha, "Evaluating the Swat Model for a Low-Gradient Forested Watershed in Coastal South Carolina," *Transactions of the ASABE*, Vol. 54, No. 6, 2011, pp. 2151-2163.

[2]   D. M. Amatya, T. J. Callahan, C. Trettin and A. Radecki-Pawlik, "Hydrologic and Water Quality Monitoring on Turkey Creek Watershed, Francis Marion National Forest, SC," *ASABE Annual International Meeting, Reno*, 2009, pp. 21-24.

[3]   D. Bosch, J. Sheridan and F. Davis, "Rainfall Characteristics and Spatial Correlation for the Georgia Coastal Plain," *Transactions of the ASAE*, Vol. 42, No. 6, 1999, pp. 1637-1644.

[4]   C. G. Garrett, V. M. Vulava, T. J. Callahan and M. L. Jones, "Groundwater-Surface Water Interactions in a Lowland Watershed: Source Contribution to Stream Flow," *Hydrological Processes*, Vol. 26, No. 21, 2012, pp. 3195-3206.

[5]   I. B. La Torre Torres, D. M. Amatya, G. Sun and T. J. Callahan, "Seasonal Rainfall-Runoff Relationships in a Lowland Forested Watershed in the Southeastern USA," *Hydrological Processes*, Vol. 25, No. 13, 2011, pp. 2032-2045.

[6]   T. M. Williams, D. M. Amatya, D. R. Hitchcock and A. E. Edwards, "Streamflow and Nutrients from a Karst Watershed with a Downstream Embayment: Chapel Branch Creek," In Press, *ASCE Journal of Hydrologic Engineering*, Posted Ahead of Print January 26, 2013.

[7]   T. J. Callahan, V. M. Vulava, M. C. Passarello and C. G. Garrett, "Estimating Groundwater Recharge in Lowland Watersheds," *Hydrological Processes*, Vol. 26, No. 19, 2012, pp. 2845-2855.

[8]   D. M. Amatya, M.K. Jha, A. E. Edwards, T. M. Williams, and D. R. Hitchcock, "SWAT-Based Streamflow and Embayment Modeling of a Karst affected Chapel Branch Watershed, SC," *Transaction of the ASABE*, Vol. 54, No. 4, 2011, pp. 11311-1323

[9]   Z. Dai, C. C. Trettin, C. Li, H. Li, G. Sun and D. M. Amatya, "Effect of Assessment Scale on Spatial and Temporal Variations in CH_4, CO_2 and N_2O Fluxes in a Forested Wetland," *Water, Air, & Soil Pollution*, Vol. 223, No. 1, 2012, pp. 253-265.

[10]   P. C. Beeson, A. M. Sadeghi, M. W. Lang, M. D. Tomer and C. S. T. Daughtry, "Sediment Delivery Estimates in Water Quality Models Altered by Resolution and Source of Topographic Data," *Journal of Environmental Quality*, 2013.

[11]   S. S. Panda, K. Burry and C. Tamblyn, "Wetland Change and Cause Recognition in Georgia Coastal Plain," *ASABE International Conference*, Dallas, 2012.

[12]   C. S. Renschler and D. C. Flanagan, "Site-Specific Decision-Making Based on RTK GPS Survey and Six Altaernative Elevation Data Sources: Soil Erosion Predictions," *Transactions of the ASAE*, Vol. 51, No. 2, 2008, pp. 413-424.

[13]   A. A. Aziz, B. L. Steward, A. Kaleita and M. Karkee, "Assessing the Effects of DEM Uncertainty on Erosion rate Estimation in an Agricultural Field," *Transactions of the ASABE*, Vol. 55, No. 3, 2012, pp. 785-798.

[14]   R. L. Perroy, B. Bookhagen, G. P. Asner and O. A. Chadwick, "Comparison of Gully Erosion Estimates Using Airborne and Ground-Based LiDAR on Santa Cruz Island, California," *Geomorphology*, Vol. 118, No. 3-4, 2010, pp. 288-300.

[15]   L. A. James, D. G. Watson and W. F. Hansen, "Using Li-DAR Data to Map Gullies and Headwater Streams under Forest Canopy: South Carolina, USA," *Catena*, Vol. 71, No. 1, 2007, pp. 132-144.

[16]   S. P. Wechsler, "Uncertainties Associated with Digital Elevation Models for Hydrologic Applications: A Review," *Hydrology and Earth System Sciences Discussions*, Vol. 11, No. 4, 2007, pp. 1481-1500.

[17]   M. W. Lang and G. W. McCarty, "LiDAR Intensity for Improved Detection of Inundation below the Forest Canopy," *Wetlands*, Vol. 29, No. 4, 2009, pp. 1166-1178.

[18]   S. V. Harder, D. M. Amatya, T. J. Callahan, C. C. Trettin and J. Hakkila, "Hydrology and Water Budget for a Forested Atlantic Coastal Plain Watershed, South Carolina," *JAWRA Journal of the American Water Resources Association*, Vol. 43, No. 3, 2007, pp. 563-575.

[19]   B. Ambroise, J. L. Perrin and D. Reutenauer, "Multi-Criterion Validation of a Semi-Distributed Conceptual Model of the Water Cycle in the Fecht Catchment (Vosges Massif, France)," *Water Resources Research*, Vol. 31, No. 6, 1995, pp. 1467-1481.

[20] J. G. Arnold, R. Srinivasan, R. S. Muttiah and J. Williams, "Large Area Hydrologic Modeling and Assessment Part I: Model Development," *JAWRA Journal of the American Water Resources Association*, Vol. 34, No. 1, 2007, pp. 73-89.

[21] Z. Dai, C. Li, C. Trettin, G. Sun, D. Amatya and H. Li, "Bi-Criteria Evaluation of the MIKE SHE Model for a Forested Watershed on the South Carolina Coastal Plain," *Hydrology and Earth System Sciences*, Vol. 14, No. 6, 2010, pp. 1033-1046.

[22] S. B. Hilliard, "Antebellum Tidewater Rice Culture in South Carolina and Georgia," In: J. R. Gibson, Ed., *European Settlement and Developmentin North America*, Uni- versity of Toronto Press, Toronto, 1978, pp. 91-115.

[23] J. K. O. Amoah, D. M. Amatya and S. Nnaji, "Quantifying Watershed Surface Depression Storage: Determination and Application in a Hydrologic Model," HYDROLOGICAL PROCESSESHydrol. Process, Wiley Online Library, 2012. wileyonlinelibrary.com

[24] PhotoScience, "Airborne Laser Terrain Mapping for Santee Experimental Forest, SC. Photo Science Geospatial Solutions," Lexington, 2007.

[25] D. M. Amatya and A. Radecki-Pawlik, "Flow Dynamics of Three Experimental Forested Watersheds in Coastal South Carolina (USA)," *ACTA Scientiarum Polonorum, Formatio Circumiectus*, Vol. 6, No. 2, 2007, pp. 3-17.

[26] D. E. Bower, C. Lowry, M. A. Lowery and N. M. Hurley, "USGS Water Resources Investigations Report WRIR 99-4015," 1999. http://sc.water.usgs.gov/publications/abstracts/wrir99-4015.html

[27] J. P. Eidson, C. M. Lacy, L. Nance, W. F. Hansen, M. A. Lowery and N. M. Hurley, Jr., "Development of a 10- and 12-Digit Hydrologic Unit Code Numbering System for South Carolina, 2005," 2005.

[28] E. B. Haley, "Field Measurements and Hydrologic Modeling of the Turkey Creek Watershed, South Carolina (SC)," Master's Thesis, Graduate School of the College of Charleston, Charleston, 2007.

[29] DeLorme, "A Topographic Map Software," 2002. www.delorme.com

[30] C. C. Trettin, D. M. Amatya, C. Kaufman, N. Levine and R. T. Morgan, "Recognizing Change in Hydrologic Functions and Pathways due to Historical Agricultural Use—Implications to Hydrologic Assessments and Modeling," *The Third Interagency Conference on Research in the Watersheds*, Estes Park, 8-11 September 2008, 5 p.

[31] J. Scurry, "Raw LiDAR Data for Berkeley County, SC, SCDNR," 2011.

[32] A. Imagery, "Quick Terrain Modeler (QTM)," Version 7.1.5, 2012.

[33] D. Civco, "ERDAS IMAGINE Remote-Sensing, Image-Processing and GIS Software," *Photogrammetric Engineering and Remote Sensing*, Vol. 60, No. 1, 1994, pp. 35-39.

[34] ESRI, "ESRI ArcGIS," Version 9.3.1., Redlands, 2009.

[35] J. D. Simley and W. J. Carswell, Jr., "The National Map—Hydrography," *US Geological Survey Fact Sheet*, Vol. 3054, No. 1, 2009, 4 p.

[36] M. Eeckhaut, J. Poesen, G. Verstraeten, V. Vanacker, J. Nyssen, J. Moeyersons, et al., "Use of LIDAR-Derived Images for Mapping Old Landslides under Forest," *Earth Surface Processes and Landforms*, Vol. 32, No. 5, 2006, pp. 754-769.

[37] M. D. Di Luzio, R. Srinivasan, J. G. Arnold and S. L. Neitsch, "Arcview Interface for SWAT2000: User's Guide-(2002)," 12 December 2012. http://www.brc.tamus.edu/swat/downloads/doc/swatav2000.pdf

[38] R. S. Kinerson, J. L. Kittle and P. B. Duda, "BASINS: Better Assessment Science Integrating Point and Nonpoint Sources," In: *Decision Support Systems for Risk-Based Management of Contaminated Sites*, A. Marcomini, G. W. Suter II and A. Critto, Eds., Springer, 2009, pp. 1-24.

[39] J. F. O'Callaghan and D. M. Mark, "The Extraction of Drainage Networks from Digital Elevation Data," *Computer Vision, Graphics, and Image Processing*, Vol. 28, No. 3, 1984, pp. 323-344.

[40] R. Jones, "Algorithms for Using a DEM for Mapping Catchment Areas of Stream Sediment Samples," *Computers & Geosciences*, Vol. 28, No. 9, 2002, pp. 1051-1060.

[41] A. R. Paz, W. Collischonn, A. Risso and C. A. B. Mendes, "Errors in River Lengths Derived from Raster Digital Elevation Models," *Computers & Geosciences*, Vol. 34, No. 11, 2008, pp. 1584-1596.

[42] J. P. Wilson, C. S. Lam and Y. Deng, "Comparison of the Performance of Flow-Routing Algorithms Used in GIS-Based Hydrologic Analysis," *Hydrological Processes*, Vol. 21, No. 8, 2007, pp. 1026-1044.

[43] K. M. Crombez, "Comparing Flow Routing Algorithms for Digital Elevation Models," *Digital Terrain Analysis (Geo428)*, Project Paper for Michigan State University, 2008, pp. 1-9.

[44] C. E. Young, "Precipitation-Runoff Relations on Small Forested Watersheds in the Coastal Plain. Study Plan Addendum No. 2," USDA Forest Service Technical Report FS-SE-1602, 1965.

[45] N. D. Gordon, T. A. McMahon, B. L. Finlayson, C. J. Gippel and R. J. Nathan, "Stream Hydrology: An Introduction for Ecologists," Wiley, Hoboken, 2004.

[46] P. Haan and R. Skaggs, "Effect of Parameter Uncertainty on DRAINMOD Predictions: I. Hydrology and Yield," *Transactions of the ASAE*, Vol. 46, No. 4, 2003, pp. 1061-1067.

[47] K. Konyha and R. Skaggs, "A Coupled, Field Hydrology: Open Channel Flow Model: Theory," *Transactions of the ASAE*, Vol. 35, No. 5, 1992, pp. 1431-1440.

[48] R. Skaggs, "A Water Management Model for Shallow Water Table Soils," Technical Report No. 134, North Carolina State University, Water Resources Research Institute of the University of North Carolina, Raleigh, 1978.

[49] J. Amoah, "A New Methodology for Estimating Water-shed-Scale Depression Storage," Ph.D. Dissertation, Department of Civil & Environmental Engineering, Florida A&M University, Tallahassee, 2008, 202 p.

[50] G. Sun, S. G. McNulty, D. M. Amatya, R. W. Skaggs, L. W. Swift, J. P. Shepard, et al., "A Comparison of the Watershed Hydrology of Coastal Forested Wetlands and the Mountainous Uplands in the Southern US," Journal of Hydrology, Vol. 263, No. 1-4, 2002, pp. 92-104.

[51] S. Laseter, "Stage-Flow Fortran Programs for Processing Discharge Data for Santee Experimental Forest Watersheds," USDA Forest Service, Coweeta Hydrologic Laboratory, Otto, 2008.

[52] D. Amatya, M. Miwa, C. Harrison, C. Trettin and G. Sun, "Hydrology and Water Quality of Two First Order Forested Watersheds in Coastal South Carolina," In: ASABE Annual International Meeting, Portland, 9-12 July 2006, 22 p.

[53] G. M. Chescheir, M. E. Lebo, D. M. Amatya, J. Hughes, J. W. Gilliam and R. W. Skaggs, "Hydrology and Water Quality of Forested Lands in Eastern North Carolina," Tech Bullet 320, North Carolina State University, Raleigh, 2008, 79 p.

[54] OHD, "Basin Delineation with a 400 m Terrain Dataset," 2011.
http://www.nws.noaa.gov/oh/hrl/gis/delineation.html

[55] W. B. Guimaraes and L. R. Bohman, "Techniques for Estimating Magnitude and Frequency of Floods in South Carolina," USGS Water-Resources Investigations Report, 1992, pp. 91-4157.

[56] J. M. Sheridan, "Peak Flow Estimates for Coastal Plain Watersheds," Transactions of the ASAE, Vol. 45, No. 5, 2002, pp. 1319-1326.

[57] H. S. Ssegane, D. M. Amatya, E. W. Tollner, Z. Dai and J. E. Nettles, "Estimation of Daily Streamflow of Southeastern Coastal Plain Watersheds by Combining Estimated Magnitude and Sequence (Tentatively Accepted)," Journal of the American Water Resources Association, In Press, 2013, pp. 1-17.

[58] D. M. Amatya, G. M. Chescheir, G. P. Fernandez, R. W. Skaggs, F. Birgand and J. Gilliam, "Lumped Parameter Models for Predicting Nitrogen Transport in Lower Coastal Plain Watersheds," Report No. 347, North Carolina Water Resources Research Institute, Raleigh, 2003.

[59] M. Beaulac and K. Reckhow, "An Examination of Land Use-Nutrient Export Relationships," Journal of the American Water Resources Bulletin, Vol. 18, No. 6, 1982, pp. 1013-1024.

[60] M. Lang, O. McDonough, G. McCarty, R. Oesterling and B. Wilen, "Enhanced Detection of Wetland-Stream Connectivity Using LiDAR," Wetlands, Vol. 32, No. 3, 2012, pp. 461-473.

15

Modelling of Environment Vulnerability to Forests Fires and Assessment by GIS Application on the Forests of Djelfa (Algeria)

Mohamed Said Guettouche, Ammar Derias
Laboratory of Geography and Territory Planning (LGAT),
University of Sciences and Technology Houari Boumediene (USTHB), Algiers, Algeria

ABSTRACT

Risk management of forest fires starts from its assessment. This assessment has been subject of several research works and many models of fire risk have been developed. One of them has been developed by ourselves, for the Mediterranean areas. However environmental vulnerability to forest fires is an important part of risk; it represents, in fact, the exposed challenge to this scourge and therefore it worths particular attention by decision makers. Thus and due to the importance of socio-economic potential of forest and the negative influence of fire on this one, we propose in this work, a model for vulnerability assessment to forest fires based on the principle of the weighted sum. Application of the proposed model suggested to use of geomatics technologies to the spatialize level of vulnerability. Within this framework, a GIS was developed and applied to the forests of Djelfa in the Saharian Atlas, as originality, it will allow the understanding of the concept of vulnerability and risk associated the steppes area scale to reach a good space control.

Keywords: Forest; Fire; Vulnerability; Model; Djelfa; GIS

1. Introduction

It's true that forest fires are difficult to identify and/or approach, the reality of the phenomenon is not easy. As many parameters are involved, especially ecological and socio-economical. The causes, frequency phenomenon and extension, of the must be searched for the vegetation structure and its environment.

Indeed, the mountainous regions of Maghreb represent high potential forest areas and are almost always in areas with high or very high density of rural population. This induces a higher risk of fire, whether in terms of hazard or in terms of vulnerability. Burnings are, increasingly, potentially important, because of the different human activities sources of ignition (barbecue, cigarette butts ...) in contact with a flammable vegetation and fuel, especially in the Algerian steppes area.

The fire risk assessment, based on historical and current data, and restitution of the results under a map form can be a remarkable contribution to forest managers in decision aid, so they can make on logical basis, all prevention policies. Cartographic degrees of risk, thus established, highlight sensitive areas at high fire risk (red areas), in which a concentration of effort and especially contingency plans are to provide objectively. Finally, we shall not forget that the fundamental purpose of the as-

sessment of fire risk is to reduce their frequency and burned areas size, through preventive measures and ensure optimal use of limited resources available to fight against fires.

In this context fire risk approach that includes our theme, as it intended to present how we assess the vulnerability to the forest fires hazard and to show the contribution of the GIS approach in that Spatialization.

2. Methodology

Risk assessment of forest fires has been subject of several research works [1-17] and many fire risk indices were established. The index that is the subject of this work is vulnerability; we have established [13] and which deserves an improvement to make it applicable in the Mediterranean areas. Also, information must be completed concerning the method and the technique used in GIS to evaluate parameters.

Based on the principle of weighted sum method, vulnerability index is designed as a model assigning each parameter a weight, depending on its socio-economic issue. It is given by the formula [13]:

$$V^{inc} = \frac{(5IP + 3IU + 2IV)}{10} \quad (1)$$

where:

IP: expresses the degree of human presence within or near the forest.

IU: expresses the degree of urbanization within or near the forest.

IV: expresses the degree of agricultural use within or near the forest.

Moreover, discussion with forestry experts of the Directorate of Forestry has found that several vulnerability criteria are omitted from this index. Indeed, this index (Equation 1), lack of parameters determining the pastoral and infrastructural potential which are characteristics of the Mediterranean forests. Then the technical evaluation of these parameters is not explicit.

Thus, and due to the importance and the degree of influence of these two issues in determining environmental vulnerability to forest fires, we propose in this work, to improve this model by adding other sub characteristic index of raised challenges and balancing the weighting. Next, we explain the method and assessment technique of various parameters by GIS.

2.1. Vulnerability Modeling to Forests Fires

To assess environmental vulnerability to forest fires, it is necessary to model each of the elements determining the vulnerability. This step is to select the parameters for each element and then using a representation model to assess vulnerability.

Parameters are the natural and socio-economic criteria of environment that will be affected by the outbreak, spread and fire intensity. These criteria are highly correlated and their combination defines vulnerability.

Based on the principle of multi-criteria analysis and adopting the method of weighted sum. This definition of vulnerability can be formalized by the following equation:

$$V^{inc} = \sum_{i=1}^{n} \lambda_i C_i \; ; \text{ with } \; \sum \lambda_i = 1 \qquad (2)$$

where:

V^{inc}: the environment vulnerability to a forests fires; C_i: the evaluation criteria and λ the weight of the criterion i $(i = 1, \cdots, n)$.

The weighting depends on the issue exposed by the criteria which is more important to human life as biological life or the economic aspect.

The modeling and evaluation process can be diagrammed in **Figure 1**.

2.2. Criteria Identification for Vulnerability and Development Equation

The notion of vulnerability is difficult to define because it includes many economic and social parameters. A distinction between economic vulnerability and human vul-

nerability can be established:

- Economic vulnerability is structural (damage to proprty, damage to houses, collective works, communications channels, etc.).
- Human vulnerability estimates of harm to those on the physical and moral (deceased, injured, missing, etc.).

Thus, vulnerability defines a degree of loss within an area affected by hazard and the environment vulnerabilty, a property or person is his ability to receive damage following an accident. This leads us to consider the vulnerability to a feared event is an estimation of what will be the seriousness of this event if occurred.

The presence of humans and their agricultural activities, animals, houses and basic infrastructure (road networks and electricity) within or near the forests are the issue whose importance determines the degree of environment vulnerability: it is the protection of human life, and agro-forestry potential, and facilities. Thus, the soo-conomic parameter is the main term in the vulnerability model.

This can be formulated mathematically by the following weighted sum equation:

$$V^{inc} = (0.3PH + 0.2PF + 0.2PA + 0.2PP + 0.1PI) \qquad (3)$$

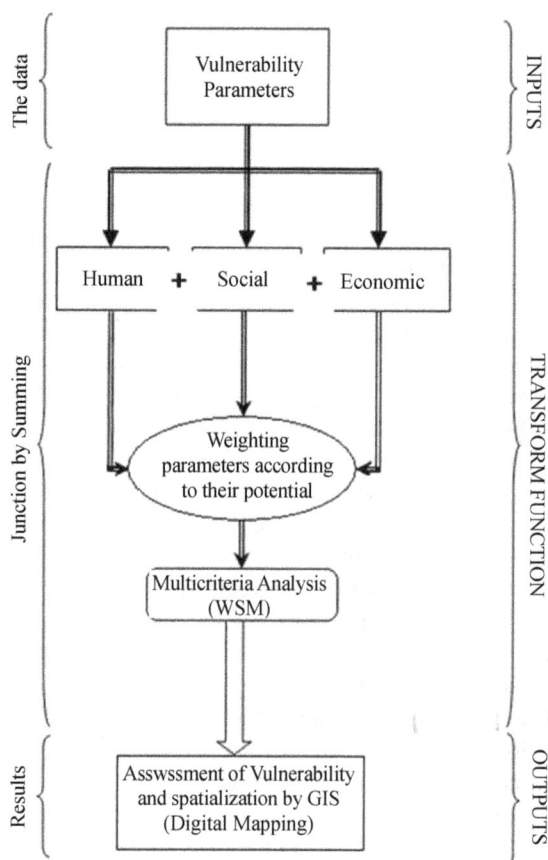

Figure 1. Process modeling and vulnerability assessment.

Modelling of Environment Vulnerability to Forests Fires and Assessment by GIS Application on the Forests of Djelfa (Algeria)

169

where:

$$PH = Np \exists SFI \qquad (4)$$

$$PF = nA/Sfu \qquad (5)$$

$$PA = Sag/SFI \qquad (6)$$

$$PP = AP/SFI \qquad (7)$$

with:

$$Ap = Dnc + Dsp \qquad (8)$$

$$Dnc = nC/Sp \qquad (9)$$

and

$$Dsp = Sp/SFI \qquad (10)$$

$$PI = So/SFI \qquad (11)$$

where:

PH: expresses the degree of human presence, it can be assessed by the number of people (Np) in a forest area of influence (Equation (4)), either within or immediately Neighborhood the forest.

PF: expresses the forestry potential is expressed here by the rate of vegetation cover. The latter is defined by the ratio of the number of trees (nA) to a unit area of 3 meters radius (Equation (5)).

PA: represents the presence of the crop activity in the influence forest area. It is defined as the ratio of the cultivated area (Sag) to the influence forest area (Equation (6)). The cultivated area is that which exists within a forest area or close it.

PP: represents the degree of pastoral activity in the influence forest area. It is defined as the ratio of the pastoral activity to the influence forest area (Equation (7)). Pastoral activity (denoted Ap) is determined by the sum of the normalized density of livestock and that of the pastoral area (Equation (8)). The normalized density of livestock is defined as the ratio of the number of head of livestock in pastoral area (Equation (9)). The normalized density of the pastoral area is the ratio of the pastoral area to forest area of influence (Equation (10)).

PI: Represents the potential for urban installations and infrastructures. It is defined by the ratio of the area occupied by these installations in the influence forest area (Equation (11)). The area occupied is defined by the sum of the areas occupied by habitations and by the networks (road, electric, other). The distance occupied by the network is referenced to a surface by taking an influence width of 25 m of both sides of the line center.

The *influence forest area* (denoted SFI) is defined by the forest plot and its close area. The neighborhood is defined by the distance of influence is which in maximum 300 meters and beginning from the boundaries of the forest plot [4]. Beyond this distance, we believe that intervention is possible for protection.

2.3. Notation Criteria and Vulnerability Assessment

Notation of the different vulnerability index and level are shown in the following **Table 1**.

2.4. Combination Matrixes

The combination between the different classes of weighted criteria (Equation (3)), one by one, indicates the level of environmental vulnerability to forest fires that we have presented under tables form (2, 3, 4 and 5).

Indeed, the weighted sum of the two classes of criteria, human potential and forestry potential, gives the degree of environmental vulnerability to forest fires (**Table 2**). This vulnerability is associated with the partial criterion, agricultural potential weighted 0.2, gives a second partial environmental vulnerability to forest fires (**Table 3**).

Table 1. Notation of vulnerability index and level.

Note	Criteria Index					Vulnerability
	PH (Nb)	PF (%)	PA (%)	PP (%)	PI (%)	Level
0	PH = 0	PF = 0	PA = 0	PP = 0	PI = 0	$V^{inc} = 0$
1	PH < 5	PF ≤ 15	PA ≤ 15	PP ≤ 15	PF ≤ 15	$0 < V^{inc} \leq 1$
2	$5 \leq PH < 10$	15 < PF ≤ 30	15 < PA ≤ 30	15 < PP ≤ 30	15 < PF ≤ 30	$1 < V^{inc} \leq 2$
3	$P \geq 10$	PF ≥ 30	PA ≥ 30	PP ≥ 30	PI ≥ 30	$V^{inc} > 2$

Table 2. Combination matrix (PH + PF = 1 er pV).

PF \ PH	0	0.3	0.6	0.9
0	**0**	**0.3**	**0.6**	**0.9**
0.2	**0.2**	**0.5**	**0.8**	**1.1**
0.4	**0.4**	**0.7**	**1.0**	**1.3**
0.6	**0.6**	**0.9**	**1.2**	**1.5**

pV: partial vulnerability.

The third level of partial vulnerability (**Table 4**) is the result of 2 pV associated with the criterion, pastoral potential weighted at 0.2. This result, combined with the criterion potential infrastructure, weighted 0.1, given the level of environmental vulnerability to forest fires (**Table 5**).

Modeling of this method using a GIS will spatialize the degree of vulnerability and create maps of synthesis.

3. Model Spatialization by GIS: Application on the Djelfa Region

The approach we have adopted, for the spatialization of environmental vulnerability to forest fires, is based on GIS [2,11-13,17].

Before giving details, it makes sense to define the area in which our spatialization will be established.

3.1. Delimitation of the Study Area

The choice of investigation area is concentrated on the district of Djelfa, in the Saharian Atlas, part of Algerian agro-sylvo-pastoral areas (**Figure 2**). This choice is based on forest landscapes diversity, and also by its central situation related to the Algerian steppe area it's different and contrasted natural and human data.

Geographically, Djelfa is a forest area in the Saharian Atlas. It is bordered in the north by the Zahrez Chott and in the south by the Saharian platform.

3.2. GIS Establishment

The satellite images, which allowed us to recognize the forest areas and to define the limits and expansions of map features necessary for the development of GIS, are those recorded by the Algerian satellite ALSAT1 in 2003, with 32 m resolution.

These images, acquired in three spectral bands (Green: 0.50 to 0.59, Red: 0.61 to 0.68 and Near Infrared: 0.79 to 0.89), were processed and analyzed by different remote sensing technical's (vegetation indices, supervised classification, etc.) to land use map in the investigation area (**Figure 3**).

The field investigation also allowed us to collect data on the forests, agricultural and pastoral potential as well as basic infrastructures and urban development. These field data were used to correct the land use map and to assess the different criteria.

On this surface based map, we digitized the road net-

Table 3. Combination Matrix (1 pV + PA = 2 pV).

PA \ 1 pV	0	0.2	0.3	0.4	0.5	0.6	0.7	0.8	0.9	1	1.1	1.2	1.3	1.5
0	0	0.2	0.3	0.4	0.5	0.6	0.7	0.8	0.9	1.0	1.1	1.2	1.3	1.5
0.2	0.2	0.4	0.5	0.6	0.7	0.8	0.9	1.0	1.1	1.2	1.3	1.4	1.5	1.7
0.4	0.4	0.6	0.7	0.8	0.9	1.0	1.1	1.2	1.3	1.4	1.5	1.6	1.7	1.9
0.6	0.6	0.8	0.9	1.0	1.1	1.2	1.3	1.4	1.5	1.6	1.7	1.8	1.9	2.1

Table 4. Combination matrix (2 pV + PP = 3 pV).

PA \ 2 pV	0	0.2	0.3	0.4	0.5	0.6	0.7	0.8	0.9	1	1.1	1.2	1.3	1.5	1.6	1.7	1.8	1.9	2.1
0	0	0.2	0.3	0.4	0.5	0.6	0.7	0.8	0.9	1.0	1.1	1.2	1.3	1.5	1.6	1.7	1.8	1.9	2.1
0.2	0.2	0.4	0.5	0.6	0.7	0.8	0.9	1.0	1.1	1.2	1.3	1.4	1.5	1.7	1.8	1.9	2.0	2.1	2.3
0.4	0.4	0.6	0.7	0.8	0.9	1.0	1.1	1.2	1.3	1.4	1.5	1.6	1.7	1.9	2.0	2.1	2.2	2.3	2.5
0.6	0.6	0.8	0.9	1.0	1.1	1.2	1.3	1.4	1.5	1.6	1.7	1.8	1.9	2.1	2.2	2.3	2.4	2.5	2.7

Table 5. Vulnerability matrix (3 pV + PI = V^{inc}).

PI \ 3 pV	0	0.2	0.3	0.4	0.5	0.6	0.7	0.8	0.9	1	1.1	1.2	1.3	1.5	1.6	1.7	1.8	1.9	2.1	2.2	2.3	2.4	2.5	2.7
0	0	0.2	0.3	0.4	0.5	0.6	0.7	0.8	0.9	1.0	1.1	1.2	1.3	1.5	1.6	1.7	1.8	1.9	2.1	2.2	2.3	2.4	2.5	2.7
0.1	0.1	0.3	0.4	0.5	0.6	0.7	0.8	0.9	1.0	1.1	1.2	1.3	1.4	1.6	1.7	1.8	1.9	2.0	2.2	2.3	2.4	2.5	2.6	2.8
0.2	0.2	0.4	0.5	0.6	0.7	0.8	0.9	1.0	1.1	1.2	1.3	1.4	1.5	1.7	1.8	1.9	2.0	2.1	2.3	2.4	2.5	2.6	2.7	2.9
0.3	0.3	0.5	0.6	0.7	0.8	0.9	1.0	1.1	1.2	1.3	1.4	1.5	1.6	1.8	1.9	2.0	2.1	2.2	2.4	2.5	2.6	2.7	2.8	3

Modelling of Environment Vulnerability to Forests Fires and Assessment by GIS Application on the Forests of Djelfa (Algeria)

171

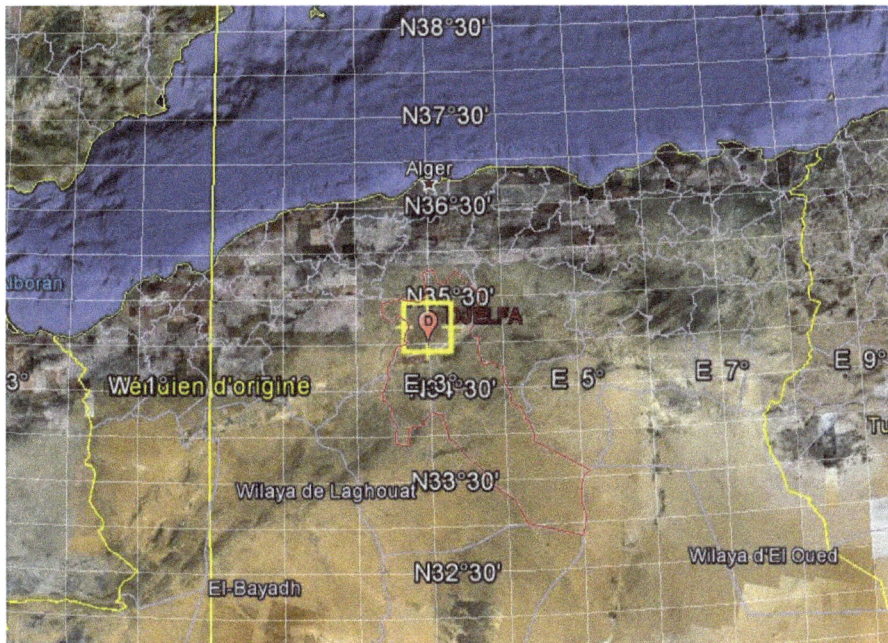

Figure 2. Location of the study area.

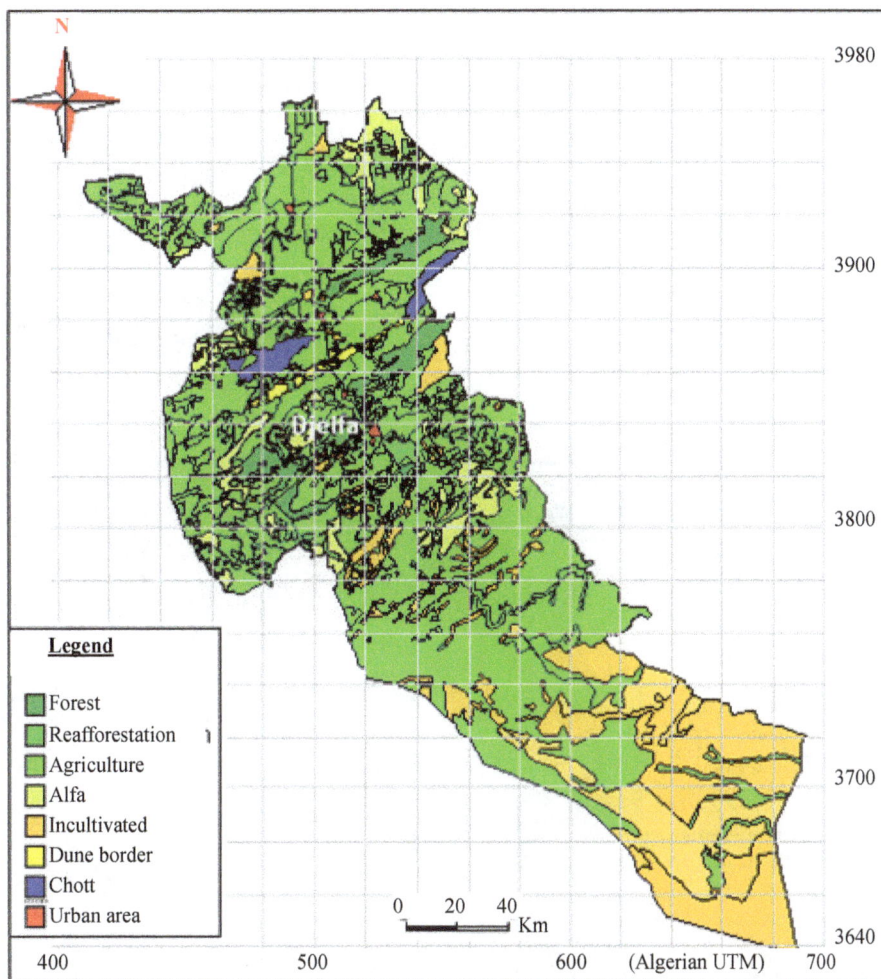

Figure 3. Land use map of Djelfa region.

work for which the influence surfaces corresponding to a radius of 50 meters have been established.

The whole data obtained by image processing or from the field investigation were compiled and implemented in GIS software (MapInfo 8) to define a model of mapping information. Indeed, a georeferenced database, organized and structured using the software, was carried out to better spatialization vulnerability to forest fires (**Figure 4**).

4. Results and Discussion

Digitalizing surface units of land use allowed us to establish the geographic database (entities) of the study area. It is used as a background graphic which we combined with the tributaries data (criteria).

Digitizing the road network has been established from the image and using the road network map of Algeria. The lines were converted to the surface by creating buffers (**Figure 5(a)**).

The buffer function was also used to create forest area of influence (**Figure 5(b)**).

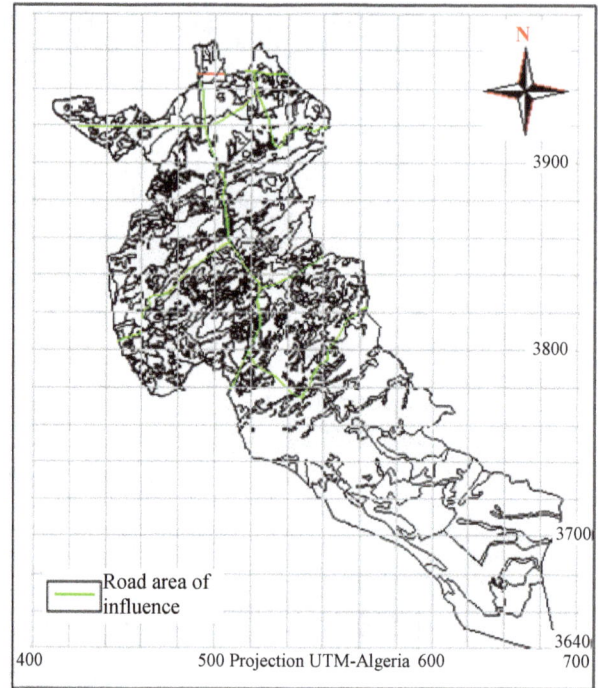

(a)

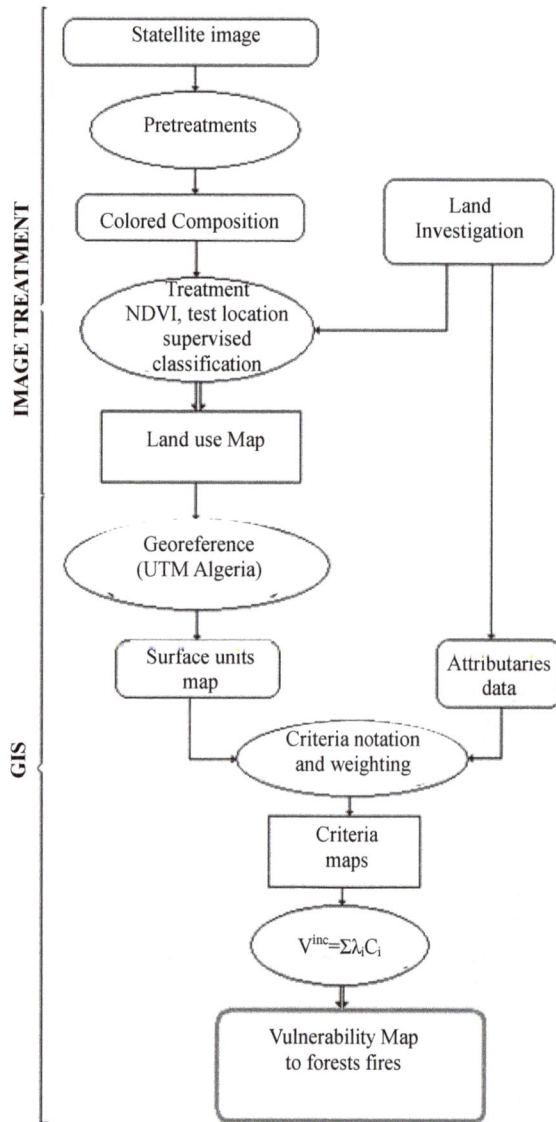

(b)

Figure 5. Creation process of influence area. (a) Map of road network and influence area; (b) Map of forest influence area.

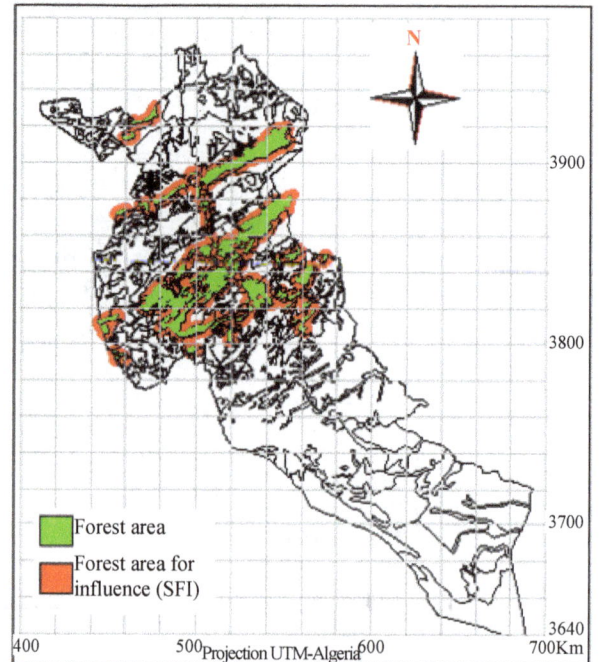

Figure 4. Diagram of vulnerability assessment and mapping process.

Modelling of Environment Vulnerability to Forests Fires and Assessment by GIS Application on the Forests of
Djelfa (Algeria)

173

The criteria were assessed and weighted according to their socioeconomic and environmental issues and potential maps of the different criteria were established (**Figures 6**).

The combination by summing the five resulting layers, (Equation (3)), were used to determine the level of environmental vulnerability to forest fires in the Djelfa region (**Figure 7**).

The map highlights the dominance of low to medium vulnerability class with an area of over 290,000 ha, (rate of 69.6%) defining a medium to low vulnerability. All these areas are occupied by not cultivate area or by Alfa. The high vulnerability class has only 7% of the total sur-

face area; it individualizes forests and their adjacent areas. This is very logical and consistent with the starting points of fires in Djelfa.

Class "not vulnerable" is less important, but well represented with an average of 14% of the area, they are in wetlands, in areas with rocky outcrops and dunes.

To validate our model, comparison between vulnerability map to forest fires was obtained, and the fires already recorded in the study area was established, helped to corroborate in most cases the results of this analysis. This comparison was used to assess the benefit limit of our model.

The superposition of vulnerability map with the sites

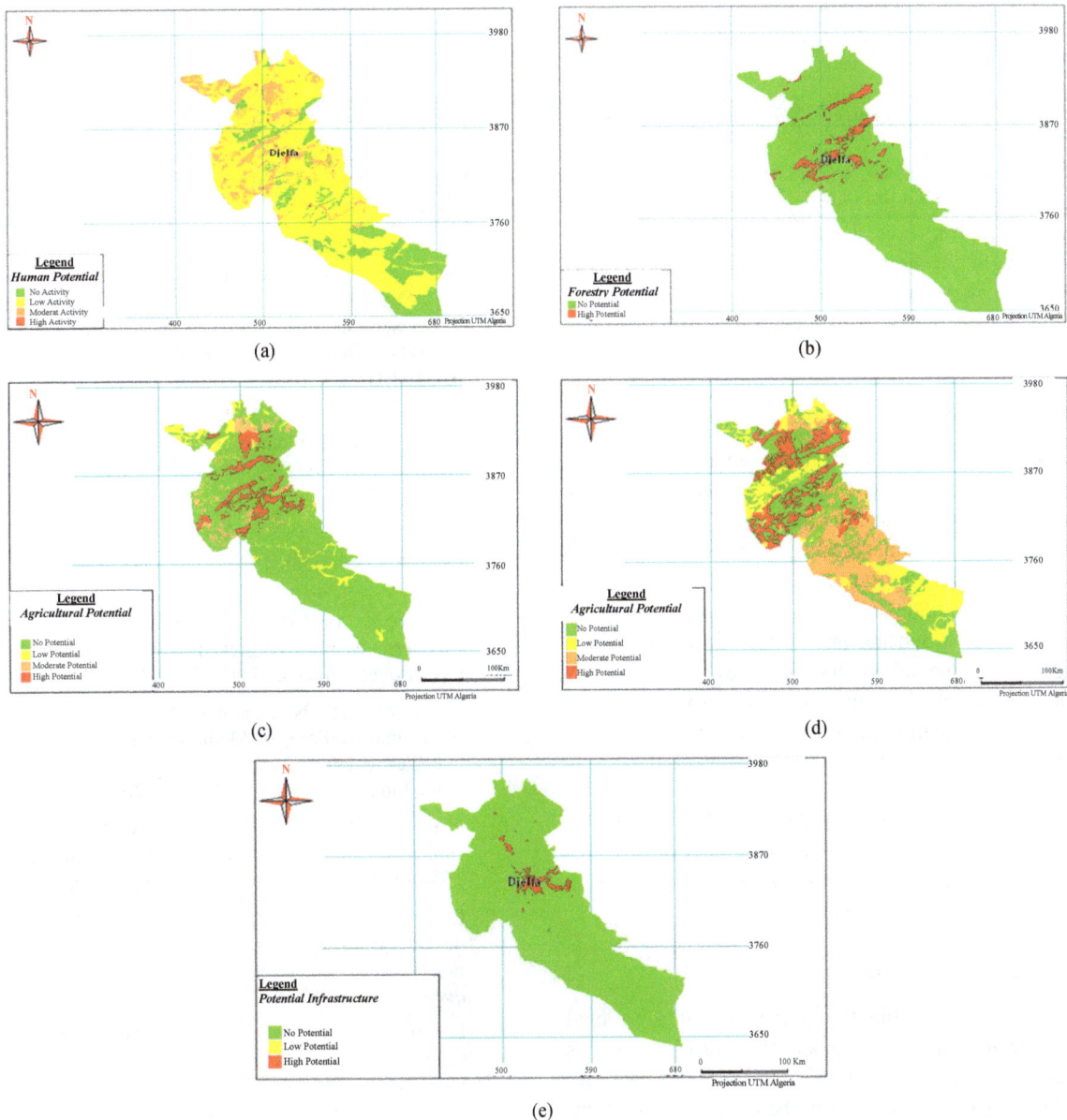

Figure 6. Results of different spatial vulnerability criteria. (a) Human potential of Djelfa region; (b) Forestry potential of Djelfa region; (c) Agricultural potential of Djelfa region; (d) Pastoral potential of Djelfa region; (e) Potential Infrastructure of Djelfa region.

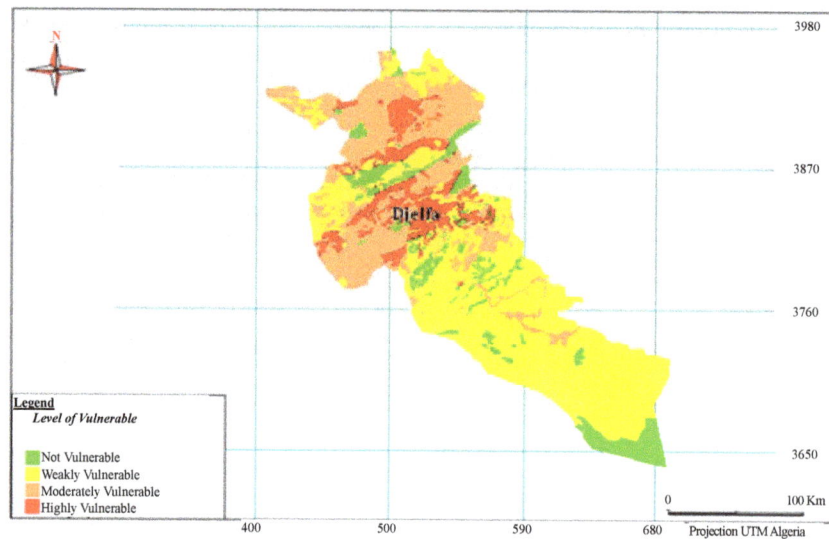

Figure 7. Map of environmental vulnerability to forests fires in Djelfa region.

already burned in the past, we allowed us to demonstrate the adequacy of the vulnerability level distribution and that of the reported fires. Therefore, the mapping model allows to highlight the areas most sensitive and to better identify and clarify priority protection areas. This map can be used for equipment installation, firewall trenching and trails establishment.

5. Conclusions

In this work, we proposed an improvement of our environmental vulnerability assessment model to forest fires, adding the issue of pastoral and human vulnerability.

Spatialization of vulnerability was established using a GIS approach. The application of the proposed model on the forest land of Djelfa region, using geomatics technologies (Remote Sensing and GIS), allows to make a map of environmental vulnerability to forest fires by combining multiple information layers took from maps and land. The operation of layers combination is used to set a map of human and socioeconomic potential of study area.

Vulnerability map is not a fight mean, but it helps foresters to make a plan and an adequate control. In addition, to manage the problem of hazardous urbanization in areas becomes possible and controllable.

6. Acknowledgements

This study comes within the framework of a national research project on modeling and the use of geomatics tools in earth sciences. We wish to thank the Forests General Directorate of Djelfa for their support, information and discussion with experts in this field.

In addition, we should be grateful to Mr. Yacine ZERARTI, translator-interpreter in University of Bou-merdes, for his help to prepare this paper in technical English language.

REFERENCES

[1] D. Alexandrian, F. Esnault and G. Calabri, "Feux de Forêt dans la Région Méditerranéenne," *Unasylva*, Vol. 50, No. 197,1999.

[2] M. Belhadj-Aissa and Y. Smara, "Application du SIG et de la Télédétection dans la Gestion des Feux de Forets en Algérie," *2nd FIG Regional Conference*, Marrakech, 2-5 December 2003.

[3] G. Carbonell, *et al.*, "Embrasement Généralisé Eclair en Feu de Forêt," European Medicines Agency, London, 2004.

[4] P. Carrega, "Le Risque d'Incendies de Forêt en Région Méditerranéenne: Compréhension et Evolution," Institut des Risques Majeurs, Montpellier, 2008.

[5] P. Carrega and N. Jeronimo, "Risque Météorologique d'Incendie de Foret et Méthodes de Spatialisation pour une Cartographie a Fine Echelle," Actes du XXème Colloque International de l'AIC, Tunis, 2007.

[6] P. Carrega, "A Meteorological Index of Forest Fire Hazard in Mediterranean France," *International Journal of Wildland Fire*, Vol. 1, No. 2, 1991, pp. 79-86.

[7] P. Carrega, "Une Formule Améliorée pour l'Estimation des Risques d'Incendie de Forêt dans les Alpes Maritimes," *Revue d'Analyse Spatiale, Quantitative et Appliquée*, No. 24, 1988, pp. 165-171

[8] E. Chuvieco, A. De Santis, D. Riaño and K. Halligan, "Simulation Approaches for Burn Severity Estimation Using Remotely Sensed Images," *Fire Ecology Special Issue*, Vol. 3, No. 1, 2007, pp. 129-152

[9] D. Tàbara, D. Saurí and R. Cerdan, "Forest Fire Risk Management and Public Participation in Changing So-

Modelling of Environment Vulnerability to Forests Fires and Assessment by GIS Application on the Forests of
Djelfa (Algeria)

175

cioenvironmental Conditions: A Case Study in a Mediterranean Region," *Risk Analysis*, Vol.23, No. 2, 2003, pp. 249-260.

[10] D. Spano, T. Georgiadis, P. Duce, F. Rossi, A. Delitala, C. Dessy and G. Bianco, "A Fire Index for Mediterranean Vegetation, Based on Micrometeorological and Ecophysiological Measurements," 2003. http://natural-hazards.jrc.it(www.docstoc.com/docs/46516849/A-fire-risk-index-for-mediterranean-vegetation-based-on)

[11] G. Faour, R. B. Kheir and A. Darwish, "Méthode Globale d'Evaluation du Risque d'Incendies de Foret Utilisant la Télédétection et les SIG: Cas du Liban," *Revue d'Télédétection*, 2006, Vol. 5, No. 4, pp. 359-377

[12] C. F. Peigneux, "Utilisation des Systèmes d'Information Géoréférée dans l'Analyse du Risque.," Université de Genève, Genève, 2003.

[13] M. S. Guettouche, A. Derias, M. Boutiba, M. A. Bounif, M. Guendouz and A. Boudella, "A Fire Risk Modelling and Spatialization by GIS. Application on the Forest of Bouzareah Clump, Algiers (Algeria)," *Journal of Geographic Information System*, Vol. 3, No. 3, 2011, pp. 247-258.

[14] M. Khader, *et al.*, "Etude du Risque Incendie a l'Aide de la Géomatique: Cas de la Forêt de Nesmoth (Algérie)," Universitat d'Alacant, Alacant, 2009.

[15] A. Mariel, "Cartographie du Niveau de Risque d'Incendie, Exemple du Massif des Maures," Centre of Agricultural Machinery, Agricultural Engineering, Water and Forests (CEMAGREF), Lyon, 1995.

[16] O. Meddour-Sahar, R. Meddour and A. Derridj, "Analyse des Feux de Forets en AlgErie sur le Temps Long 1876-2007," 2008. www.ciheam.org

[17] A. Missoumi and K. Tadjerouni, "SIG et Imagerie Alsat1 pour la Cartographie du Risque d'Incendie de Forêt," *2nd FIG Regional Conference*, Marrakech, 2-5 December 2003, p. 14.

16

GIS as an Efficient Tool to Manage Educational Services and Infrastructure in Kuwait

Khalid Al-Rasheed[1], Hamdy I. El-Gamily[2,3]
[1]Civil Engineering Department, College of Engineering, Kuwait University, Kuwait City, Kuwait
[2]Geoinformatics Center, Kuwait Institute for Scientific Research (KISR), Kuwait City, Kuwait
[3]National Authority for Remote Sensing and Space Sciences (NARSS), Cairo, Egypt

ABSTRACT

The State of Kuwait Ministry of Education (MoE) has clearly defined land use standards for the location of public schools, and an inventory of reserved lands for future facilities. Unless, there is a geographical efficient tool to manage and plan the education system in a rapidly developing country such as Kuwait there will be huge deficit in such services. Geographic Information Systems (GIS) was used to fill in this gap and effectively evaluate and analyze their facilities and unoccupied lands to ensure they continue to meet the population and future needs of Kuwaiti students. This paper utilized the GIS to inventory, map, and analyze MoE facilities and unoccupied land reservations with a goal of improved planning and decision making. Unfortunately, the initial spatial analysis of the data showed huge percent of districts that have no schools failing to meet the minimum standard of the MoE including kindergartens, primary schools, intermediate schools for girls, intermediate schools for boys, secondary schools for girls and secondary schools for boys at 72%, 71%, 48%, 43%, 54%, and 55% respectively. Such critical results will enable the decision makers to prioritize the immediate action of relocation the schools or widen the services and accessibility. Moreover, the analysis of the data showed a critical and immediate need to reserve land for five districts where they are heavily populated and lacked reserved land. However, based on long term land use plans, there is an urgent need to relocate some land and reserve others to meet the future urbanization plans and population growth.

Keywords: Kuwait; Education; GIS; Infrastructure; Land Use Planning; Land Reservation

1. Introduction

Overpopulation growth in developing countries always put excessive pressure on government for facility management. Key element is the education and allocation of the school within the urbanized areas to meet the community needs. Unless there is a tool to help decision makers for relocating of schools according to the standardized criterion, it will be mysteries planning. The Kuwait school system includes a diverse range of students and inventory of schools. The Kuwait Ministry of Education (MoE) is the body responsible for administering public schools including, special needs, religious schools, and adult literacy education [1]. MoE also oversees and regulates the private schools operating within Kuwait. MoE property holdings include school facilities and administration facilities, as well as unoccupied land parcels that have been reserved throughout the Kuwait urban area for future facilities to meet the future needs of over population growth.

To foster the use of new geographic information tools for effective planning of the school within the Kuwaiti populated area, the Kuwait National Educational Atlas Project for Services and Infrastructure (NEASI)—phase 1 was initiated. This project was a challenge to support the decision and policy makers in MoE with inventorying, mapping, managing, and analyzing existing school facilities, and land use planning for the unoccupied lands that reserved for future school facilities. MoE had defined land use standards for the placement of public schools, and an inventory of reserved lands for future facilities. The NEASI phase 1 project was a two-year joint mission conducted by MoE and the Kuwait Institute for Scientific Research (KISR) to: inventory, map, and analyze MoE facilities and unoccupied land reserved for future education construction.

Prior to the NEASI project, MoE lacked the ability to evaluate its entire inventory of facilities and unoccupied lands against their established criteria in a spatial context. The NEASI project identified five functions that Geographic Information Systems (GIS) technology could improve: 1) document and map MoE land holdings and their uses, including those that are unoccupied and/or awaiting development, school sites and administrative sites; 2) document and organize buildings and mainte-nance activities data; 3) integrate student and staff re-cords (non-spatial) with geographical data sets; 4) de-velop educational indicators and educational land use planning tools to improve decision making in Kuwait. This paper focuses on the land use analysis and planning aspect of the project.

GIS school mapping has much been used in the last two decades for educational planning [2]. GIS database provides comprehensive framework and organization of spatial and non-spatial data to efficiently help decision makers and planners. It apparently provides a mapping tool for the relationships between the distribution of schools and the school age population within the populated areas [3]. It is an efficient tool in managing and planning the accessibility to the educational schools.

In this paper, challenge was to develop GIS system for the MoE for planning and a decision making tool for im-mediate and future planning of the educational zones in Kuwait.

2. Study Area Description

The State of Kuwait is a small country located on the Arabian Gulf, and is bordered by Saudi Arabia and Iraq. Urban settlement is concentrated along the southern edge of the Kuwait Bay and along the western edge of the Arabian Gulf. An estimated three (3) million people live in Kuwait with about one third are at school age [1].

The administrative hierarchy of Kuwait is starting form the largest scale of governorate through districts to the smallest area of blocks. The Kuwait Educational Zones are formed by blocks. The study area is made up of all contiguous districts and blocks in Kuwait with an urban designation. As such, the districts of Kabd and Failaka Island were not included in the analysis.

3. Materials and Methods

As the case of many developing countries, the availabil-ity of geographical data for Kuwait is limited and there is not yet a national GIS database. Therefore, the major political boundaries of governorates, districts, and blocks were developed. This was using analog map sheets that were converted into GIS format. However, the detailed scale of parcel boundaries and street centerlines were ob-tained electronically from the Kuwait Municipality and the Kuwait Ministry of Public Works respectively. The map projected and alignment of both datasets were hand-led using the GIS package.

The remaining data required for analysis, such as de-mographical data and schools, was developed from avai-lable analog and digital sources, and then manual field data collection was conducted to complete and verify the features and attributes. This included capturing all of the individual buildings, and parcels owned by MoE. As MoE had multiple, sometimes conflicting, and missing data sources it was necessary to field verification of the school building and parcels. Records were maintained at individual school facilities, and they were also kept at the MoE administrative offices. School facility, student, and teacher records were developed at the building level, and land use/land holdings were developed at the parcel le-vel.

Prior of developing the GIS database user needs as-sessment process was carried out to identify the spatial and non-spatial data needs of the MoE for managing and planning the educational facilities. This task focused on data needs analysis and GIS requirements analysis to identify applications that will be built to utilize the geo-database [4-6]. **Figure 1** provides a block diagram illus-trating the project development methodology to inven-tory, map and analyze MoE facilities and infrastructure.

ArcGIS package was used to develop the GIS database and publish on web-based services. The overall architect-ure of the developed system is shown in **Figure 2** [7]. In addition to the SDSS for site suitability and education indicators, custom NEASI tools and applications were developed based on the requirements analysis. It was customized to provide users with easy to understand in-terface [7-12]. The NEASI applications and customiza-tions were developed using: C#.net 2008 Version 3.5, ArcGIS Server, and ArcObjects API. The ArcGIS Server ADF template was used to configure the map and related components. The system is developed to provide various service tools including:

- Tools and functions to explore, and display informa-tion available in the geodatabase.
- The unoccupied land tool supports the user in locating available unoccupied lands that belong to MoE, based on administrative reference such as governorates, dis-tricts, and urban blocks.
- Tool for map production to generate an easy costum-ed map layouts including map title, map scale, map legend and north direction arrow.
- Search tool to search and locate schools based on school names, school stages, and/or administrative units.
- Documentation and reporting tool to provide life link of the school records. The uploaded documents can be retrieved by other users if they have permission to see and download such documents.
- Tools for updating the attribute or spatial data sets from the field using the web application.

The GIS database was validated on site using series of data capture sheets prepared for each district. The data capture sheets mapped and classified all MoE properties. The field data team visited each site to verify, update or

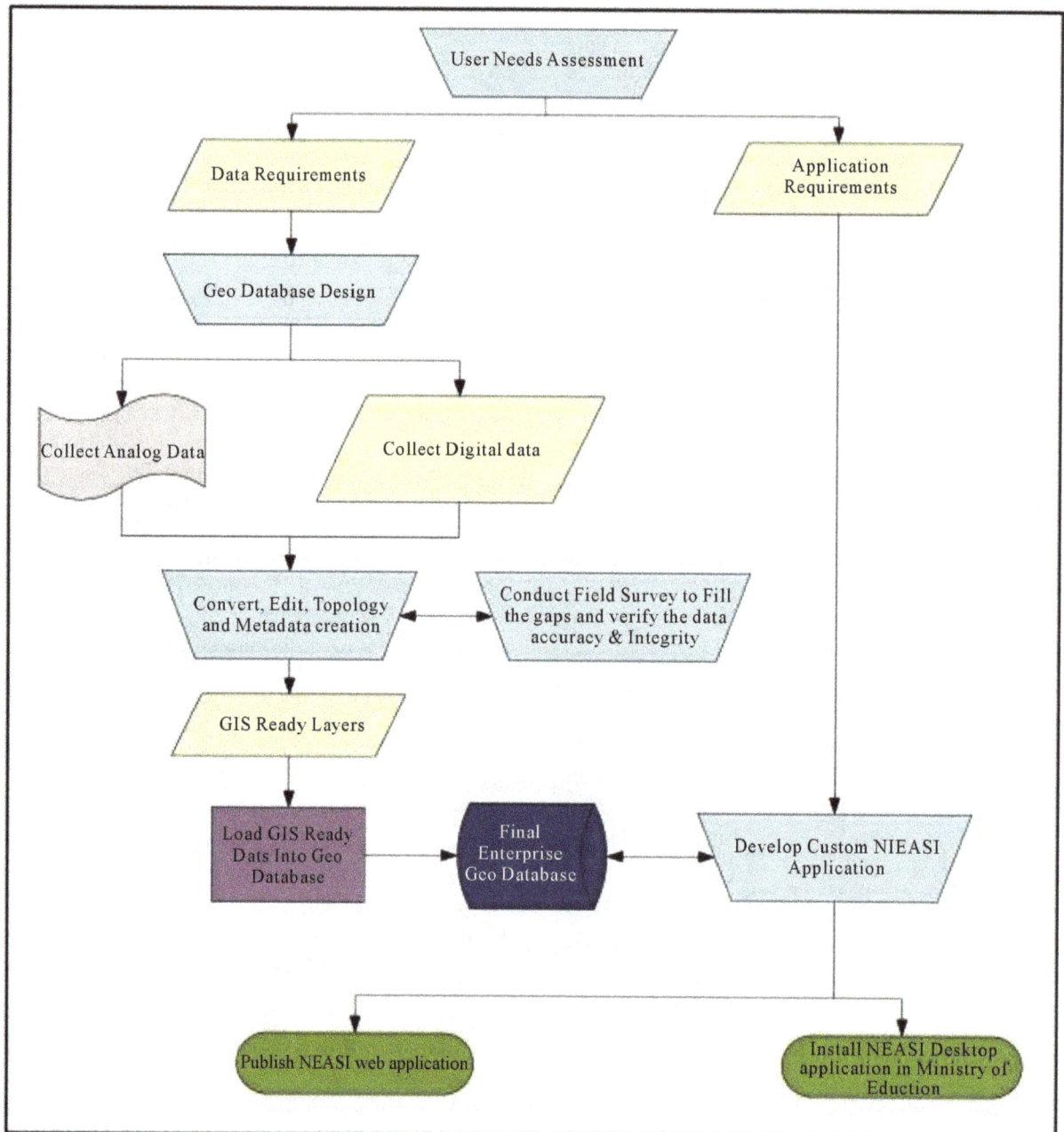

Figure 1. NEASI block diagram of the GIS processes.

collect new descriptions of the properties. The benefit of this task was two-fold because it resulted in high accuracy data and provided a training opportunity for MoE personnel. **Figure 3** shows an example of a data capture sheet that prepared and used by KISR project team and MoE personnel to collect data.

4. Results

The GIS analysis was firstly to evaluate the locations of existing public schools to determine if the existing schools were located consistent with the MoE general

placement standards. The analysis was conducted for public kindergarten, primary, intermediate, and secondary schools. The general placement standards for school stages are as follows:

- Schools should be located inside urban/residential districts.
- Each block should contain one (and only one) kindergarten, and one (and only one) primary school.
- Each district should contain one (and only one) intermediate girl's school, and one (and only one) intermediate boy's school.

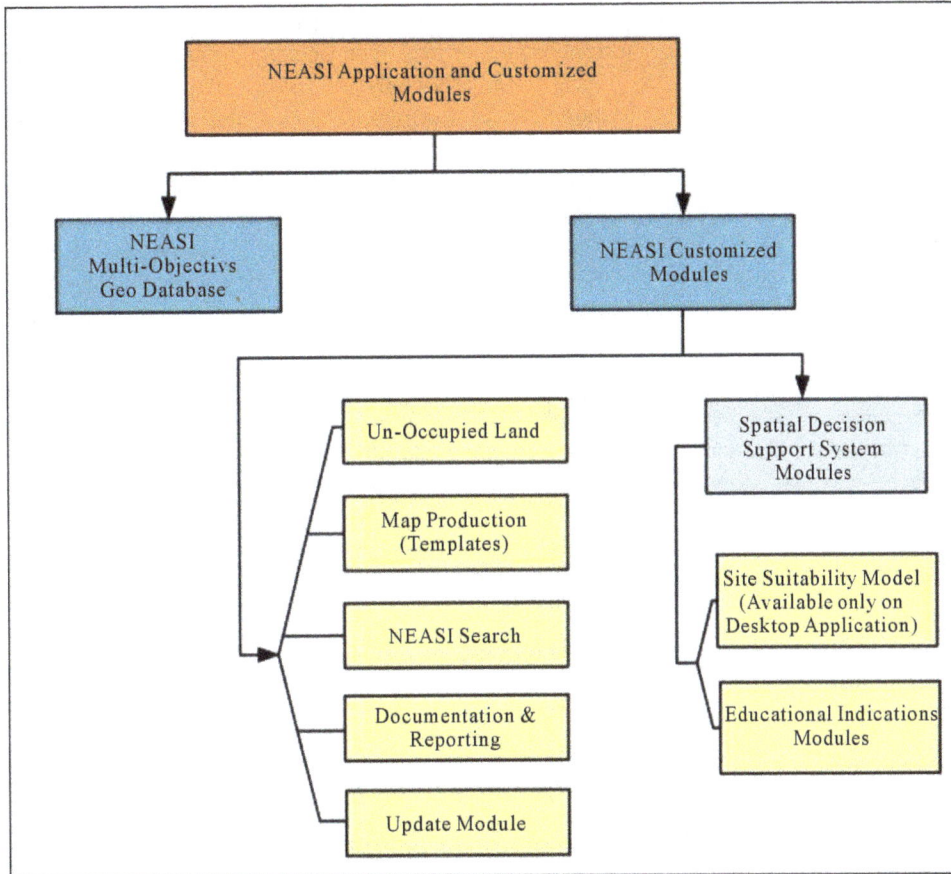

Figure 2. The architecture of the GIS system of MoE.

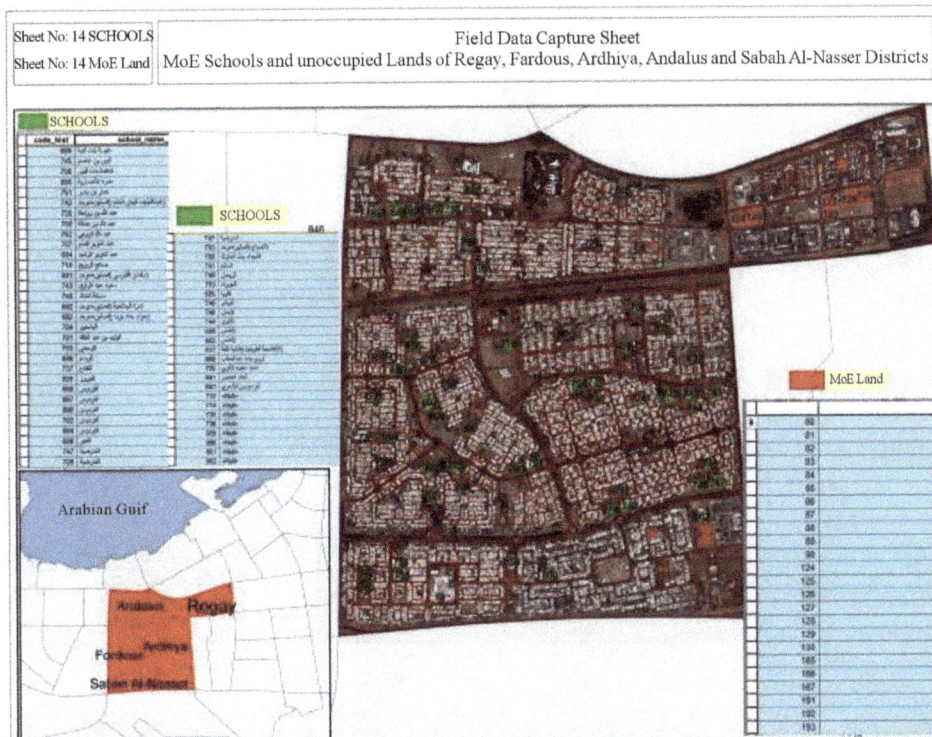

Figure 3. Example of the field data capture sheet.

- Each district should contain one (and only one) secondary girl's school, and one (and only one) seconddary boy's school.

MoE mapped and attributed 815 public schools in which 212 are kindergartens, 259 primary schools, 207 intermediate schools, and 137 secondary schools. The spatial visualization of these schools within the territory of each block aided decision makers to analyze the distribution of the schools according to the MoE standard.

4.1. Kindergartens

Actually, there are about 707 blocks in the Kuwaiti urbanized area. The general standard of both the kindergartens and primary schools is one school per block. The GIS analysis of these two types of schools showed effect-

tively the gaps where blocks are out of education services. The actual placement of kindergartens revealed that only 28% of urbanized blocks contain kindergartens; in which 26% meet the MoE standards and contain only one kindergarten and 2% contain 2 kindergartens per block, and 1 block contained 3 kindergartens. Conversely, the analysis alarmed seriously that 72% of all urban blocks do not contain any kindergartens. Such results show the effectiveness of GIS analysis to alarm the decision makers and policy planners to take an immediate action to resolve this problem. **Table 1** lists the relationship between the number of blocks and distribution number of kindergartens. **Figure 4** provides the spatial distribution of the kindergartens schools within the educational zones in Kuwait.

Table 1. Distribution of public kindergartens.

Kindergartens per block			
No. of schools per block	No. of blocks	Total number of blocks	%
0	509	707	72
1	185	707	26
2	12	707	2
3	1	707	0

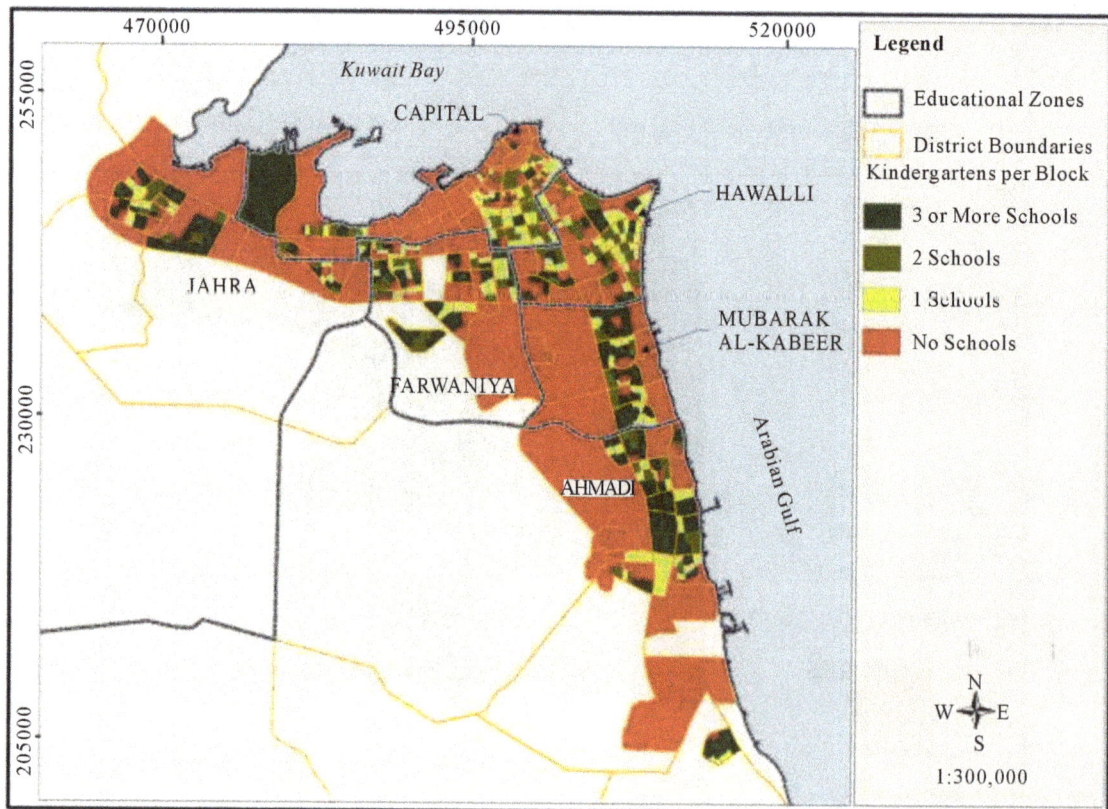

Figure 4. Kuwait public kindergartens.

4.2. Primary Schools

The spatial analysis of the distribution of primary schools per block was nearly similar to the kindergartens. 71% of the urban blocks have no primary schools. However only 29% of the urban blocks have primary schools; in which 23% has just one school per block. More impressively, 44 blocks were recorded with more than one school up to 5 schools per block in a single block. **Table 2** lists the statistical records of these schools per urban blocks. It is apparently that GIS analysis has shown how much miss-planning of primary and kindergartens schools location along the populated community (**Figure 5**). This effectively could help MoE decision makers to take an imme-

diate action towards the services and distribution of primary schools and the urgent needs for either establishing new schools or relocation of others to provide complete services and easy accessibility to the community.

4.3. Intermediate Schools

The Kuwaiti educational system for intermediate school is established to be for gender segregation, which means that there are girls and boys schools. The basic standard is one intermediate school per district per gender. Analytical study of the girl's intermediate school has clearly shown that about 48% of the districts had no girl's intermediate school. The rest of about 52% has variable

Table 2. Distribution of public primary schools.

Primary schools per block			
No. of schools per block	No. of blocks	Total number of blocks	%
0	503	707	71
1	160	707	23
2	41	707	6
3	2	707	0
5	1	707	0

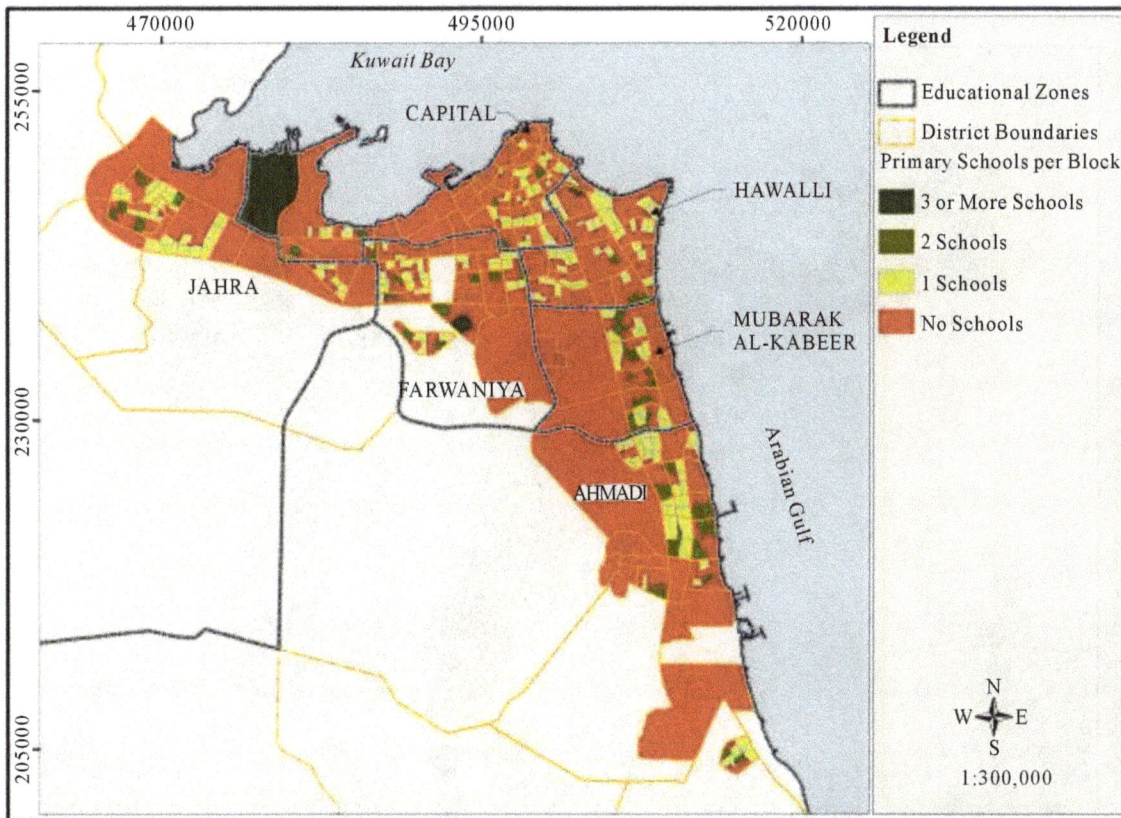

Figure 5. Kuwait public primary schools.

number one or more school. 25% of the districts met the MoE standards and had only one school, 20% of the districts had 2 schools, 6% had 3 schools, and only 1% had 6 girl's intermediate schools. **Table 3** lists the statistic of the intermediate girl's school. However, **Figure 6** shows the spatial distribution of these kinds of schools.

Complementary to the intermediate school system is the boy's intermediate schools, which are distributed to accommodate the boys of the intermediate school age. Better than the girl's school, only 43% of the districts had no school for boys and 57% had one school or more. Similar to the girls, 25% of the districts met the MoE requirements and had only 1 school. However, 32% had more than 1 school for boys distributed as 19% with 2 schools per district, 6% had 3 schools per district, 6%

had 4 schools per district, and there was a single district with 5 boy's intermediate schools. **Table 4** lists the statistics of the intermediate boy's schools within the Kuwait districts. **Figure 7** shows the graphical distribution of these schools.

4.4. Secondary Schools

Secondary school is the final level that qualifies students to the university and similar to the intermediate schools, the Kuwait educational system is based on segregation of secondary school per district per gender. The geographical analysis of the girl's secondary schools shows that high percentage of 54% districts that have no girl's secondary school. Only 47% of the districts have one or

Table 3. Distribution of public girl's intermediate schools.

Girl's intermediate schools per district			
No. of schools per district	No. of districts	Total number of districts	%
0	59	121	48
1	30	121	25
2	24	121	20
3	7	121	6
6	1	121	1

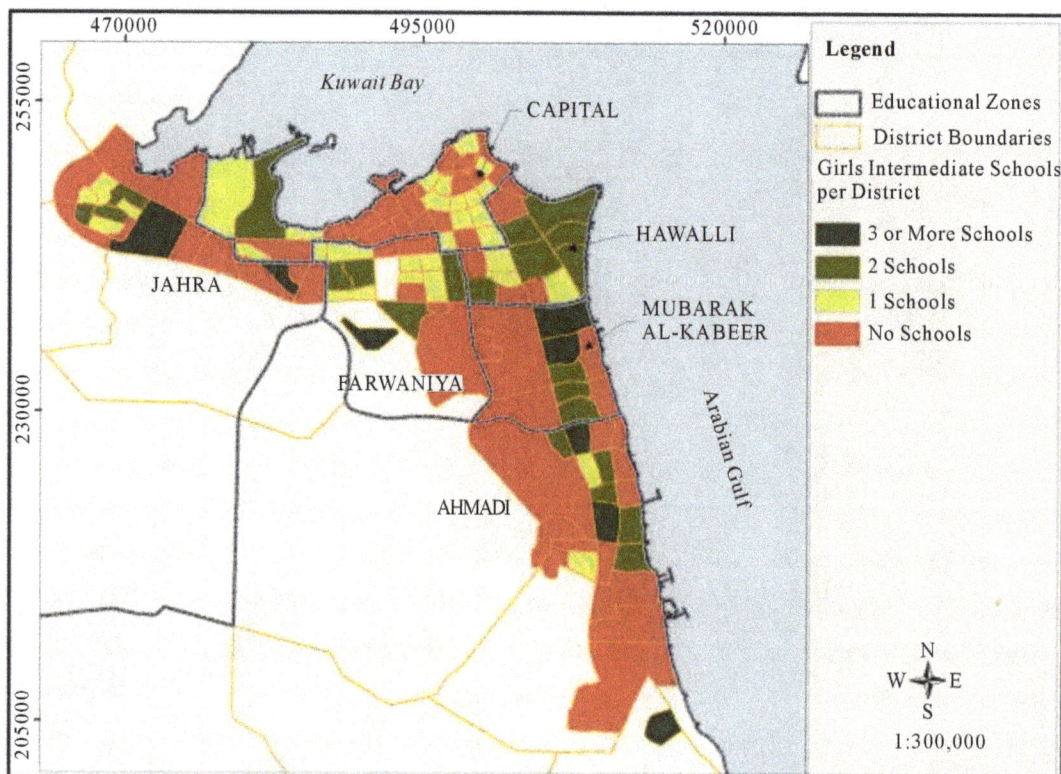

Figure 6. Kuwait public girl's intermediate schools.

Table 4. Distribution of public boy's intermediate schools.

Boy's intermediate schools per district			
No. of schools per district	No. of districts	Total number of districts	%
0	52	121	43
1	30	121	25
2	23	121	19
3	8	121	6
4	7	121	6
5	1	121	1

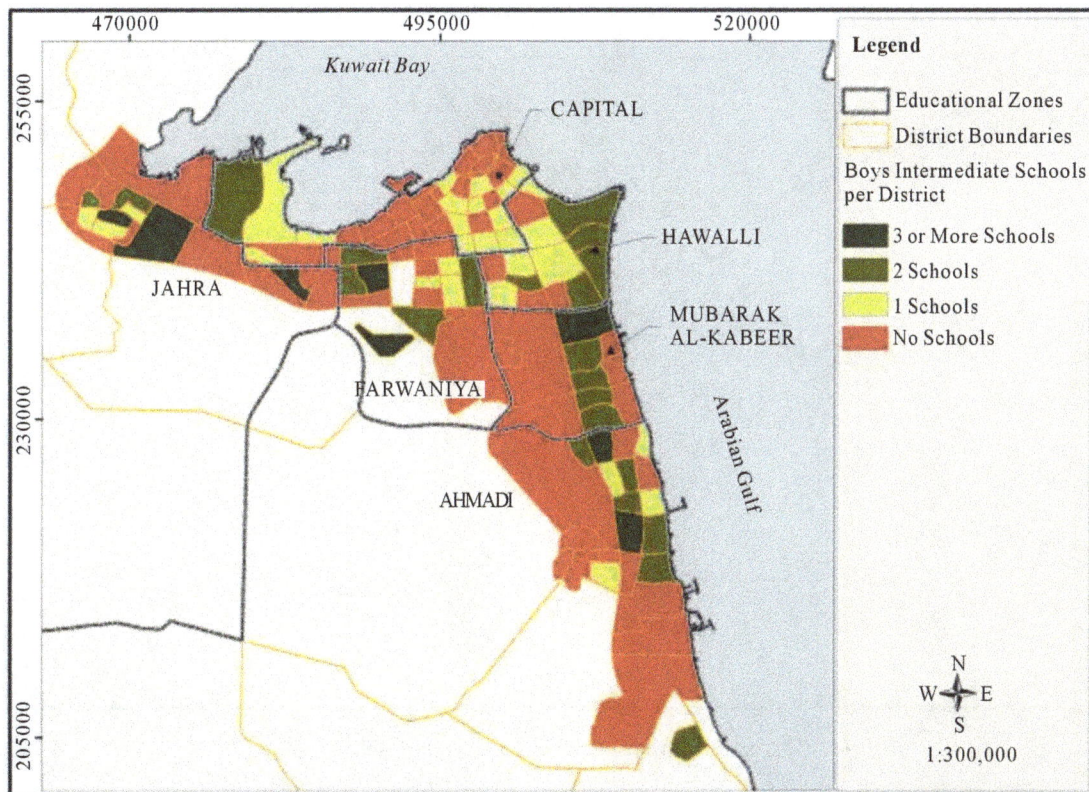

Figure 7. Kuwait public boy's intermediate schools.

more school in which 33% have met the general standards of 1 school per district. The rest of 13% was shown that 12% have 3 schools and 1% have 3 girl's secondary schools. **Table 5** lists the statistics of the girl's secondary school. **Figure 8** presents the graphical distribution of these schools along the Kuwaiti urbanized districts.

Complementary to the girl's secondary school system is the boy's secondary schools which are distributed to accommodate the boys of the secondary school age. Similar to the girls, about 55% of the districts have no schools. The remaining 45% of the districts have one or more schools, in which 39% met the standard of 1 boy's secondary school per district and only 6% have 2 schools.

None of the districts contained more than 2 boy's seconddary schools. **Table 6** lists the statistics of the boy's secondary schools. **Figure 9** presents the spatial distribution of this type of schools within the Kuwait districts.

5. Future Planning

MoE has an inventory of 268 unoccupied parcels in reserve for future educational facilities to meet the future demands for population growth. These parcels are distributed throughout the 707 urban blocks in Kuwait. **Table 7** provides a summary of the relationship between the existing public school and the land reserved for future

Table 5. Distribution of public girl's secondary schools.

Girl's secondary schools per district			
No. of schools per district	No. of districts	Total number of districts	%
0	65	121	54
1	40	121	33
2	15	121	12
3	1	121	1

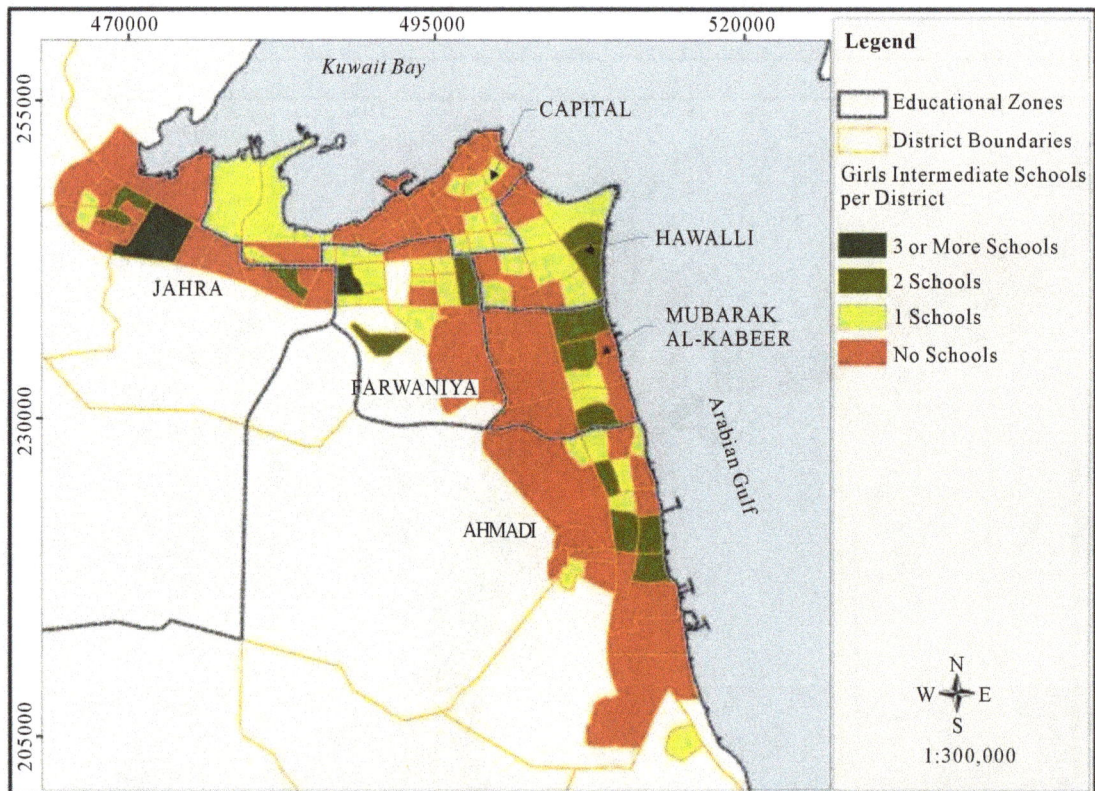

Figure 8. Kuwait public girl's secondary schools.

Table 6. Distribution of public boy's secondary schools.

Boy's Secondary Schools per District			
No. of schools per district	No. of districts	Total number of districts	%
0	66	121	55
1	47	121	39
2	8	121	6

school facilities. 25% of all urban blocks in Kuwait have at least one parcel in reserve for future educational facilities. This includes blocks that already have one or more schools and blocks that have no schools. Conversely, there are 75% of all urban blocks in Kuwait that have no land in reserve for future educational facilities. The data available for analysis did not distinguish residential from commercial and industrial classes, or population at the block level, within the urbanized area. This is may be one of the main reasons that increase the percentage of districts and blocks that have not schools of different stages.

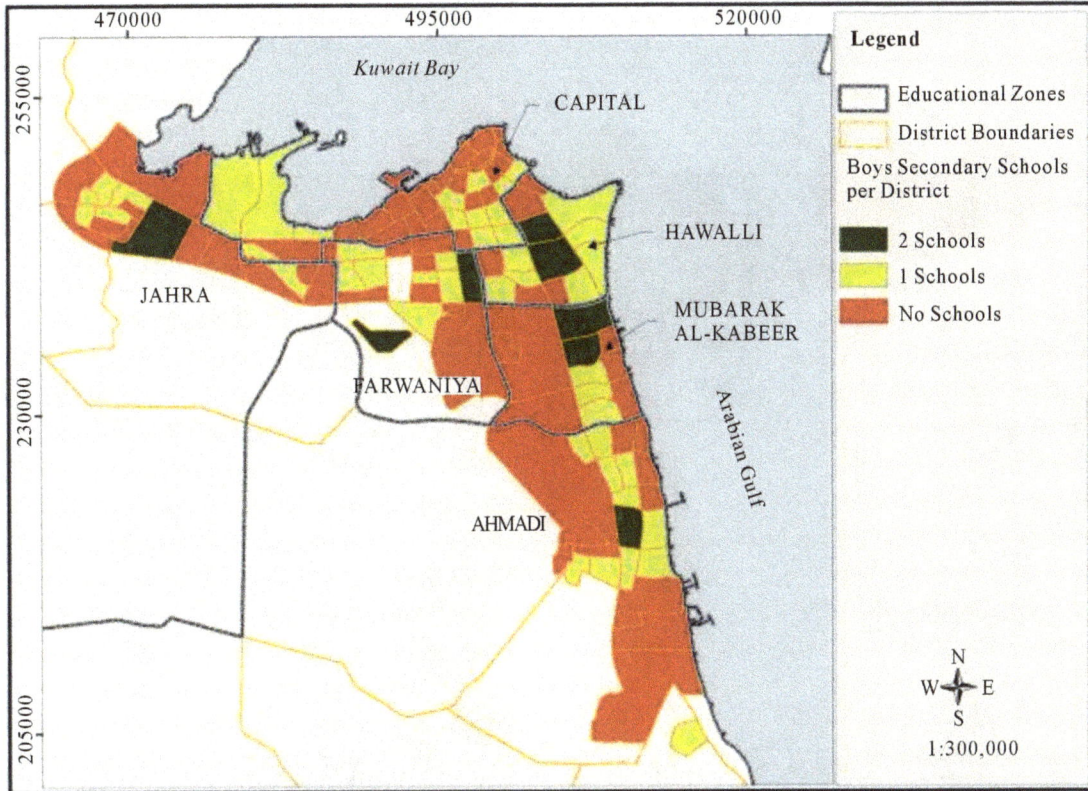

Figure 9. Kuwait public boy's secondary schools.

Table 7. Summary of unoccupied lands.

Status	No. blocks	Total blocks	% Each	Sum
Block has school and has reserved land	91	707	13%	25%
Block does not have school and has reserved land	86	707	12%	
Block has school and does not have reserved land	262	707	37%	75%
Block does not have school and does not have reserved land	268	707	38%	

GIS has enabled an improved understanding for the immediate needs of MoE to indentify urban residential blocks which had no unoccupied land in reserve. This, indeed, spatially prioritizes to the education policy makers to develop an action plan for the immediate requirements as well both medium and long term. **Figure 10** presents the spatial distribution of these reserved parcels along the Kuwait urban area.

6. Conclusions

Kuwait is a developing country that has a rapid urbanization growth rate of 2.1, which requires an efficient planning mechanism for infrastructure to meet such rapid growth. MoE has faced such challenge and made a commendable job of managing and planning educational land uses until now, using the tools and information avai-

lable. GIS technology was efficiently used to locate the number of schools at all levels per blocks and therefore to clearly evaluate the districts that failed to meet the MoE standard of single school per block. Unfortunately, 72% of the districts lack the existence of kindergarten schools, which require an urgent plan to resolve such critical problem. This kind of school at such younger age might require re-planning of allocating more than one school per block to ease the accessibility; only 12% achieved such high standard. This was also the case of the primary school where about 71% of the districts lack to host single primary school.

However, more detailed information is needed about future development, especially residential to more accurately plan for educational land reservations. If the detailed land use designations of the Kuwait Master Plan data were converted to GIS and incorporated into the

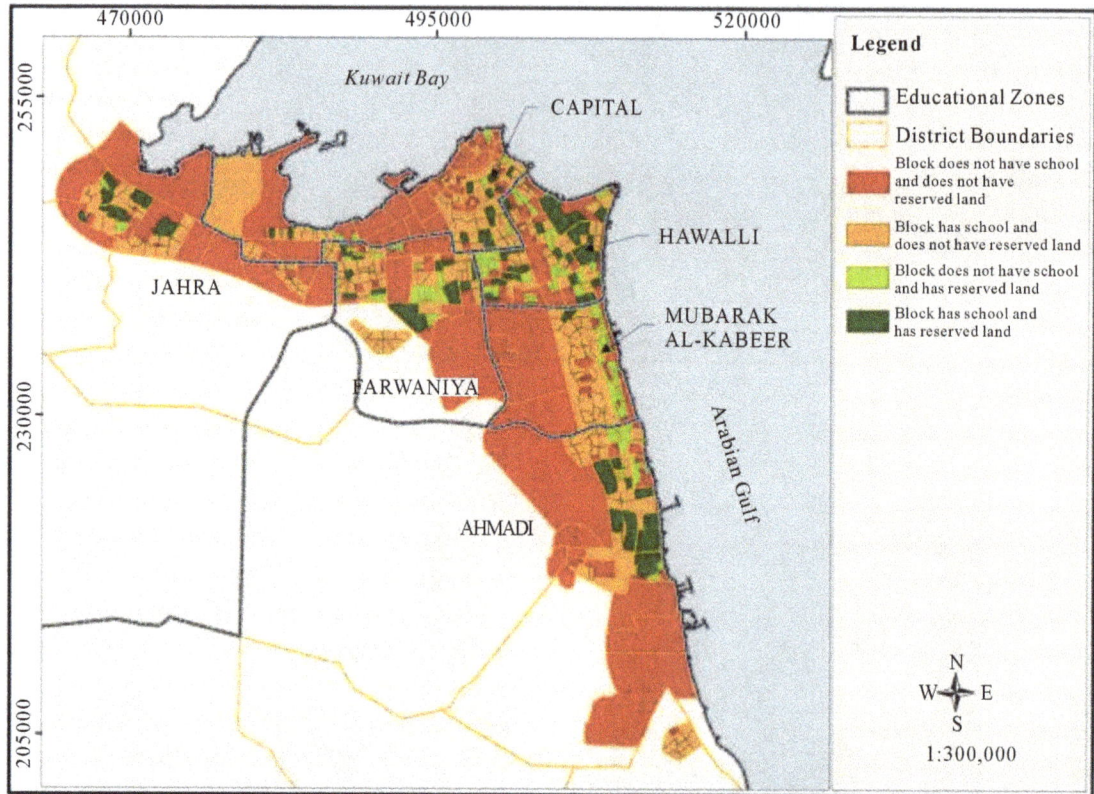

Figure 10. Distribution of unoccupied land reserves.

analysis GIS tools, the model would be enhanced to propose ultimate solution in a time dimension. For example, the existing model has shown a critical need to reserve additional land in residential blocks in Kuwait for pre-existing development and to secure additional lands for the blocks and districts designated for future residential development. As more GIS data is developed and disseminated for Kuwait, the ability to model and analyze educational land use will be improved

7. Acknowledgements

The authors would like to express their appreciation and gratitude to the Ministry of Education (MoE) and Kuwait Institute for Scientific Research (KISR) for jointly funding the NEASI (SP001S) project. Further, we would like to thank the members of steering and technical committees for their continued support. Special thanks are due to the project team from MoE and KISR Geoinformatics Center.

REFERENCES

[1] Central Intelligence Agency, "World Fact Book," 2012. https://www.cia.gov/library/publications/the-world-factbook/geos//ku.html

[2] D. J. Maguire, M. F. Goodchild and D. W. Rhind, "Geographical Information Systems: Principles and Applications," Longman, Harlow, 1991.

[3] S. J. Hite, "GIS-Generated School Mapping Materials of Two Counties in Hungary Prepared for Françoise Caillods," International Institute for Educational Planning, Paris, 2006.

[4] H. W. Calkins, "Local Government GIS Development Guides," 1996. http://www.archives.nysed.gov/a/records/mr_pubGIS03.shtml

[5] R. Laurini and D. Thompson, "Fundamentals of Spatial Information Systems," Academic Press, London, 1992.

[6] I. Heywood, S. Cornelius and S. Carver, "An Introduction to Geographical Information Systems (Third Edition)," Pearson Education Limited, Harlow, 2006.

[7] H. El-Gamily, A. Al-Othman and S. Al-Fulaij, "The National Educational Atlas for Services and Infrastructure-Phase 1," Final Report, Kuwait Institute for Scientific Research, Safat, 2011.

[8] R. J. Shavelson, L. M. McDonnell and J. O. Rand, "What Are Educational Indicators and Indicator Systems?" *Practical Assessment, Research and Evaluation*, Vol. 2, No. 11, 1991.

[9] C. Acedo, M. Uemura, C. J. Thomas, Y. Nagashima and K. L. Ngow, "Education Indicators for East Asia and Pacific," World Bank Report, Washington DC, 1999.

[10] European Commission Education and Culture DG Brussels, "Key Education Indicators on Social Inclusion and Efficiency," Final Report, 2005-4751/001-001 EDU-ETU, Brussels, 2006.

[11] UNESCO Institute for Statistics, "Education Indicators Technical Guidelines," UNESCO, Paris, 2009.

[12] I. Attfield, M. Tamiru, B. Parolin and A. DeGrauwe, "Improving Micro-Planning in Education through a Geographical Information System: Studies on Ethiopia and Palestine," UNESCO Publishing, Paris, 2002.

17

Geospatial Evaluation for Ecological Watershed Management: A Case Study of Some Chesapeake Bay Sub-Watersheds in Maryland USA

Isoken T. Aighewi, Osarodion K. Nosakhare
Benedict College, Columbia, South Carolina, USA

ABSTRACT

Geospatial technology is increasingly being used for various applications in environmental management as the need for sustainable development becomes more evident in today's rapidly-developing world. As a decision tool, Geographic Information system (GIS) and Global positioning System (GPS) can support major decisions dealing with natural phenomena distributed in space and time. Such is the case for land use/cover known to impact ecosystems health in very direct ways. Our study examined one such application in managing land use of some sub-watersheds in the eastern Shore of Maryland, USA. We conducted a 20-year historical land use/cover evaluation using Landsat-TM remotely sensed images and GIS analysis and water monitoring data acquired during the period by Maryland Department of Natural Resources, including sewage discharge of some municipalities in the area. The results not only showed general trends in land use patterns, but also detailed dynamics of land use-land cover classes, impact on water quality, as well as other useful information for guiding both terrestrial and aquatic ecosystems management decisions of the sub-watersheds. The use of this technology for evaluating trends in land use/cover on a decade-by-decade basis is recommended as standard practice for managing ecosystem health on a sustainable basis.

Keywords: Geospatial; Land Use; Water Quality; Remote Sensing; Nutrients; Watershed; GIS

1. Introduction

Information has always been the cornerstone of effective decisions [1]. Consequently, most scientists and environmental managers in the US and many developed countries of the world now fully embrace geospatial technology (Geographic information systems, remote sensing and global positioning system) for the study of the environment, reporting on environmental phenomena, and modeling how the environment is responding to natural and man-made factors. Modeling land use/cover in relation to ecosystems is among such uses. Land use change has been known to be one of the most ubiquitous anthropogenic influences on global ecosystems [2]. Land cover patterns have also changed dramatically during the last century especially in North America with these historic changes leaving persistent legacies; similarly, the amount of land converted to urban and agricultural uses and the spatial arrangement of riparian habitats are useful indicators of the status of riverine ecosystems of the present times [3]. It has been reported that Change occurring on land affects water quality and thus the ecological health of the aquatic ecosystems; and such has been correlated with the degradation of biological, chemical and physical properties of streams within the Chesapeake Bay watershed [4]. The Lower Eastern Shore watershed and Coastal Bays of Maryland-subsets of the Chesapeake Bay watershed, have experienced rapid urbanization in the last decade with the increase in real estate development and roads as obvious indicators. Jantz *et al.*, [5] observed a 61% increase in developed land within the Chesapeake Bay watershed between 1990 and 2000, with most (64%) of the new development occurring on agricultural lands and grasslands, while 33% occurred on forested lands. A decade ago, the US Environmental Protection Agency [6] reported that urbanization was threatening Maryland streams and that if the rate of urban sprawl continued, more streams will likely degrade. In that year, about 16% of Maryland's land area was urban and was expected to grow to 21% in the next 25 years,

Geospatial Evaluation for Ecological Watershed Management: A Case Study of Some Chesapeake Bay
Sub-Watersheds in Maryland USA

189

while only 42% of the state was forested.

A Coastal Change Analysis Program (C-CAP) was initiated by the National Oceanic and Atmospheric Administration (NOAA) for monitoring LULC changes of the coastal regions of the US on a 5-year basis starting from 1996 [7]; and is still currently investigating techniques for analyzing land cover change data trends. Similarly, the Multi-Resolution Land Characteristics Consortium (MRCL) and the National Land Cover Data (NLCD) also exists for 1973, 1992, and 2001 [8]. However, no published analysis of long term trends in land use/cover exist-particularly in view of the rapidly changing demography as well as the agronomic and poultry industries of the coastal Eastern Shore of Maryland since the 1980's.

The state of US surface water systems has also been receiving major attention due to point and non-point source pollution from anthropogenic sources. For example, a survey of some US surface waters [9] showed that about 44% of the assessed stream miles, 64% of assessed lake acres, and 30% of the assessed bay and estuarine square miles were not clean enough to support uses such as fishing and swimming. The leading causes of impairment-pathogens, mercury, nutrients, and organic enrichment/low dissolved oxygen are from sources such as atmospheric deposition, agriculture, hydrologic modifications, and unknown or unspecified sources. Although the Clean Water Act (CWA) was intended to restore and maintain the chemical, physical, and biological integrity of the nation's surface waters [10], this goal has thus continued to pose a major challenge for water quality compliance, particularly due to contribution for non-point sources.

Significant relationships between land use-land cover (LULC) and water quality have been well documented over the years [11-18]. Agriculture, urban activity and industrialization are major sources of non-point pollution that contribute significant amounts of Phosphorus (P) and Nitrogen (N) to surface waters in the United States [19]. Urbanization has also been identified as a major threat to Maryland streams [6], and that a continuation of this trend could bring more streams into degraded status. This report classified 46% of all streams in Maryland in poor biological health conditions. It has been predicted that non-point source pollution will increase in the future if current land use-land cover practices continue [19]. It has been reported [17] that the extent of urban lands and its proximity to streams was the most important factor in predicting N and P concentrations in stream water. Also, lakes with highly forested catchments had lower levels of lead and chlorine and were less prone to eutrophication than lakes in non-forested catchments [20]. Urbanization and population growth usually lead to increased volume of wastewater, requiring treatment before discharge into surface waters; this invariably results in higher volumes

of sewage effluents (usually containing high P and N) and thus increased point source pollution. The most deleterious effects of sewage discharge into coastal environment are eutrophication [21]. Sewage effluent discharged into surface water can also result in significant effects on marine biota [22], leading to changes in abundance, biomass and diversity of the organisms. While several studies exist for relating cause-and-effect of point and non-point pollution on water quality, studies that incorporate long-term trends are few. Such studies could provide better and more holistic insight into factors influencing surface water quality; and could potentially provide more precise and useful information for decisions on land use and water system management at the watershed and landscape levels.

The objectives of this study therefore were to 1) apply geospatial techniques for evaluating the historical Land use-land cover (LULC) trends in the lower Eastern Shore watersheds of Maryland over a 20 year period; and 2) evaluate the influence of historical land use-land cover changes and sewage loading on surface water quality of some lower Eastern shore watersheds.

2. Study Area

This study was conducted in the lower Eastern Shore watershed and coastal bays of Maryland situated between longitudes 74°59'15.2"W and 76°17'5.6"W and latitudes 37°54'12.4"N and 38°53'10.7"N. It is located between the Atlantic Ocean and the Chesapeake Bay and drains approximately 5596.69 km^2 in Wicomico, Somerset and Worcester counties; and some portions of Caroline and Dorchester counties.

Major land use in the area includes cropland, forestry, pasture and urban; and water bodies include estuary, river/stream and wetlands. The area which is less than 100 feet above sea level includes a total of 23 sub watersheds [23] (see **Figure 1**). The major economic activities in the lower eastern shore are poultry production and grain farming-corn, soybeans and barley, including other important economic activities such as fishery and tourism.

3. Data and Methods

To capture historical land use/cover changes, remotely sensed satellite data were sourced from the United States Geological Survey's Center for Earth Resources Observation Sciences (USGS-EROS). To this end, *Landsat-TM* satellite data of Maryland (Path 14, rows 33 and 34) for 1986 (May 6), 1996 (May 1) and 2006 (April 27) were obtained in GeoTiff format. These images were of 10 days temporal variability (1986-2006) due to cloud cover and were systematically corrected; the data were of very high acquisition quality and had been geo-referenced and atmospherically corrected. The reflective bands 1 - 5 and

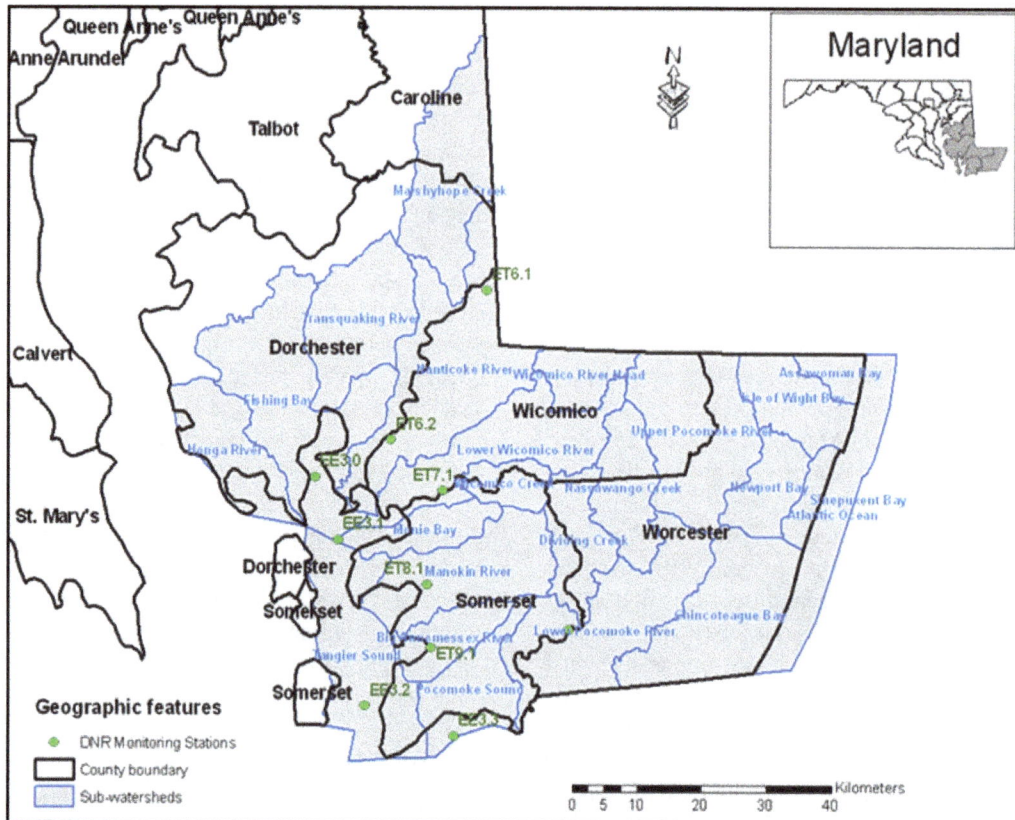

Figure 1. Lower eastern shore of Maryland study sites.

7 were of pixel size of 30 m, while the thermal band 6 was 60 m. All maps were projected using NAD83 UTM Zone 18.

Historical (1986-2006) water quality data of the lower Eastern Shore watersheds monitored by Maryland's Department of Natural Resources (MD-DNR) was acquired from the US Environmental Protection Agency Chesapeake Bay Program [24]. The scope was limited to assessment of physical and chemical water quality parameters such as Total phosphorus (TP), Total nitrogen (TN), total suspended solids (TSS), chlorophyll-a (CHLA), Secchi disc depth (SECCI), dissolved oxygen (DO), pH (PH), specific conductivity (SPCOND), salinity (SAL) and water temperature (WTEMP). Water quality data were pre-processed by taking the means for each month. Representative months for each season were taken to avoid bias from missing data. In this regard, the month of January was taken to represent winter, April to represent spring and July and November to represent summer and fall respectively. Only sites with continuous water quality monitoring data for 1986-2006 were included in the analyses. In order to validate the water quality data, water samples were randomly collected from GPS-guided sampling sites in July, 2006 and analyzed both *in-situ* and in the laboratory using the same standard procedures for the historic data. Because no significant differences

existed, they were excluded from the final analysis. Similarly, the historical water quality data for the coastal bays were rather sporadic and thus was excluded from the final analysis.

LULC classification for the study area was done in Environment for Visualizing Images (*ENVI* 4.5) acquired from ITT Visual Information Solutions [25]. Bands 7, 4 and 2 were selected for supervised classification using Mahalanobis distance method after several trials. The reference group or regions of interest selection was guided by aerial photos, Google Earth and ground-truthing and personal knowledge of the study area. LULC classification system of Anderson *et al.*, [26] was used for classification. It classifies land use-land cover into nine major categories: Urban or built up land, Crop/ Agricultural Land, Rangeland, Forest Land, Water, Wetland, Barren Land, Tundra, Perennial Snow or Ice at the level 1. However, rangeland, tundra and perennial snow land cover types are absent in the study location and were eliminated from the classification scheme. The classified images were exported into a GIS Environment ArcGIS 9.2 [27] where spatial analysis was completed. The 1986 *Landsat* images for rows 33 and 34 were merged and each sub watershed was masked and extracted. Areas of various land use in each watershed were quantified by multiplying the number of pixels for each land use by the spatial resolution (30 m × 30 m) of the *Landsat* images from which the LULC data was derived. Land use changes at 10-year intervals (1986-1996 and 1996-2006), and 20-year interval (1986-2006) were derived by overlaying the respective LULC maps for each interval. The overall change map was produced by overlaying the LULC map of the end of the study period (2006) over the initial map of 1986.

The Kruskal-Wallis one-way non-parametric analysis of variance test and the post hoc paired was employed to evaluate the how significant the land use/cover changed between each time intervals for the same area.

4. Results and Discussions

4.1. Land Use-Land Cover Dynamics

Figure 2 shows the general trends in extent of the major land use/cover in two decades. Whereas there were increasing trend with respect to the extent of urban lands, forest lands and surface water cover, barren lands, wetlands and croplands deceased during the 20-year study period.

The land use/cover of the study site in 1986, 2006 and the aggregate change during the period are shown in **Figures 3(a)-(c)** respectively; the latter was derived from map algebra techniques in Arc-GIS. During this period, forestlands and area covered by water increased by 8.5% and 10% respectively, while urban land increase by 121.8%. However, there was a net loss of agricultural lands (19.6%), wetlands (21.3%) and barren lands (51.3%) within the same period.

All the sub watersheds in the study area experienced an increase in urban land use between 1986 and 2006 (**Table 1**) except the narrow coastal bays bordering the Atlantic Ocean and consisting of basically water (99.6%) and beaches (0.2%). The largest net gain in urban land occurred in the Lower Wicomico sub-watershed. Urban land increased by 18.26 km² in the Lower Wicomico River sub watershed during the study period (11.37%).

Table 1. Changes in urban land in the lower eastern shore sub-watersheds (1986-2006).

Subwatersheds	1986 (km²)	1996 (km²)	2006 (km²)	Net Change (km²)
Marshyhope Creek	7.89	13.92	20.11	12.22
Big Annemessex River	3.12	1.82	6.20	3.08
Nanticoke River	8.06	16.36	23.48	15.42
Transquaking River	3.66	7.58	13.39	9.73
Fishing Bay	4.21	5.16	13.11	8.90
Wicomico River Head	6.41	10.26	14.23	7.82
Upper Pocomoke River	5.46	12.29	16.72	11.26
Lower Wicomico River	18.43	24.10	36.69	18.26
Honga River	2.02	2.09	4.03	2.01
Nassawango Creek	1.73	3.30	5.11	3.38
Dividing Creek	0.90	2.14	4.78	3.88
Wicomico Creek	1.22	2.48	4.69	3.47
Monie Bay	0.53	1.55	2.83	2.30
Manokin River	3.70	8.39	13.50	9.80
Lower Pocomoke River	10.67	13.70	22.34	11.67
Tangier Sound	3.39	2.85	4.83	1.44
Pocomoke Sound	4.30	3.06	7.48	3.18
Assawoman Bay	5.60	4.48	6.02	0.42
Isle of Wight Bay	10.09	12.32	16.89	6.81
Atlantic Ocean	1.22	1.72	0.79	-0.43
Newport Bay	4.65	4.52	7.61	2.95
Sinepuxent Bay	3.85	3.36	4.42	0.57
Chincoteague Bay	4.27	3.91	6.68	2.42
Sum	115.37	161.36	255.94	140.57
Mean	5.02	7.02	11.13	6.11
SD	4.02	5.95	8.60	4.57
SE	0.84	1.24	1.79	0.95

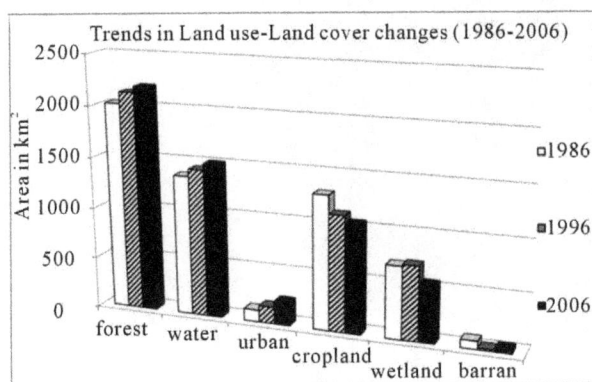

Figure 2. Trends in land use-land cover changes from 1986-2006.

Using the Kruskal-Wallis test and the post hoc paired comparisons, the lower Eastern Shore sub watersheds increased significantly ($p < 0.05$, H = 9.87, n = 23) in urban land area between 1986 and 2006 after a total of 227.29 km² of other land uses were converted to urban lands. However, changes between 1986-1996 and 1996-2006 intervals were not significant.

Most of the gains in urban land (56% or 127.20 km²) occurred on agricultural lands with the greatest change

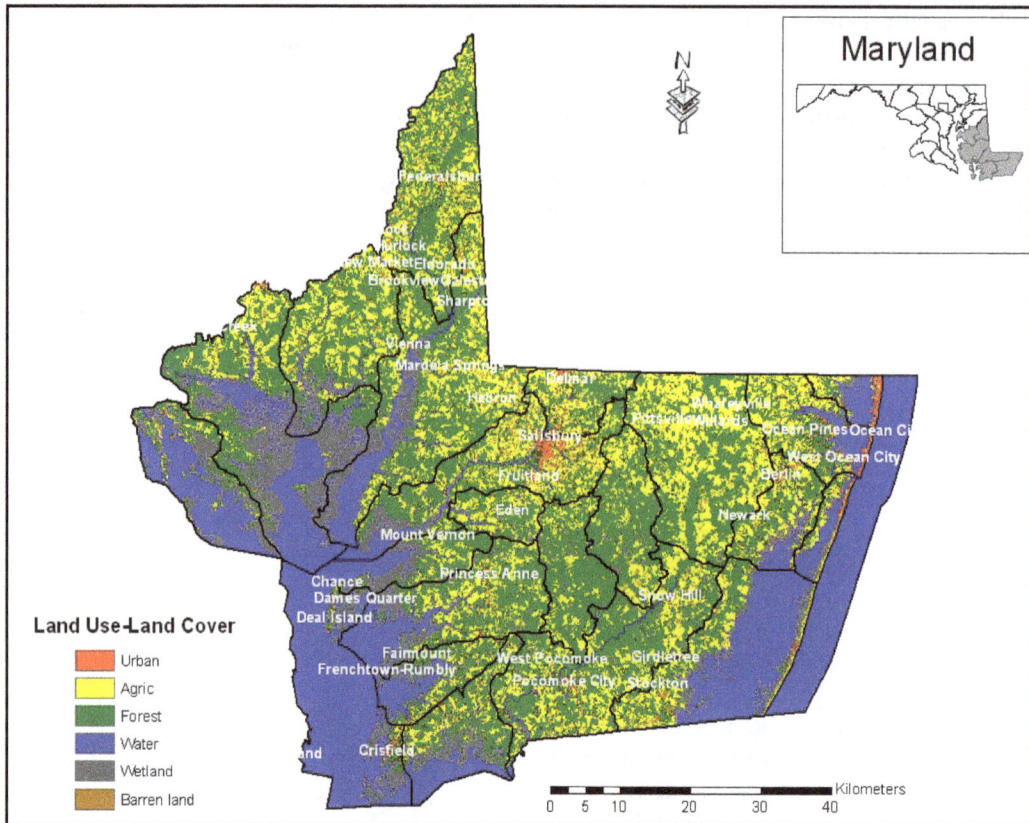

(a)

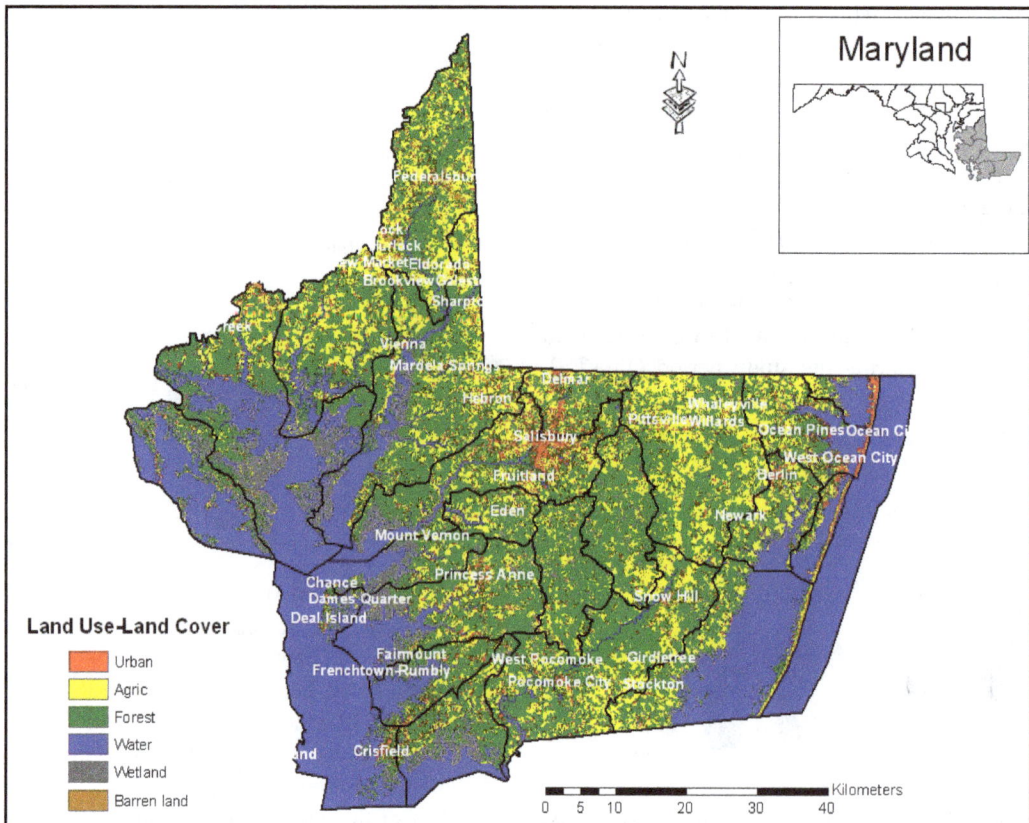

(b)

Geospatial Evaluation for Ecological Watershed Management: A Case Study of Some Chesapeake Bay
Sub-Watersheds in Maryland USA

193

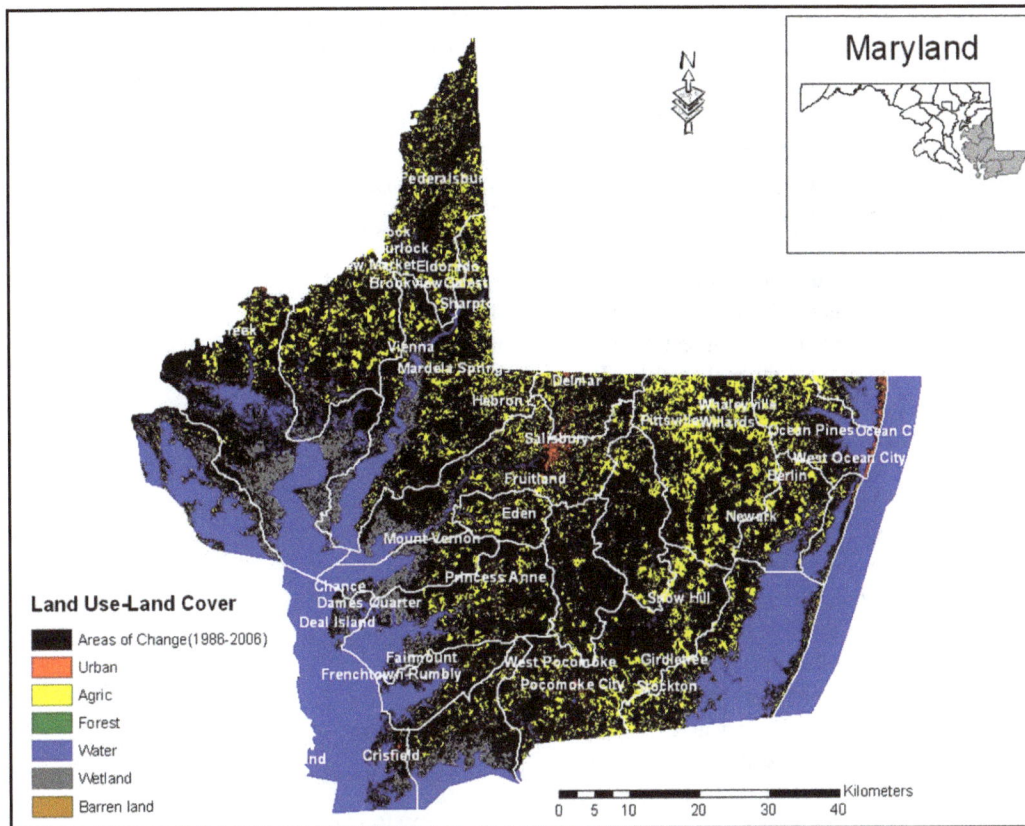

(c)

Figure 3. (a) Land use/cover of Maryland eastern shore sub-watersheds 1986; (b) Land use/cover of Maryland eastern shore sub-watersheds 2006; (c) Land use change in Maryland eastern shore sub-watersheds (1986-2006).

occurring in the Lower Wicomico River sub watershed (18.3 km^2), Nanticoke River (13.1 km^2), Upper Pocomoke River (11 km^2), Marshyhope Creek (10.2 km^2) and Lower Pocomoke River (10.5 km^2) sub watersheds. This trend is not unexpected considering the fact that residential developments and commercial growth attendant to increasing world population growth are not only occurring in urban centers but also in previously small towns and lands previously used for agricultural purposes [28]. About 33% (75.17 km^2) of urban land growth occurred on forested lands and the most significant loss to urbanization also occurred in the Lower Wicomico (8.1 km^2), Marshyhope Creek (8.1 km^2), Nanticoke River (8.0 km^2) and Lower Pocomoke River (7.1 km^2) sub watersheds. This trend is similar to that reported in 2004 [5] where a 61% increase in developed lands within the Chesapeake Bay watershed from 1990-2000 were observed and were attributed mostly to new urban development (64%) occurring on agricultural lands and grasslands, while 33% occurred on forested lands. The National Oceanic and Atmospheric Administration (NOAA) data also showed a similar trend for the lower Eastern Shore between 1996 and 2005. A total loss of approximately 12.65 km^2 of wetlands to urbanization occurred, while 8.19 km^2 of

barren land were converted to urban lands during the 20-year study period.

The total populations of the lower Eastern Shore counties (Dorchester, Somerset, and Wicomico and Worcester counties) were approximately 155,708 in 1986, 176,905 in 1996 and 198,155 in 2006. At an average annual growth rate of 0.98%, the projected population for 2030 was 249,700 and will be double the population in 1970 [29]. Therefore, the gains in urban land can be attributed to the changes in the demography of the major cities/towns in the sub watersheds. For example, Salisbury, in the Lower Wicomico River sub watershed has been experiencing a rapid growth in housing development due to increase in population from approximately 16,850 in 1986 to 27,172 in 2006—a 61% increase (Maryland Department of Planning, 2006). Such growth in urban lands create impervious surfaces and thus reduce infiltration, and increase nutrient, sediment and other pollutant loadings into the aquatic ecosystems which lowers quality of the surface waters [30-32]. Urbanization in Marshyhope Creek (12.22 km^2) can be attributed to population growth in Federalsburg and Hurlock while population growth in towns such as Hebron, Vienna, Mardela Springs and Sharptown is responsible for the 15.42 km^2 increase in urban land use observed in Nanticoke River sub watershed. Population growth in Pocomoke City and Snow Hill are mainly responsible for the 11.67 km^2 increase in urban land use in Lower Pocomoke River sub watershed. A combined urban land gain of 7.23 km^2 in Isle of Wight and Assawoman Bay is due to population growth in Ocean City—a popular tourist city in the Eastern Shore of Maryland.

There was a net loss of about 256.16 km^2 (19.6%) of cropland land in the lower Eastern Shore sub watersheds between 1986 and 2006 in general. The largest net loss of agricultural lands occurred in the Lower Wicomico River sub watershed (**Table 2**). In this sub watershed, Agricultural land decreased from 108.62 to 74.92 km^2 from 1986-2006. However, there were no significant differences in agricultural land use between 1986 and 1996, 1996 and 2006 or 1986 and 2006 (p < 0.70, H = 0.83, n = 23). This finding was corroborated with the Census of Agriculture conducted by the United States Department of Agriculture every five years that showed a decline in the use of land for crop production [33]. Agricultural lands in Maryland decreased consistently over the years from 686,964 ha in 1987 to 553,324 ha in 2007, representing a 19.5% loss. However, the poultry industry in Maryland has been on a steady rise. Broilers and other meat-type chickens sold have increased from 257,070,110 in 1987 to 296,373,113 in 2007 (up by 13.3% between 1987 and 2007).

The greatest loss of agricultural lands occurred in the Lower Wicomico River sub watershed followed by Nanticoke River and Upper Pocomoke River sub watersheds. Agricultural lands lost 127.20 km^2 to urban sprawl. A large land mass (457.31 km^2) of agricultural lands was converted to forested lands during the study period. Nanticoke River (58.4 km^2), Upper Pocomoke River (49.9 km^2), Marshyhope Creek (42.5 km^2), Lower Wicomico River (39.2 km^2), Lower Pocomoke River (41.1 km^2) and Transquaking River (36.2 km^2) sub-watersheds experienced large changes from agricultural land to forested land. Forest lands were also lost to agriculture (301.61 km^2) between 1986 and 2006 in the Lower Eastern Shore with most of those changes occurring in Marshyhope Creek (40.5 km^2), Nanticoke River (40.8 km^2), Upper Pocomoke River (33.3 km^2),

In the Lower Pocomoke River (18.3 km^2) and Transquaking River (27.6 km^2) subwatershed, Agricultural land to wetland change was 54.83 km^2 while an approximate 21.59 km^2 of agricultural land in 1986 became barren in 2006. Marshyhope Creek and Nanticoke River sub watersheds have lost 7.5 km^2 and 4.6 km^2 of agricultural lands respectively to barren lands. About 636.89 km^2 (about 47% of agricultural lands) remained unaltered during the same period especially in the Upper Pocomoke River (109.3 km^2) and the Nanticoke River sub

Table 2. Changes in agricultural lands in the lower eastern shore sub-watersheds (1986-2006).

Subwatersheds	1986 (km^2)	1996 (km^2)	2006 (km^2)	Net Change (km^2)
Marshyhope Creek	124.26	140.98	112.39	−11.87
Big Annemessex River	18.78	19.98	11.33	−7.45
Nanticoke River	155.52	136.02	126.69	−28.83
Transquaking River	96.3	97.22	81.1	−15.20
Fishing Bay	58.8	51.2	37.49	−21.31
Wicomico River Head	42.46	34.64	32.21	−10.25
Upper Pocomoke River	177.28	111.74	149.31	−27.97
Lower Wicomico River	108.62	92.45	74.92	−33.70
Honga River	11.42	9.09	6.89	−4.53
Nassawango Creek	49.6	38.98	40.05	−9.55
Dividing Creek	30.4	30.61	27.81	−2.59
Wicomico Creek	28.34	23.89	22.4	−5.94
Monie Bay	15.05	14.38	10.84	−4.21
Manokin River	58.82	48.63	45.05	−13.77
Lower Pocomoke River	121.82	113.81	110.39	−11.43
Tangier Sound	3.82	5.93	2.84	−0.98
Pocomoke Sound	29.83	26.78	19	−10.83
Assawoman Bay	10.63	7.32	7.76	−2.87
Isle of Wight Bay	61.08	35.56	45.06	−16.02
Atlantic Ocean	0.30	0.32	0.08	−0.22
Newport Bay	43.17	28.11	33.98	−9.19
Sinepuxent Bay	10.40	7.41	6.63	−3.77
Chincoteague Bay	50.25	46.49	46.56	−3.69
Sum	1306.94	1121.53	1050.78	−256.16
Mean	56.82	48.76	45.69	−11.14
SD	50.26	43.75	43.07	−7.18
SE	10.48	9.12	8.98	−1.50

watersheds (75.2 km^2). While a total of 250.16 km^2 agricultural land was lost to other land uses during the study period, an average of 11.14 ± 1.50 km^2 were lost per sub watershed.

Very substantial areas of forest lands (1447.29 km^2 or 71%) remained unaltered between 1986 and 2006 for Lower Pocomoke River, Upper Pocomoke River, Nanticoke River, Fishing Bay and Marshyhope Creek. However, whereas forest land occupied 2022.6 km^2 in 1986, only 2193.9 km^2 remained in 2006 (**Figure 4**), with a net

Geospatial Evaluation for Ecological Watershed Management: A Case Study of Some Chesapeake Bay
Sub-Watersheds in Maryland USA

195

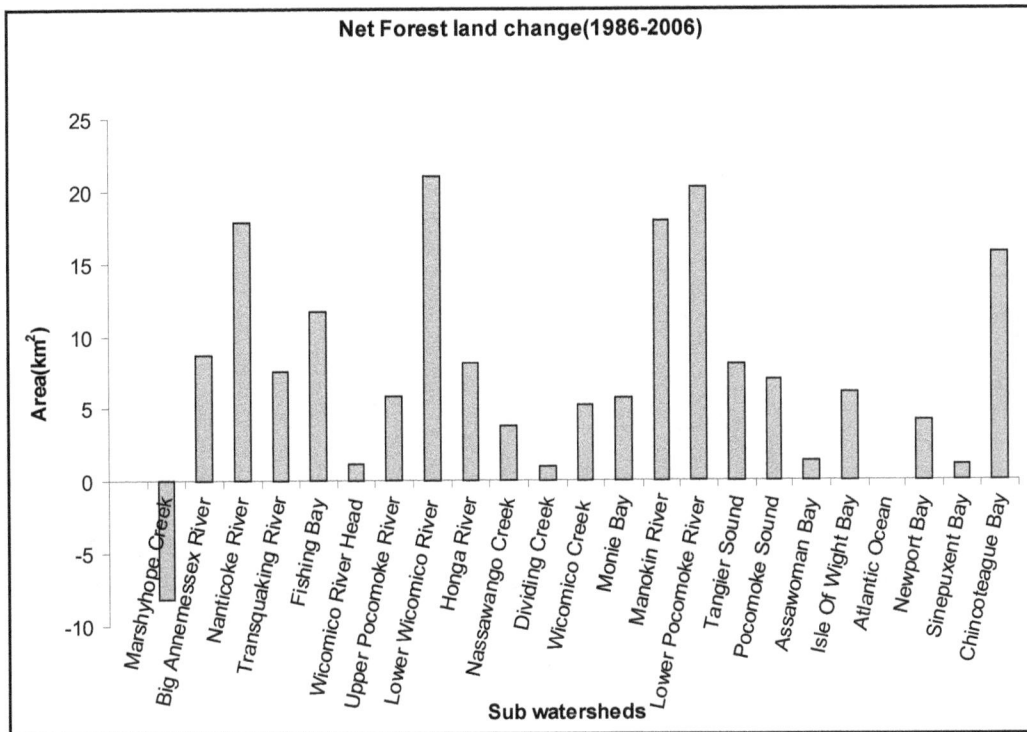

Figure 4. Forest land change in Maryland eastern shore (1986-2006).

gain of 171.27 km^2 (8.5%) during the 20-year period. The recent U.S Census of Agriculture [32] reported increase in woodlands and pastures in Maryland in general. However, this change was not significant between all-time intervals investigated (p < 0.90, H = 0.28, n = 23). Although forested lands increased in most sub-watersheds, Marshyhope Creek actually experienced a net loss of 8. This result is corroborated by the NOAA land use-land cover data which also showed an increase in forested lands for the Lower Eastern Shore between 1996 and 2005 [7]. Increase in forested land in the Lower Eastern Shore can be attributed to natural forest regrowth and Maryland Forest Conservation Act enacted in 1991. This Act stipulates that, "gaining approval of the required Forest Conservation Plan (development of more than one acre) may require long term protection of included priority areas or planting/replanting (afforestation or reforestation) a sensitive area off-site" [34]. The largest net gain in forest land (21.07 km^2) during the study period was in the Lower Wicomico sub watershed. Gains in forest land from other land uses decreased in the following order: Agricultural land to forest (457.31 km^2), wetland to forest (204.45 km^2), urban to forest (38.48 km^2), water to forest (27.53 km^2) and barren land to forest (19.38 km^2).

There was a net gain of about 135.89 km^2 (10%) of areas covered by water in the lower Eastern Shore sub watersheds between 1986 and 2006 (**Figure 5**). During the study period, however, 154.61 km^2 of wetlands became inundated. Most of this inundation occurred in Fishing Bay (55.77 km^2) situated at the edge of Chesapeake Bay in Dorchester county. Conversely, only 31.44 km^2 of water covered areas in 1986 became wetlands in 2006 with Fishing Bay also experiencing the most change of 6.4 km^2. Some forested lands (32.2 km^2) were also inundated by water during the study period, while 9.6 km^2 of agricultural lands also became inundated with water. About 6.6 km^2 of urban lands similarly became inundated by water while a total change from barren land to water in the lower Eastern Shore was 1.63 km^2 during the study period. The increase in water cover in the Lower Eastern Shore of Maryland is due in part to sea-level rise-perhaps a global warming effect on the estuarine tributaries of the Chesapeake Bay which empties into the Atlantic Ocean. This rise is indicated by the decrease in the extent of wetlands and salt marshes (by 22%) through submergence, and barren land (which decreased by 2%) during the study period. Hilbert [35] reported a similar trend in the Grand Bay National Estuarine Research Reserve area of Mississippi in the Northern coast of Gulf of Mexico from 1974 to 2001.

Loss of wetlands was observed in 17 out of the 23 sub watersheds in the study area (**Table 3**). In general, there was a 23% net loss (150.02 km^2) of wetlands from 1986-2006 in the study area. But these losses were not significant (p < 0.50, H = 1.80, n = 23) for the periods between 1986-1996, 1996-2006 or 1986-2006. Similar results were observed in NOAA land use-land cover data

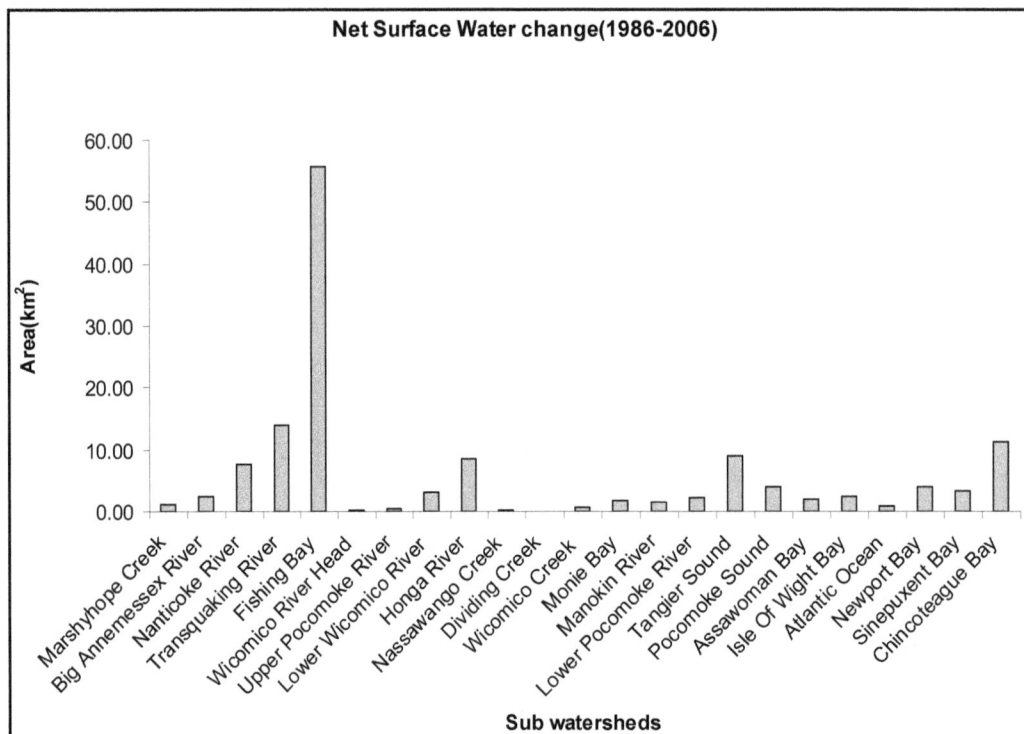

Figure 5. Net surface water cover in Maryland eastern shore sub-watersheds (1986-2006).

which also showed loss of wetlands for the lower Eastern Shore between 1996 and 2005 [7]. The vital ecological functions of wetlands—such as water quality improvement/preservation, fish and wildlife habitats, reduction of flood damage, shoreline erosion protection etc., make their decline of great ecological concern. The Congaree Bottomland Hardwood Swamp in South Carolina was estimated to remove pollutants equivalent to that removed annually by a $5 million waste water treatment plant [18]. The decreasing extent of wetlands in the lower Eastern Shore has the potential to compromise several ecological services.

Our results also indicate that during the study period, 154.61 km^2 of wetlands became covered by water. Most of this wetland inundation by water (55.77 km^2 representing 41%) occurred in Fishing Bay where the largest net loss of 53.16 km^2 (35%) of wetlands occurred. Nutria (*Myocastor coypus*) has been a primary force in accelerating wetland loss in the Black Water basin—where Fishing Bay is located—as well as other sub watersheds in Maryland. Nutria feed on marsh vegetation, expose the mud and thereby predispose marshes to erosion. Consequently, the marsh surface sinks and the vegetation is lost to flooding. Large area of marsh lands in the Black Water National Wildlife Refuge (BNWR) within the same watershed have been reported lost to Nutria [36]. Although, this destructive rodent has been eradicated from BNWR [37], 53% of the remaining marshes in BNWR is considered unhealthy and is likely to be lost in the future [35].

It is obvious therefore that the activities of these non-native rodent species on the wetlands may have contributed in part to the decreasing wetlands and increasing extent of water cover in the lower eastern shore of Maryland. Reciprocally, only 31.44 km^2 of areas covered by water in 1986 have become wetlands in 2006 with Fishing Bay also experiencing the most change of 6.4 km^2. This indicates, fluctuations between areas covered by water and wetland in Fishing Bay but with more wetlands becoming flooded. Of the 704.25 km^2 of wetlands in 1986, only 299.56 km^2 of wetlands (which represents about 43%) have remained unaltered between 1986 and 2006 especially in Fishing Bay (79.3 km^2), Nanticoke River (35.4 km^2), Honga River (29.6 km^2) and Manokin River (25.9 km^2) sub watersheds. Approximately 204.45 km^2 of wetlands became forested between 1986 and 2006. Most of this change has occurred in the Lower Pocomoke River (25 km^2), Manokin River (23.7 km^2) watersheds as well as Fishing Bay (21.6 km^2). A total of 34.09 km^2 of wetlands was lost to agricultural land use in the study region between 1986 and 2006. Small areas of wetlands also changed to agricultural land in the Lower Pocomoke River (5.4 km^2), Manokin River (3.4 km^2) and Fishing Bay (3.0 km^2). About 12.65 km^2 of wetlands became urbanized in 2006.

There was a decrease in barren lands (51.3%) in all the sub watersheds (**Table 4**) except for Marshyhope Creek (with a net gain of 2.22 km^2), and Sinepuxent Bay (1.05 km^2) while Assawoman Bay experienced no net gain.

Geospatial Evaluation for Ecological Watershed Management: A Case Study of Some Chesapeake Bay
Sub-Watersheds in Maryland USA

197

Table 3. Changes in wetlands in the lower eastern shore sub-watersheds (1986-2006).

Subwatersheds	1986 (km^2)	1996 (km^2)	2006 (km^2)	Net Change (km^2)
Marshyhope Creek	1.1	14.18	5.51	4.41
Big Annemessex River	23.1	18.98	16.46	−6.64
Nanticoke River	60.32	61.28	52.6	−7.72
Transquaking River	32.82	28.63	19.74	−13.08
Fishing Bay	163.85	106.17	110.69	−53.16
Wicomico River Head	0.65	4.93	3.47	2.82
Upper Pocomoke River	5.08	62.02	18.7	13.62
Lower Wicomico River	31.81	37.64	31.13	−0.68
Honga River	56.02	45.63	42.34	−13.68
Nassawango Creek	12.66	26.35	17.62	4.96
Dividing Creek	16.58	16.2	14.88	−1.70
Wicomico Creek	7.58	8.19	5.15	−2.43
Monie Bay	31.01	28.61	25.64	−5.37
Manokin River	60.45	52.98	46.96	−13.49
Lower Pocomoke River	36.69	32.53	20.29	−16.40
Tangier Sound	40.47	35.01	23.04	−17.43
Pocomoke Sound	41.3	36.74	38.5	−2.80
Assawoman Bay	6.48	7.67	5.43	−1.05
Isle of Wight Bay	7.77	22.71	9.61	1.85
Atlantic Ocean	0.04	0.16	0.01	−0.04
Newport Bay	14.30	24.41	14.54	0.24
Sinepuxent Bay	6.99	7.13	4.64	−2.35
Chincoteague Bay	47.18	41.45	27.29	−19.89
Sum	704.25	719.60	554.23	−150.02
Mean	30.62	31.29	24.10	−6.52
SD	35.09	23.82	23.76	−11.33
SE	7.32	4.97	4.96	−2.36

Table 4. Changes in barren lands in the lower eastern shore sub-watersheds (1986-2006).

Subwatersheds	1986 (km^2)	1996 (km^2)	2006 (km^2)	Net Change (km^2)
Marshyhope Creek	10.62	2.29	12.84	2.22
Big Annemessex River	0.14	0.05	0.03	−0.11
Nanticoke River	9.05	1.65	6.7	−2.35
Transquaking River	5.87	0.59	2.83	−3.04
Fishing Bay	2.09	0.05	0.21	−1.88
Wicomico River Head	3.02	0.46	1.33	−1.69
Upper Pocomoke River	4.24	0.80	1.08	−3.16
Lower Wicomico River	10.61	0.71	2.61	−8.00
Honga River	0.41	0.00	0.01	−0.40
Nassawango Creek	3.2	0.06	0.5	−2.70
Dividing Creek	0.78	0.09	0.15	−0.63
Wicomico Creek	1.39	0.34	0.23	−1.16
Monie Bay	0.19	0.01	0.03	−0.16
Manokin River	2.26	0.09	0.15	−2.11
Lower Pocomoke River	6.79	0.64	0.43	−6.36
Tangier Sound	0.08	0.07	0.02	−0.06
Pocomoke Sound	0.58	0.05	0.04	−0.54
Assawoman Bay	0.61	0.13	0.65	0.04
Isle Of Wight Bay	2.38	0.28	1.06	−1.32
Atlantic Ocean	1.14	2.02	0.11	−1.04
Newport Bay	2.40	0.50	0.24	−2.17
Sinepuxent Bay	2.35	2.78	3.40	1.05
Chincoteague Bay	10.01	3.56	4.02	−5.99
Sum	80.22	17.22	38.67	−41.55
Mean	3.49	0.75	1.68	−1.81
SD	3.55	1.00	2.96	−0.59
SE	0.74	0.21	0.62	−0.12

Barren lands, which are mainly beaches in Assawoman Bay and Isle of Wight Bay where Ocean City (a popular tourist city) is located have experienced increase in water covered areas by 2.07 km^2 in Assawoman Bay and 2.56 km^2 in Isle of Wight Bay. This may be due to the rising sea level-a trend which has also been reported by Hilbert [34] in the Grand Bay National Estuarine Research Reserve area of Mississippi in the Northern coast of Gulf of Mexico from 1974-2001. The lower Eastern Shore sub watersheds recorded a significant net loss of about 41.55 km^2 of barren lands between 1986 and 2006 ($p < 0.05$, H = 13.28, n = 23). Barren land significantly ($p < 0.05$) decreased from 80.22 km^2 in 1986 to 17.22 km^2 in 1996; however, changes between 1996 and 2006 was not significant.

About 43.20 km^2 of barren land was converted to agricultural land while 8.19 km^2 of barren land became urbanized. As of 2006, approximately 19.38 km^2 of barren lands became forested within the 20-year period, with the largest changes occurring in the Lower Wicomico

River (2.7 km^2), Marshyhope Creek (2.4 km^2) and Nanticoke River (2.0 km^2) sub watersheds. Only 6.68 km^2 of barren land remained unaltered especially in the Marshyhope Creek (1.6 km^2).

4.2. Land Use and Water Quality and Nutrient Loading

A stepwise regression analysis of the land use and water quality/nutrient variables and the resulting correlation matrix is shown in **Table 5**. A significantly positive correlations was observed between forest land cover and agricultural land use ($r^2 = 0.95$); urban land use and forest land cover ($r^2 = 0.72$); Total N and Total P levels ($r^2 = 0.68$). However, significantly negative correlation was observed between Secchi depth (SECCHI) and Total P; as well as the latter and total suspended solids (TSS). No significant correlations were observed between the other combinations of variables evaluated. Land use-land cover, wastewater treatment plants sewage load, water quality and climatic data were combined and used to develop regression models. However, none was possible for Total N-obviously because of the ubiquitous nature of N in the environment.

We observed a significant ($p < 0.05$) decrease in Total Phosphorus (TP) concentration for the water systems during the period investigated (**Figure 6**) ranged from to 0.093 mg/l in 1986 to 0.044 mg/l in 2006. However,

some location variations were observed e.g., the Lower Pocomoke River and the Nanticoke River recorded the highest mean TP of 0.121 ± 0.100 mg/l and 0.083 ± 0.05 mg/l respectively. These surface waters are within watersheds made up of 24.5% and 27% agricultural lands respectively; consequently the crop fields may have been contributing substantially to the phosphorus loading of the water systems-particularly where conventional cropping methods involving inorganic fertilizers or poultry manure are used, coupled with sewage discharge to Nanticoke River and Lower Pocomoke River from Sharptown and Snowhill wastewater treatment plants. Lower

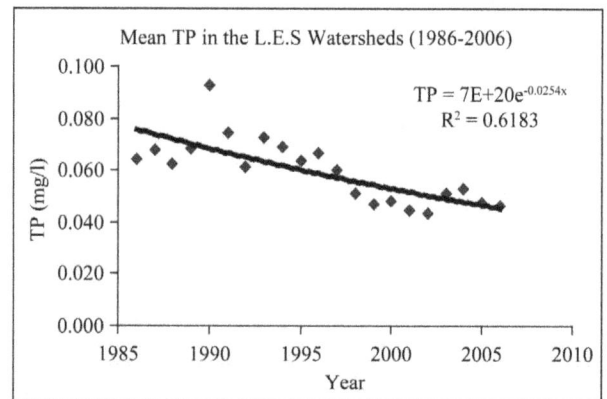

Figure 6. Mean total phosphorus of Maryland eastern shore surface waters (1986-2006).

Table 5. Pearson correlation of log-transformed data.

	URBAN	AGRIC	FOREST	WATER	BARREN	TN	TP	TSS	SALINITY	SPCOND
URBAN							−0.22			
AGRIC	0.69									
FOREST	0.72	0.95								
WATER		−0.41								
WETLAND		0.37	0.45	0.54						
BARREN	0.6	0.8	0.68				−0.22			
TN										
TP	−0.22	−0.22			−0.21	0.68				
TSS						0.51	0.64			
CHLA						0.41	0.41	0.37		
SALINITY						−0.68	−0.48	−0.48		
SECCHI					0.26	−0.52	−0.71	−0.74		
WWTPTN	0.54	0.28					0.24	−0.28		
WWTPTP	0.55	0.29					−0.22	−0.26		
TIME	0.4				−0.29					
PRECI pH									−0.21	0.83

Geospatial Evaluation for Ecological Watershed Management: A Case Study of Some Chesapeake Bay Sub-Watersheds in Maryland USA

199

levels of TP at Tangier Sound and Monie Bay sites also correlated with small areas of agricultural lands.

Urban lands increased at the expense of agricultural lands and barren lands which decreased by 9.7% and 10.2% respectively from 1986-2006. Increasing urban lands increases the extent of impervious surfaces from buildings, roads and runoff that are known to accelerate nutrients, sediments and chemical loadings into the aquatic systems [30,31]. However, an analysis of the water quality monitoring data from several stations in the stream networks did not adequately support this trend as the total phosphorus decreased generally during the same period.

On the other hand, decrease in agricultural lands and phosphorus input through point sources (Waste Treatment plants) during the same period may have influenced the general trends observed-perhaps as a result of better compliance to the National Pollutant Discharge Elimination System (NPDES) rules (for point source) in the last two decades; or more efficient use of fertilizers and soil conditioners (from non-point source).

Total Nitrogen (TN) levels varied within a narrow range during the 20-year study period, (1.00 - 1.71 mg/l). In general, no discernible trends were observed for Nitrogen levels in the surface waters as can be seen in **Figure 7**. On the other hand, point source discharge of nitrogen from the waste treatment plant increased significantly during the period (**Figure 8**). The increase is due to the steady growth in population and thus more waste being produced and processed by the various treatment plants in the area. For specific sampling sites however, some differences were observed. For example, Nanticoke River and the Lower Pocomoke River sub-watersheds had the highest TN with mean N of 3.218 ± 1.2 mg/l and 1.912 ± 0.7 mg/l respectively. TN increased significantly ($p < 0.05$) at the Nanticoke River, while a decrease was observed at the Lower Pocomoke River. Agricultural land decreased by 6% in Nanticoke River and decreased by 2% in Lower Pocomoke River sub watershed, but there was a substantial increase in croplands (24.5% and 27.0% respectively) in 2006.

Increased nutrient availability in surface water systems due to anthropogenic causes could lead to eutrophication and increase in chlorophyll-*a* levels in coastal waters resulting from increased phytoplankton biomass. Chlorophyll-*a* level provides a useful proxy indicator of the amount of nutrients incorporated into phytoplankton biomass.

Dissolved Oxygen (DO) affects aquatic life [22]. In this study, DO levels showed a significant decrease from 1986-2006 in general (**Figure 9**). However, there were site variations such as the Lower Wicomico River watershed where increases in urban population resulted in increased nutrient loading rates from Salisbury, Fruitland

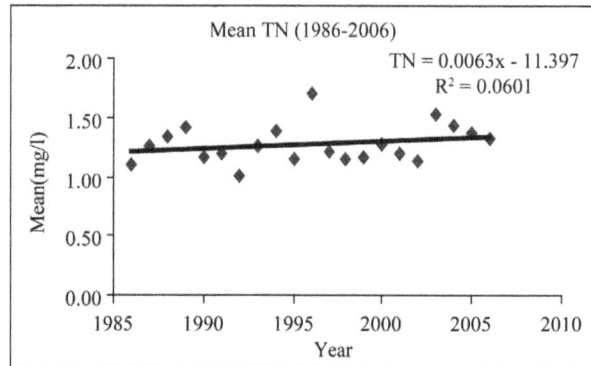

Figure 7. Mean total N in maryland eastern shore surface water (1986-2012).

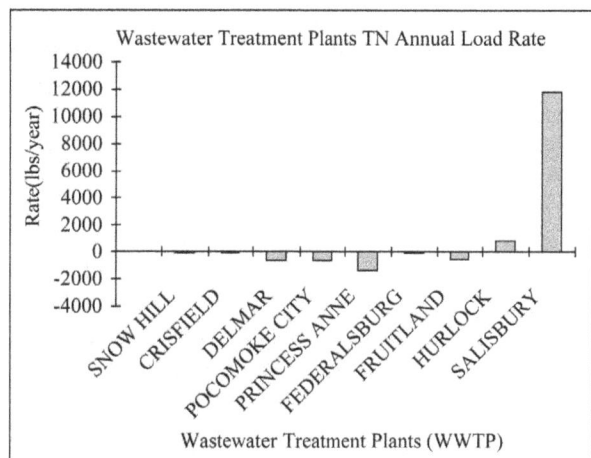

Figure 8. Mean annual nitrogen loading from wastewater treatment plants.

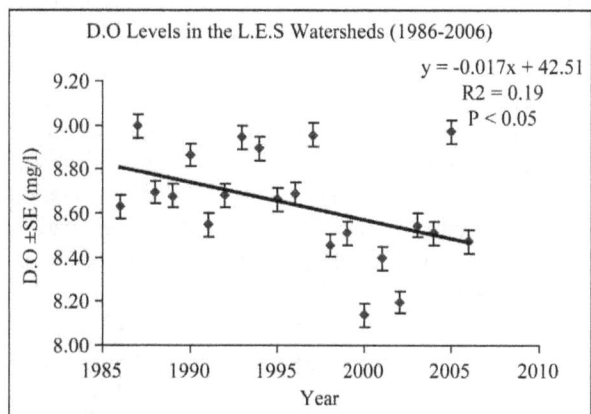

Figure 9. Dissolved oxygen levels in the lower eastern shore watersheds (1986-2006).

and Delmar wastewater treatment plants that empty into the Wicomico River and thus low DO levels during the period into the water system. On the other hand, the Annemessex River, Fishing bay and Monie bay had relatively higher DO due to fewer wastewater discharge into their surface water systems. The mean annual TP discharged into surface water was highest (44379.4 kg/year) at the Salisbury WWTP; however there was a decrease in the rate of TP discharged into the Wicomico River (475 kg/year) from Salisbury WWTP during the study period. TP decreased from an average of 19,727 kg/year in 1986 to 8835 kg/year in 1992 and remained stable from 1993 to 2006. Other WWTP with significant (p < 0.05, n = 20) decreasing TP load rates are Crisfield, Delmar, Princess Anne and Federalsburg. TP loading rates have however been increasing significant (p < 0.05) at Hurlock (197 kg/year).

5. Conclusions

Urban land increased followed by surface water and forest lands in the Lower Eastern Shore watersheds from 1986-2006. However, there were net losses of crop/agricultural lands, wetlands and barren lands during the two decades of this study. Most of the urban land gain occurred on crop/agricultural land while a third occurred on forested land. The largest gains in urban land as well as loss of crop/agricultural, forest and barren lands occurred in Lower Wicomico River sub-watershed. Net area covered by water increased by 135.9 km² from 1986-2006 for all sub watersheds and 154.61 km² of wetlands was inundated or covered by water. Most of such coverage (41%) occurred in Fishing Bay, in Dorchester County, and was attributed to the rising sea level as these tributaries of the Chesapeake Bay empties into the Atlantic Ocean. 17 out of 23 sub watersheds in the lower Eastern Shore experienced decreased wetlands areas from 1986-2006. The net area of wetlands lost was 150 km² especially in Fishing Bay (35%) and threatens the Blackwater National Wildlife Refuge (BNWR) located there. This signals change in the coastal ecology attributable in part to global climate change and the consequent sea level rise as well as wetland subsidence due to the destructive feeding activities of *Myocastor coypu* (Nutria)-a non native rodent species which feeds on marsh vegetation. Declining wetlands have serious ecological implications with respect to the various ecological services it provides: notably habitat loss for shellfish and waterfowls, flood buffer and wastes filter. It is also envisaged that the change in salinity of the brackish water would have major implications on the biota and could affect the biodiversity of the wetlands and surface water alike. Excessive concentration of P is the most common cause of eutrophication in freshwater lakes, reservoirs, streams and in headwaters of estuarine systems [37]. However, in this study, both Nitrogen and Phosphorus did not show a similar trend-despite the enormous nutrient loading from the wastewater treatment plants in the sub-watersheds. Whereas Phosphorus trend for the study location showed a decrease during the period, the trend was mostly attributable to the declining crop/agricultural land use-rather

than the decrease in P loading from wastewater treatment plants. Obviously, the increase in urban land use and the resultant increase in impervious surface area and runoff may have contributed other pollutants rather than P. Although there were no discernible trends in TN in general, the Lower Wicomico River which receives very heavy N load annually from three wastewater treatment plants (Delmar, Fruitland and Salisbury) showed a generally increasing trend-albeit local. This result is supported by a Hawaiian coastal waters study [21] where the discharge of effluent from two wastewater treatment plants did not significantly impact water quality parameters outside the zone of initial dilution. The Lower Wicomico River also received the highest Phosphorus load-though with a declining trend like other; this may probably reflect more compliance to National Pollutant Discharge Elimination System (NPDES) rules over the years. This site is apparently being impacted by both urban land use and waste treatment plant discharge resulting from the growing human population (particularly, Salisbury) in the watershed. There was also a general decrease in the dissolved oxygen levels in the surface waters during the period due to increasing Nitrogen load from several wastewater treatment plants in the area with implications for eutrophication.

The geospatial technology employed in this work has demonstrated the versatility of GIS/Remote sensing for quantifying past changes in land use/cover with respect to identifying precisely where changes occurred (Change detections) in the use of the lands that can guide urban and regional planners. Furthermore, these techniques have revealed the gradual effect of the climate change in the rising sea-level. However, it is suggested that a comprehensive historical land use/land cover be done every decade in order to detect significant changes as was observed in this study. Furthermore, developing nations are encouraged to take advantage of the remotely sensed Landsat data-which exist for the entire world for analyzing past and present trends in land use and cover for improving environmental and urban planning in general.

REFERENCES

[1] J. K. Berry, "Map Analysis: Understanding Spatial Patterns and Relationships," 2007.
http://www.innovativegis.com/basis/MapAnalysis

[2] V. H. Dale, S. Brown, R. Haeuber, N. T. Hobbs, N. Huntly, R. J. Naiman, W. E. Riebsame, M. G. Turner and T. Valone, "Ecological Principles and Guidelines for Managing the Use of Land," *Ecological Applications*, Vol. 10, No. 3, 2000, pp. 639-670.

[3] S. E. Gergel, M. G. Turner, J. R. Miller, J. M. Melack and E. H. Stanley, "Landscape Indicators of Human Impacts to Riverine Systems," *Aquatic Science*, Vol. 64, No. 2, 2002, pp. 118-128.

[4] M. A. Palmer, G. E. Moglen, N. E. Bockstael, S. Brooks, J. E. Pizuto, C. Wiegand and K. Van Ness, "The Ecological Consequences of Changing Land Use for Running Waters, with a Case Study of Urbanizing Watersheds in Maryland," *Human Population and Freshwater Resources*, Yale University Press, New Haven, 2002, pp. 85-113.

[5] P. Jantz, S. Goetz and C. A. Jantz, "Urbanization and the Loss of Resource Lands within the Chesapeake Bay Watershed," *Environmental Management*, Vol. 36, No. 6, 2004, pp. 808-825.

[6] D. M. Boward, P. F. Kazyak, S. A. Stranko, M. K. Hurd and T. P. Prochaska (US Environmental Protection Agency) "From the Mountains to the Sea: The State of Maryland's Freshwater Streams," *EPA Report* 903-R-99-023, Maryland Department of Natural Resources, Annapolis, 1999, 64 p.

[7] National Oceanic and Atmospheric Administration, "Coastal Services Centre," *Land Cover Analysis: Northeast Land Cover*, 2007.
http://www.csc.noaa.gov/crs/lca/northeast.html

[8] MRCC, "Multi-Resolution Land Characteristics Consortium," National Land Cover Database, 2008.
http://www.mrlc.gov/index.php

[9] US Environmental Protection Agency, Office of Wastewater Management, "National Water Quality Inventory: Report to Congress," *Reporting Cycle*, EPA 841-R-08-00, 2004.

[10] US Environmental Protection Agency, "Clean Water Act, Major Environmental Laws," 2002.
http://www.epa.gov/region5/water/pdf/ecwa.pdf

[11] M. A. Palmer, G. E. Moglen, N. E. Bockstael, S. Brooks, J. E. Pizzuto, C. Wiegand and K. Van Ness, "The Ecological Consequences of Changing Land Use for Running Waters: The Suburban Maryland Case," *Yale Bulletin of Environmental Science*, Vol. 107, 2002, pp. 85-113.

[12] P. Basnyat, L. D. Teeter, K. M. Fynn and B. G. Lockaby, "Relationships between Landscape Characteristics and No-Point Inputs to Coastal Estuaries," *Environmental Management*, Vol. 23, No. 4, 1999, pp. 539-549.

[13] J. S. Harding, E. F. Benfield, P. V. Bolstad, G. S. Helfman and E. B. D. Jones, "Stream Biodiversity: The Ghost of Land Use Past," *Proceedings of the National Academy of Sciences of the USA*, Vol. 95, No. 25, 1998, pp. 14834-14847.

[14] N. E. Roth, J. D. Allan and D. E. Ericson, "Landscape Influences on Stream Biotic Integrity Assessed at Multiple Spatial Scales," *Landscape Ecology*, Vol. 11, No. 3, 1996, pp. 141-156.

[15] J. Omernik, A. Abernathy and L. Male, "Stream Nutrient Levels and Proximity of Agricultural and Forest Land to Streams: Some Relationships," *Journal of Soil and Water Conservation*, Vol. 36, No. 4, 1981, pp. 227-231.

[16] L. Osborne and M. Wiley, "Empirical Relationships between Land Use/Cover Patterns and Stream Water Quality in an Agricultural Catchment," *Journal of Environmental Management*, Vol. 26, 1988, pp. 9-27.

[17] J. Karr and I. Schlosser, "Water Resources and the Land-Water Interface," *Science*, Vol. 201, No. 4352, 1978, pp. 229-234.

[18] US Environmental Protection Agency, "Watersheds Academy: Wetland Functions and Values," 2008.
http://www.epa.gov/watertrain/wetlands/index.htm

[19] S. R. Carpenter, N. F. Caraco, D. L. Correll, R. W. Howarth, A. N. Sharpley and V. H. Smith, "Nonpoint Pollution of Surface Waters with Phosphorus and Nitrogen," *Ecological Applications*, Vol. 8, No. 3, 1998, pp. 559-568.

[20] N. E. Detenbeck, C. M. Elonen, D. L. Taylor, L. E. Anderson, T. M. Jicha and S. L. Batterman, "Region, Landscape and Scaleeffects on Lake Superior Tributary Water Quality," *Journal of the American Water Resources Association*, Vol. 40, No. 3, 2004, pp. 705-720.

[21] R. Parnell, "The Effects of Sewage Discharge on Water Quality and Phytoplankton of Hawaiian Coastal Waters," *Marine Environmental Research*, Vol. 55, No. 4, 2002, pp. 293-311.

[22] M. S. Adam, J. L. Stauber, M. E. Binet, R. Molloy and D. Gregory, "Toxicity of a Secondary-Treated Sewage Effluent to Marine Biota in Bass Strait, Australia: Development of Action Trigger Values for a Toxicity Monitoring Program," *Marine Pollution Bulletin*, Vol. 57, No. 6-12, 2008, pp. 587-598.

[23] US Environmental Protection Agency, "Surf Your Watershed," 2009.
http://cfpub.epa.gov/surf/locate/hucperstate_search.cfm?statepostal=MD

[24] US Environmental Protection Agency, "Chesapeake Bay Program," Historic Water Quality Data, Annapolis, 2007.

[25] ITTVIS, "Environment for Visualizing Images," *ENVI*, 2008.

[26] J. R. Anderson, E. E. Hardy, J. T. Roach and R. T. Witmer, "A Land Use and Land Cover Classification System for Use with the Remote Sensor Data," *Geological Survey Professional Paper* 964. *A Revision of the Land Use Classification System as Presented in US Geological Survey Circular*, 1976, p. 671.

[27] Environmental Systems Research Institute, ArcGIS9.2, Redlands, 2007.

[28] T. N. Carlson and S. T. Arthur, "The Impact of Land Use-Land Cover Changes Due to Urbanization on Surface Microclimate and Hydrology: A Satellite Perspective," *Global and Planetary Change*, Vol. 25, No. 1-2, 2000, pp. 49-65.

[29] Maryland Department of Planning, "Historical and Projected Total Population for Maryland's Jurisdictions," Planning Data Services, 2006.
http://www.mdp.state.md.us/msdc/popproj/TOTPOP_PROJ06.pdf

[30] Z. Tang, B. A. Engel, B. C. Pijanowski and K. J. Lim, "Forecasting Land Use Change and Its Environmental Impact at a Watershed Scale," *Journal of Environmental Management*, Vol. 76, No. 1, 2005, pp. 35-45.

[31] M. A. Mallin, S. H. Ensign, M. R. McIver, G. C. Shank, and P. K. Fowler, "Demographic, Landscape, and Meteorological Factors Controlling the Microbial Pollution of Coastal Waters," *Hydrobiologia*, Vol. 460, No. 1-3, 2001, pp. 185-193.

[32] M. A. Van Buren, W. E. Watt, J. Marsalek and B. Anderson, "Thermal Enhancement of Storm Water Runoff by Paved Surfaces," *Water Research*, Vol. 34, No. 4, 2000, pp. 1359-1371.

[33] United States Department of Agriculture, 2009. http://www.agcensus.usda.gov./Publications/2007/Full_R eport/Volume_1,_Chapter_1_State_Level/Maryland/st24 _1_001_001.pdf

[34] Maryland Department of Natural Resources, "Forest Service: Urban and Community Forestry," Forest Conservation Act, 1995. http://www.dnr.state.md.us/forests/programapps/newFCA .asp

[35] W. K. Hilbert, "Land Cover Change within the Grand Bay National Estuarine Research Reserve: 1974-2001," *Journal of Coastal Research*, Vol. 22, No. 6, 2006, pp. 1552-1557.

[36] US Fish and Wildlife Service, "Blackwater National Wildlife Refuge: Nutria and Blackwater Refuge," 2009. http://www.fws.gov/blackwater/nutriafact.html#damage

[37] D. I. Correli, "Phosphorus: A Rate Limiting Nutrient in Surface Waters," *Poultry Science*, Vol. 78, No. 5, 1998, pp. 674-682.

Permissions

The contributors of this book come from diverse backgrounds, making this book a truly international effort. This book will bring forth new frontiers with its revolutionizing research information and detailed analysis of the nascent developments around the world.

We would like to thank all the contributing authors for lending their expertise to make the book truly unique. They have played a crucial role in the development of this book. Without their invaluable contributions this book wouldn't have been possible. They have made vital efforts to compile up to date information on the varied aspects of this subject to make this book a valuable addition to the collection of many professionals and students.

This book was conceptualized with the vision of imparting up-to-date information and advanced data in this field. To ensure the same, a matchless editorial board was set up. Every individual on the board went through rigorous rounds of assessment to prove their worth. After which they invested a large part of their time researching and compiling the most relevant data for our readers. Conferences and sessions were held from time to time between the editorial board and the contributing authors to present the data in the most comprehensible form. The editorial team has worked tirelessly to provide valuable and valid information to help people across the globe.

Every chapter published in this book has been scrutinized by our experts. Their significance has been extensively debated. The topics covered herein carry significant findings which will fuel the growth of the discipline. They may even be implemented as practical applications or may be referred to as a beginning point for another development. Chapters in this book were first published by Scientific Research Publishing Inc.; hereby published with permission under the Creative Commons Attribution License or equivalent.

The editorial board has been involved in producing this book since its inception. They have spent rigorous hours researching and exploring the diverse topics which have resulted in the successful publishing of this book. They have passed on their knowledge of decades through this book. To expedite this challenging task, the publisher supported the team at every step. A small team of assistant editors was also appointed to further simplify the editing procedure and attain best results for the readers.

Our editorial team has been hand-picked from every corner of the world. Their multi-ethnicity adds dynamic inputs to the discussions which result in innovative outcomes. These outcomes are then further discussed with the researchers and contributors who give their valuable feedback and opinion regarding the same. The feedback is then collaborated with the researches and they are edited in a comprehensive manner to aid the understanding of the subject.

Apart from the editorial board, the designing team has also invested a significant amount of their time in understanding the subject and creating the most relevant covers. They scrutinized every image to scout for the most suitable representation of the subject and create an appropriate cover for the book.

The publishing team has been involved in this book since its early stages. They were actively engaged in every process, be it collecting the data, connecting with the contributors or procuring relevant information. The team has been an ardent support to the editorial, designing and production team. Their endless efforts to recruit the best for this project, has resulted in the accomplishment of this book. They are a veteran in the field of academics and their pool of knowledge is as vast as their experience in printing. Their expertise and guidance has proved useful at every step. Their uncompromising quality standards have made this book an exceptional effort. Their encouragement from time to time has been an inspiration for everyone.

The publisher and the editorial board hope that this book will prove to be a valuable piece of knowledge for researchers, students, practitioners and scholars across the globe.

List of Contributors

Carolina Rojas
Department of Geography, Faculty of Architecture, Urbanism and Geography, University of Concepción, Center for Sustainable Urban Development—Chile, Concepción, Chile

Joan Pino
CREAF (Center for Ecological Research and Forestry Applications), Autonomous University of Barcelona, Barcelona, Spain

Iván Muñiz
Department of Applied Economics, Autonomous University of Barcelona, Barcelona, Spain

Ivan Frigerio, Stefano Roverato and Mattia De Amicis
Department of Earth and Environmental Sciences, University of Milano Bicocca, Milan, Italy

Amar Guettouche and Farid Kaoua
Faculty of Civil Engineering, University of Sciences and Technology Houari Boumediene, Algiers, Algeria

Hala A. Effat
Environmental Studies and Land Use Department, National Authority for Remote Sensing and Space Sciences, NARSS, Cairo, Egypt

Mohamed N. Hegazy
Division of Geological Applications and Mineral Resources, National Authority for Remote Sensing and Space Sciences, NARSS, Cairo, Egypt

Antonio Miguel Martínez-Graña and Jose Luis Goy
Geology Department, External Geodynamics Area, Sciences Faculty, University of Salamanca, Salamanca, Spain

Caridad Zazo
Section Geology, National Museum of Natural Sciences, Madrid, Spain

Fatwa Ramdani
Institute of Geography, Department of Earth Science, Graduate School of Science, Tohoku University, Sendai, Japan

Giuliana Lauro
Department of Industrial and Information Engineering, Second University of Naples, Aversa, Italy

Khalid Al-Ahmadi and Atiq Al-Dossari
Space Research Institute, King Abdulaziz City for Science and Technology, Riyadh, Saudi Arabia

Ali Al-Zahrani
King Faisal Specialist Hospital and Research Centre, Riyadh, Saudi Arabia

Kenneth R. Sheehan
Water Systems Analysis Group, University of New Hampshire, Durham, USA

Stuart A. Welsh
US Geological Survey, West Virginia Cooperative Fish and Wildlife Research Unit, Morgantown, USA

Ragab Khalil
Landscape Architecture Department, KAU, Jeddah, Saudi Arabia
Civil Engineering Department, Assiut University, Assiut, Egypt

Mehdi Mekni
Department of Math Science and Technology, University of Minnesota, Crookston, USA

Verônica Wilma Bezerra Azevedo and Chigueru Tiba
Department of Nuclear Energy, Federal University of Pernambuco, Recife, Brazil

Youness Kharchaf, Hassan Rhinane, Abdelhadi Kaoukaya and Abdelhamid Fadil
Geosciences Laboratory, Faculty of Sciences-Ain Chock, Hassan II University, Casablanca, Morocco

Devendra Amatya and Carl Trettin
USDA Forest Service, Cordesville, USA

Herbert Ssegane
University of Georgia, Athens, USA

Sudhanshu Panda
Gainesville State College, Oakwood, USA

Mohamed Said Guettouche and Ammar Derias
Laboratory of Geography and Territory Planning (LGAT), University of Sciences and Technology Houari Boumediene (USTHB), Algiers, Algeria

Hamdy I. El-Gamily
Geoinformatics Center, Kuwait Institute for Scientific Research (KISR), Kuwait City, Kuwait
National Authority for Remote Sensing and Space Sciences (NARSS), Cairo, Egypt

Khalid Al-Rasheed
Civil Engineering Department, College of Engineering, Kuwait University, Kuwait City, Kuwait

Isoken T. Aighewi and Osarodion K. Nosakhare
Benedict College, Columbia, South Carolina, USA